POST-CONSUMER WASTE RECYCLING AND OPTIMAL PRODUCTION

Edited by **Enri Damanhuri**

Post-Consumer Waste Recycling and Optimal Production
Edited by Enri Damanhuri

Published by InTech
Janeza Trdine 9, 51000 Rijeka, Croatia

Publishing Process Manager Romina Skomersic

Technical Editor Teodora Smiljanic

Cover Designer InTech Design Team

Additional hard copies can be obtained from orders@intechopen.com

Post-Consumer Waste Recycling and Optimal Production, Edited by Enri Damanhuri
p. cm.
ISBN 978-953-51-0632-6

For EU product safety concerns:
IN TECH d.o.o., Prolaz Marije Krucifikse Kozulić 3, 51000 Rijeka, Croatia,
info@intechopen.com or visit our website at intechopen.com.

Meet the editor

Dr Enri Damanhuri is Professor at the Department of Environmental Engineering, Faculty of Civil and Environmental Engineering, Institut Teknologi Bandung. He graduated from the Department of Sanitary Engineering ITB, in 1975. In 1987, Dr Damanhuri obtained his degree of Docteur Ingenieur from the University of Paris VII, France, in Chemistry of Pollution. His doctoral thesis was on the role of leachate recirculation on the kinetic of refuse degradation at landfill. His research is focused on alleviating problems associated with solid and hazardous waste management in Indonesia. He also works as a consultant in the area of technology and management of solid and hazardous waste. Dr Damanhuri has been involved in preparing various regulation concepts related to solid and hazardous waste management and technology for Indonesia.

Contents

Preface

Depletion of mineral resources is one of the major problems in the world. We have consumed available resources as if they were unlimited, and at the same time we generate a huge quantity of waste, thereby contaminating the Earth. The amount of waste in our society reflects how much resources we consume. We, as modern society, create more products and packaging which is visually appealing but cannot be easily *decomposed*. Energy demands, resource limitations, and environmental pollution are closely linked. Available resources and the environmental capacity to absorb waste are limited, and consequently may threaten the existence of a sustainable society. It is a well-known fact that resource recovery and recycling benefit the environment.

Resource recovery and recycling offer great potential to enhance resource management and reduce waste disposal. They will prolong the lifespan of landfills, and restrain natural resource depletion to ensure sustainable development of resource-intensive industries. Resource recovery and waste recycling are very important practices for community and industrial activities, and they contribute to the solution of the problems. As the population of the world increases, recycling is becoming increasingly more important. This approach provides effective solutions to cope with depletion of natural resources and waste generation caused by the mass production and mass consumption of the present highly civilized social system.

Waste recycling is a process for reusing materials, either directly, indirectly, or through energy utilization. Recycling is the process of separating, collecting and remanufacturing or converting used or waste products into new materials. It involves a series of steps to produce new products; turns materials that would otherwise become waste into valuable resources. The term "waste recycling" in its simplest sense means the return of a discarded material or article to the same product system, such as the return of waste paper to make new paper. The recycling of materials has the potential to reduce greenhouse gas emission, energy consumption, and other environmental impacts due to the substitution of virgin materials with recycled materials. The emissions from extraction and manufacturing of products from virgin materials can be minimized.

This book is concerned with several aspects of waste materials recycling, and is divided into three sections. The first section addresses the current problem of household waste generation in Romania, followed by an explanation of roles of stakeholders, both informal and formal sectors, in post-consumer waste activities. Case studies in Brazil, Indonesia and Japan are presented. Many countries, such as Japan, encourage their residents to use group collection programs, and some municipalities even provide financial support. The second section discusses the analysis tools for recycling system, particularly the concept of dual traceable ownership system, the economic aspects of recycling, recycling of mobile phones, design for e-waste recycling deposit system in China and reverse logistics model post-consumers for PET in Brazil. The third section of this book focuses on the recycling process and optimal production. It consists of modeling of recycling in LCA, optimal production decision in the closed-loop supply chain, multi-step of products approach based on ADSM, material and energy flow in iron and steel enterprise, the energy consumption of recycling processes, and mechanism to reduce the size of different materials in recycling industries.

I hope that this book will convey both the need and means for recycling and resource conservation activities to a wide readership, at both academician and professional level, and contribute to the creation of a sound material-cycle society.

As an editor of this book, I would like to thank all the authors for their considerable contribution. I also express my gratitude to Ms. Romina Skomersic, the Publishing Process Manager at InTech, who did an excellent job at encouraging us to do our tasks within the deadlines. Without her this book would not have been published.

Enri Damanhuri
Department of Enviromental Engineering,
Faculty of Civil and Enviromental Engineering,
Institute of Technology Bandung, Bandung,
Indonesia

Section 1

Post-Consumer Waste and Recycling

1

Assessing the Efficiency of a Proposed Project in Waste Management

Scorţar Lucia-Monica
Babeş-Bolyai University, Cluj-Napoca
Romania

1. Introduction

This paper addresses a current and very important problem, namely household waste generation that, on the one hand, affects the environment and human health and on the other hand, it reflects how inefficient the usage of natural resources by the society is.

Nature is a good human environment, more or less altered by people, and mankind is part of it. Environment means all the elements that surround us, humans often leaving their footprint on them, mostly in a destructive way; hence the need of interventions from the state authorities for nature protection and conservation. Leniently watching people's actions that destroy nature and the environment, would mean a distortion of their own sense of conservation, a degradation of personal and community life.

If industrialization and urbanization are activities more than necessary in human social development, humans also have to find alternatives to the process of altering the environment, in order to preserve a clean environment so that life can exist and be as more beautiful.

It can be noted by everyone that even where a single human lives, he makes some changes to the environment and, implicitly, creates waste, especially household waste. The larger the community, the bigger the waste quantity is, so that the existence of uncontrolled, accumulated household creates major problems that must be solved urgently and permanently. As a result, they must be managed so as not to burden to suffocation the community, thus imposing waste selective collection, reuse, recycling and treatment, and finally storing waste remains.

In the current waste management strategies, the trend is an integrated system based on prevention of waste, minimizing waste quantity, recycling and reusing of waste, treatment with a large number of technology, and ultimately, waste remains disposal, also taking care of population and environment health ("waste management hierarchy").

At the beginning of XXI century, we believe that not us, but generally people do not have the required education to treat waste problem as something serious that can influence decisively our future existence.Technical creativity in waste recovery did not have the same pace as the creativity in developing new products and adopting and implementing the technologies for achieving them. We believe that, viewed as an actual and future businesses, waste recovery will find its required technical capacity.

Once the waste was produced, the best way to reduce or eliminate its negative impact on the environment is its recovery. Recovering household waste fractions requires a separation between recoverable components. Since the separation of mixed waste is already done at great expense and often with insufficient results, useful materials must be collected by collection systems and recovered before mixing them with other parts of the waste.

This paper is proposing a new investment project in household waste management for a group of associated communities in both urban and rural areas, forming Comunitatea Urbană Arieş, judeţul Cluj (CUA).

Comunitatea Urbană Arieş is an association of local authorities from the area Arieş-Turda-Câmpia Turzii with public institution vocation and attributions for cooperation created through the free will act expressed by the participating Local Councils, according to their legal prerogatives, with the current regulations and the regulations of the European Union. It is appointed by the member councils to perform for them and on their behalf public services of common interest.

The fundamental criterion for joining and participating in the Comunitatea Urbană Arieş is for each Local Council the efficient access to common resources and their rational and integrated management, in order to protect the environment (Comunitatea Urbană Arieş, Statutul Asociaţiei, Turda, 2005).

The proposed waste management scheme is analyzed based on the principle of population access to sanitation services, under which public authorities are responsible for organizing public services of the community so that all members have equal access to these services.

The analysis showed that the effectiveness of the appropriate scheme for this area cannot be conceived outside a recovery circuit for useful material contained in household waste (paper/cardboard, glass, metal, plastics, including the capitalization of organic waste).

This project demonstrates a sustainable approach to waste management, proposing investments in the purchase of containers for selective collection, transport facilities, sorting, composting, recycling, organic waste incineration and final ecological disposal.

This project is designed to serve Comunitatea Urbană Arieş (117.780 inhabitants and 2.975 businesses) which is composed of two urban (Turda and Câmpia Turzii) and nine surrounding rural areas (Mihai Viteazu, Călăraşi, Frata, Aiton, Luna, Petreştii de Jos, Săndulești, Tritenii de Jos, Viişoara).

We believe that the modern system of sustainable management of waste proposed for Comunitatea Urbană Arieş will contribute to reducing the amount of waste stored in this area, as this is a suitable system which treats the main fraction of household waste to protect the environment.

2. Goal, targets, location

Waste management situation in the studied area cannot be extended for too long, as uncontrolled waste accumulations are affecting both environmental health and that of the population. The proposed investment project includes activities related to the hierarchy in terms of waste management (prevention, collection, recovery and disposal), as there is no proper system of waste management in this area.

This project concerns both actions in collection, transport, recycling and composting useful materials of household waste and a further step in remaining waste incineration, in an incinerator equipped with two modules. Thus, only the ashes resulting from incineration are finally stored.

These actions that should be provided in an integrated waste management system are phased as follows:

1. Waste processing at the source (in the apartment, household, institution),
2. Providing selective waste collection service (collection container, where the public service responsibility begins),
3. Waste transport with special vehicles,
4. Interim storage of waste (transfer station located between Câmpia-Turzii and Turda),
5. Recovery of selected waste fractions will be done by special institutions, and mixed collected waste will be treated in the sort-treatment station located on the former chemical plant Turda platform
6. Operating a waste incineration plant for the unused remaining waste from the sorting station located between Turda and Câmpia-Turzii
7. Transporting of the resulting ash to the ecological county landfill that will be opened in the village Feleacu.

In this scheme, the sorted waste will be recovered through their direct sale to recyclers. In this way, incomes can be obtained that will allow support of operating expenses with direct impact for the sanitation tariffs. Based on this scenario, we will forecast costs and revenues involved in the functioning of this waste management scheme proposed in Comunitatea Urbană Arieş and we will determine the efficiency of this system.

Applying this investment project is aimed at: empowering people in terms of practical activities for the selective collection of waste, increasing the coverage of sanitation services at bearable price, reducing quantities of waste and recovery of useful materials contained in waste, the final goal of these actions being a healthy and clean environment.

a. Non-selective collection of waste for the envisaged project consists of:
   - For the urban areas: Collecting waste from households includes: systematic structure of the current collection routes, extending these routes and have them served by the sanitation department, which will be reorganized in accordance with local laws and regulations of sanitation. Waste collection service will be extended.

The operating principle is collection at the collection points, respectively, each household will have pre-collection containers and collection points will be grouped as follows: in block of flats areas collection points will be arranged for each 200 apartments, and in houses areas the collection vehicles will travel from door to door to collect the waste.

   - For the rural areas: Collection and transport vehicles will retrieve mixed waste from door to door, covering the rural area. Social and administrative areas (administrative units, commercial units, schools and kindergartens) will benefit of the purchase of a number of pre-collection containers with a capacity of 1,1 m^3, used for urban areas only.

b. Selective collection of recoverable waste consists of:

- For the urban areas: In order to selectively collect waste, we propose organizing selective waste collection points, each having four containers for recoverable waste. We propose the collection of four fractions of recoverable waste: glass (in a 2-compartment container), metal, paper/cardboard and plastic. In houses areas, for every 100 households there will be arranged a point of collection, and in blocks of flats areas, for each 200 apartments, a collection point will be arranged.
- For the rural areas: For the selective waste collection, collection points will be arranged in each village. Selective collection points in villages are designed to serve 150 rural households, because of the lower waste ratio generated as opposed to the urban areas. Collection points structure is identical to those in urban areas.

3. Interpretation and analysis data

The poroposed waste management scheme is sized for a service area that includes 117.780 people and 2.975 companies, dispersed in two urban and nine rural areas, forming Comunitatea Urbană Arieş. (Studiu de Oportunitate privind delegarea gestiunii serviciului de salubritate în municipiul Turda, GPA Business Consulting, August 2006).

After the analysis and the description of the proposed waste management scheme, the object estimates of the investment components have been set and also a general estimate of the proposed management scheme has been developed, as follows:

Object estimate	Investment components	Value of the objects estimate (with VAT)
Object estimate 1	Setting up 388 collection points	3.069.386 euro
Object estimate 2	Purchasing transportation means	773.500 euro
Object estimate 3	Setting up the transfer station	450.466 euro
Object estimate 4	Setting up the sorting/treating station	4.838.724 euro
Object estimate 5	Purchasing and distributing collection containers to the population	555.603 euro
Object estimate 6	Setting up the incineration station	4.941.957 euro

Tabel 1. The object estimates of the investment components

The general total sum of the investment including VAT is situated at the value of 14.629.636 Euro. This value, although it seems a high value, compared to the usual schemes promoted in our country cover those segments of the management scheme which are not usually included in the budgets of the projects which need financing, such as the distribution of collection containers to the population, including the rural area, as well as the expenditures for setting up a modern sorting/treating station.

Determining the precision of the total investment value depends on both information sources and how to deepen the calculations during the project's execution design. The value of the object estimate is obtained by summing the values of the works categories that make the object. The value of these works categories is estimated on the base of the works quantities and their prices.

In order to achieve this investment objective, after drawing the general estimate, the total investment value of 14.629.636 euro resulted, obtaining the following values for categories of work:

- Expenses for obtaining the land: they are considered 0, because the land on which the investment objectives are placed belong to the public domain and will be made available to the project by the beneficiary City Council;
- Expenses for land planning (enbankments): 2.542.000 euro;
- Expenses for utilities necessary for the objective (wiring, plumbing, heating, telecommunications equipment): 73.640 euro;
- Expenses for design and technical assistance: 12.980 euro;
- Expenses for basic investment (construction and installation, equipment, assembling the machineries): 11.484.672 euro;
- Other expenses (site organization, fees, taxes, extraordinary expenses): 178.320 euro;
- Expenses for technological tests and delivering to the beneficiary (including training of operating personnel): 53.690 euro.

Household waste generated by Comunitatea Urbană Arieş, in 2009, was 45.417 tons distributed as follows:

- In the urban environment there have been generated 27.170 tons;
- In the rural areas there have been generated 5.127 tons;
- The amount generated by businesses was 13.127 tons. Based on these data from local authorities and taking into account the waste generation index for urban and rural areas (according to County Plan for Waste Management), a forecast has been done for the quantities of household waste generated in this area. Generation indicators are calculated both for municipal waste and for household waste, based on the generated quantity and on the number of persons served. Projected quantities of recoverable materials (paper, glass, metals, plastics) from individuals and businesses in the Comunitatea Urbană Arieş were determined by multiplying the amounts recovered by the number of individuals and businesses served.

The quantity of household waste treated in the treatment station is calculated as the difference between household waste generated and waste recovered. Biodegradable waste occupies a large proportion of the total amount of household waste generated by the studied population, namely:

- 61% biodegradable waste is generated in urban areas;
- 55% biodegradable waste is generated in rural areas.

This waste fraction will be treated in the sort-treatment station, the compost beeing properly capitalized. Final waste is waste that remains after the fraction of biodegradable waste was recovered from household waste for treatment. This fraction will be stored safely on the green ramp Feleacu.

The water content of biodegradable waste to be processed is considered 30%. A 70% of dry matter is recovered as compost used for agricultural areas and the remaining 30% is considered final residue being deposited.

Next, we consider as useful the forecast of the total amount of household waste deposited at the landfill, by aggregating the quantities of waste that remains after treatment (waste treatment) and the amount of residue resulting from the composting process.

Analysis horizon of the proposed project for Comunitatea Urbană Arieş is 21 years, the recommended duration for waste management projects being between 20 to 30 years. The analysis horizon of the project (economic lifetime) is the time for which it is expected to obtain the project's benefits.

Waste incineration involves burning waste in special facilities called incinerators, which ensure high combustion temperatures that determine the neutralization of waste using containment and gas purification equipments.

The majority of modern solid waste incinerators produce less particulate and gaseous pollutants than their predecessors, which had few environmental controls on air emissions by regulatory bodies world-wide, multi stage pollution control systems are becoming more common. The operation of the combustion process plays an important role in the formation of some pollutants. Carbon monoxide, nitrogen oxides, hydrocarbons and other volatile organic compound emissions can be minimised by optimising the combustion process. The combustion of waste can produce trace quanties of dioxins and furans.(McDougall et al., 2001)

Reintroducing the useful materials contained in household waste in the manufacturing process yields clear advantages. Recovered materials have been previously purified and processed so that their use in manufacturing activity involves a cleaner environment and less energy consumption.

As a result of waste incineration, only 20% of the input incineration waste will be landfilled, in the form of unrecoverable slag and ash, this deposited refuse requiring storage expenses provided in the following table. The following expenses are required for processing the waste in the incineration plant:

- Expenditure on fuel for the flame holder: 60 euros per ton of waste burned;
- Environmental monitoring expenses: approx. 75.000 euros annually;
- Maintenance and operating costs (salaries, overhead, materials, maintenance, repairs)

Starting with 2013, vehicles and equipment maintenance is estimated at 168.000 euro / year due to higher degree of wear, while in the first four years it is estimated that the maintenance will be about half, i.e. 84.000 euro.

Implementing this project will create 67 jobs on indefinite term (stations heads, electromechanical maintenance workers, administrative staff, drivers, etc..) and expected salary level for the first year of operation is 501.840 euro.

- Used waters purification expenditure: about 1,2 euros per ton of processed waste;
- Gas cleaning costs: replacing filter cartridges, about 13 euros per ton of processed waste;
- Expenditure with environmental taxes: about 5 euros per ton of processed waste (environment fund contributions, local eco-taxes, stock registration system of green certificates for CO2 emissions, eco-audit costs etc.).

Depreciation expenses (investment value/number of years of operation): 247.098 euro.

Year	Generated household waste	Recovered waste	Household waste for treating	Biodegradable waste	Eliminated waste	Water from biodegradable waste	Dry substances from biodeg. waste	Compost	Residue from composting	Total landfilled waste
2009	45.417,46	13.754,93	31.662,53	18.364,27	13.298,26	5.509,28	12.854,99	8.998,49	3.856,50	17.154,76
2010	45.811,92	14.399,35	31.412,58	18.219,29	13.193,28	5.465,79	12.753,51	8.927,45	3.826,05	17.019,33
2011	46.209,83	15.073,95	31.135,87	18.058,80	13.077,07	5.417,64	12.641,16	8.848,81	3.792,35	16.869,41
2012	46.611,19	15.780,17	30.831,02	17.881,99	12.949,03	5.364,60	12.517,40	8.762,18	3.755,22	16.704,25
2013	47.016,05	16.519,47	30.496,58	17.688,02	12.808,57	5.306,41	12.381,61	8.667,13	3.714,48	16.523,05
2014	47.424,44	17.293,41	30.131,03	17.476,00	12.655,03	5.242,80	12.233,20	8.563,24	3.669,96	16.324,99
2015	47.836,39	18.103,60	29.732,78	17.245,01	12.487,77	5.173,50	12.071,51	8.450,06	3.621,45	16.109,22
2016	48.251,92	18.951,76	29.300,16	16.994,09	12.306,07	5.098,23	11.895,86	8.327,11	3.568,76	15.874,83
2017	48.671,07	19.839,65	28.831,42	16.722,22	12.109,20	5.016,67	11.705,56	8.193,89	3.511,67	15.620,86
2018	49.093,87	20.769,14	28.324,73	16.428,34	11.896,39	4.928,50	11.499,84	8.049,89	3.449,95	15.346,34
2019	49.520,35	21.742,17	27.778,18	16.111,34	11.666,84	4.833,40	11.277,94	7.894,56	3.383,38	15.050,22
2020	49.950,55	22.760,79	27.189,76	15.770,06	11.419,70	4.731,02	11.039,04	7.727,33	3.311,71	14.731,41
2021	50.384,49	23.827,13	26.557,36	15.403,27	11.154,09	4.620,98	10.782,29	7.547,60	3.234,69	14.388,78
2022	50.822,21	24.943,43	25.878,78	15.009,69	10.869,09	4.502,91	10.506,78	7.354,75	3.152,04	14.021,12
2023	51.263,75	26.112,03	25.151,71	14.587,99	10.563,72	4.376,40	10.211,60	7.148,12	3.063,48	13.627,20
2024	51.709,13	27.335,38	24.373,74	14.136,77	10.236,97	4.241,03	9.895,74	6.927,02	2.968,72	13.205,69
2025	52.158,39	28.616,05	23.542,34	13.654,56	9.887,78	4.096,37	9.558,19	6.690,73	2.867,46	12.755,24
2026	52.611,56	29.956,71	22.654,85	13.139,81	9.515,04	3.941,94	9.197,87	6.438,51	2.759,36	12.274,40
2027	53.068,68	31.360,18	21.708,50	12.590,93	9.117,57	3.777,28	8.813,65	6.169,56	2.644,10	11.761,67
2028	53.529,78	32.829,40	20.700,38	12.006,22	8.694,16	3.601,87	8.404,35	5.883,05	2.521,31	11.215,46
2029	53.994,90	34.367,46	19.627,44	11.383,91	8.243,52	3.415,17	7.968,74	5.578,12	2.390,62	10.634,15

Table 2. **Prediction of the waste quatities (ton)**

Year	Landfilling costs	Overhead costs	Fuel costs, vehicle/ equipments maintenance	Depreciation expenditure	Salaries	Flame holder fuel expenditure	Used waters/ gas purification	Environment taxes/ Environment monitoring	TOTAL EXPENSES
2009	140.669,01	52.200,00	212.128,00	677.387,00	501.840,00	1.029.285,45	243.597,56	160.773,79	3.017.940,81
2010	149.770,14	54.196,80	215.450,74	677.387,00	547.005,60	1.021.160,06	241.674,55	159.310,96	3.065.955,84
2011	158.572,50	54.874,14	218.232,27	677.387,00	596.236,10	1.012.164,87	239.545,69	158.810,97	3.115.073,92
2012	167.042,48	55.590,85	220.743,99	677.387,00	649.897,35	1.002.254,88	237.200,32	158.525,26	3.167.066,69
2013	175.144,32	56.310,62	334.521,39	677.387,00	708.388,12	991.382,95	234.627,30	158.310,99	3.333.591,22
2014	182.839,94	57.021,39	336.695,38	338.694,00	772.143,05	979.499,66	231.814,92	158.168,14	3.053.404,73
2015	190.088,81	57.642,62	338.629,29	338.694,00	841.635,92	966.553,28	228.750,94	157.953,86	3.115.398,10
2016	196.847,85	58.318,27	340.601,87	338.694,00	917.383,15	952.489,58	225.422,53	157.953,87	3.181.988,53
2017	203.071,22	59.068,63	342.613,91	338.694,00	999.947,64	937.251,78	221.816,26	157.953,88	3.253.424,90
2018	208.710,22	59.834,01	344.666,19	338.694,00	1.089.942,92	920.780,39	217.918,03	157.953,89	3.330.134,60
2019	213.713,10	60.614,69	346.759,51	338.694,00	1.188.037,79	903.013,08	213.713,10	157.953,90	3.412.653,49
2020	218.024,86	61.410,98	348.894,70	338.694,00	1.294.961,19	883.884,59	209.186,02	157.953,91	3.501.570,53
2021	221.587,14	62.223,20	351.072,60	338.694,00	1.411.507,69	863.326,53	204.320,61	157.953,92	3.597.532,80
2022	224.337,94	63.051,66	353.294,05	338.694,00	1.538.543,39	841.267,29	199.099,93	157.953,93	3.701.251,01
2023	226.211,48	63.896,70	355.559,93	338.694,00	1.677.012,29	817.631,86	193.506,21	157.953,94	3.813.505,59
2024	227.137,94	64.758,63	357.871,13	338.694,00	1.827.943,40	792.341,66	187.520,86	157.953,95	3.935.153,24
2025	227.043,27	65.637,80	360.228,55	338.694,00	1.992.458,30	765.314,41	181.124,41	157.953,96	4.067.134,10
2026	225.848,93	66.534,56	362.633,12	338.694,00	2.171.779,55	736.463,92	174.296,46	157.953,97	4.210.479,68
2027	223.471,64	67.449,25	365.085,78	338.694,00	2.367.239,71	705.699,91	167.015,65	157.953,98	4.366.321,41
2028	219.823,10	68.382,24	367.587,50	338.694,00	2.580.291,28	672.927,86	159.259,59	157.953,99	4.535.900,03
2029	214.809,74	69.333,88	370.139,25	338.694,00	2.812.517,50	638.048,74	151.004,87	157.954,00	4.720.575,85

Table 3. Prediction of the total expenditure of the operation of the proposed project (Euro)

The ascertainment of the depreciation was achieved by applying depreciation rates on the input value of assets. It is included in operating expenses. Depreciation for buildings / facilities, equipment and machinery was calculated by applying depreciation rates on the value of the depreciable asset. The depreciation rates is given by:

$$C\,a(\%) = \frac{100}{t} \tag{1}$$

where:

Ca - depreciation rate, expressed as a percentage;
t - the normal duration of the asset, expressed in years.

The results for depreciation rates for construction / facilities, equipment and machinery are: 11%, 20% and 5%.

Storage fee costs were obtained by multiplying the total quantity of stored waste by the presumed value of storage fee. In order to safely store the final waste, a storage fee is considered that is applied to each ton of deposited waste. This fee is expected to grow in order to determine the local authorities to take measures to reduce the quantities sent to storage thus encouraging their sale.

This waste management plan proposed for Comunitatea Urbană Arieş involves, in addition to costs resulting from the incineration explained above, also the operating expenses for the first waste management scheme in the area (proposed scheme from volume no. 27 E/October/2009 of the Transylvanian Review of Administrative Sciences journal).

In the integrated concepts of waste management, the incineration process can be an important component.

The reintroduction of separately collected organic waste fractions in the flow of substances, we can obtain an apparent easing of the waste deposits from the researched villages and towns, and the recoverable heat fraction is optimized in terms of its composition and its thermal power.

The purpose of incineration is:

- minimizing waste quantities;
- destroying dangerous biodegradable components;
- transforming waste in inert (inactive) material;
- recovery of energy contained in the incinerated waste.

All waste incinerators must meet the objectives of European and national legislation. In parallel, incinerators must meet the conditions regarding energy recovery from waste, i.e. heat recovery and other forms of energy derived from waste incineration. Incinerators should be operated so that the final ashes can be sorted to recover recyclable metals and to be used in construction, with no impact on soil and groundwater.

Every civilized country must implement and use the waste recovery infrastructure to protect the environment and life itself. Environmental benefits of recycling are much more effective than any other action to protect the environment. Recycling conserves energy. Much less energy is needed to transform recycled materials into new products, compared with the production of raw materials.

Following the implementation of such a project it is very important to consider the social and environmental benefits:

- the improvement of water, soil and air quality in the area where a waste management project has been implemented;
- increase of life expectancy due to limitation of pollution;
- local development generated by the project (increasing land prices, development of tourism and investor-friendliness, new employment opportunities etc.);
- awareness-raising among the general public as to the resources contained in household waste.

Year	Income from electricity sale	Income from green certificate stock exchange	Income from by-products recovery	Total incineration income	Generated energy (MWh)	Number of green certificates
2009	205.857,09	1.157.946,14	85.773,79	1.449.577.01	8.577,38	25.732,14
2010	187.212,68	1.148.805,06	85.096,67	1.421.114,41	8.509,67	25.529,00
2011	185.563,56	1.138.685,48	84.347,07	1.408.596,11	8.434,71	25.304,12
2012	183.746,73	1.127.536,74	83.521,24	1.394.804,71	8.352,12	25.056,37
2013	181.753,54	1.115.305,81	82.615,25	1.379.674,60	8.261,52	24.784,57
2014	179.574,94	1.101.937,12	81.624,97	1.363.137,03	8.162,50	24.487,49
2015	177.201,43	1.087.372,44	80.546,11	1.345.119,98	8.054,61	24.163,83
2016	174.623,09	1.071.550,78	79.374,13	1.325.548,00	7.937,41	23.812,24
2017	171.829,49	1.054.408,26	78.104,32	1.304.342,06	7.810,43	23.431,29
2018	168.809,74	1.035.877,94	76.731,70	1.281.419,37	7.673,17	23.019,51
2019	165.552,40	1.015.889,72	75.251,09	1.256.693,21	7.525,11	22.575,33
2020	162.045,51	994.370,16	73.657,05	1.230.072,72	7.365,70	22.097,11
2021	158.276,53	971.242,35	71.943,88	1.201.462,75	7.194,39	21.583,16
2022	154.232,34	946.425,70	70.105,61	1.170.763,65	7.010,56	21.031,68
2023	149.899,17	919.835,84	68.135,99	1.137.871,00	6.813,60	20.440,80
2024	145.262,64	891.384,37	66.028,47	1.102.675,48	6.602,85	19.808,54
2025	140.307,64	860.978,71	63.776,20	1.065.062,55	6.377,62	19.132,86
2026	135.018,38	828.521,90	61.371,99	1.024.912,28	6.137,20	18.411,60
2027	129.378,32	793.912,40	58.808,33	982.099,04	5.880,83	17.642,50
2028	123.370,11	757.043,84	56.077,32	936.491,27	5.607,73	16.823,20
2029	116.975,60	717.804,83	53.170,73	887.951,16	5.317,07	15.951,22

Table 4. Prediction of the total income obtained from the waste incineration process (Euro)

To determine the total revenue resulting from the operation of this waste management scheme, with the revenues from exploitation of recoverable materials (paper/cardboard, glass, plastic, metal and organic waste) and the income from the sanitation charges, revenues from the incineration process will be taken into account, as follows:

- income from sale of electricity: about 12 euros per ton of waste processed; one tonne of waste processed = 0,5 MW recovered energy;
- income from the green certificate stock of exchange: cca 45 Euro per green certificate; each MWh of recovered energy yields 2 green certificates;
- income resulting the recovery of by-products resulting from incineration: about 5 euros per ton of processed waste (mineral acids, furnace slag used in construction materials industry, etc.).

Year	Income from materials recovery	Population service charge	Companies service charge	Incineration income	TOTAL INCOME
2009	964.547,68	1.413.360,00	234.410,71	1.449.577,01	4.061.895,41
2010	1.002.840,13	1.440.120,56	239.567,75	1.421.114,41	4.103.642,85
2011	1.042.866,71	1.467.387,80	244.838,24	1.408.596,11	4.163.688,86
2012	1.084.708,15	1.495.171,32	250.224,68	1.394.804,71	4.224.908,86
2013	1.128.448,95	1.523.480,90	255.729,62	1.379.674,60	4.287.334,07
2014	1.174.177,57	1.552.326,48	261.355,68	1.363.137,03	4.350.996,75
2015	1.221.986,59	1.581.718,23	267.105,50	1.345.119,98	4.415.930,30
2016	1.271.972,95	1.611.666,49	272.981,82	1.325.548,00	4.482.169,26
2017	1.324.238,11	1.642.181,78	278.987,42	1.304.342,06	4.549.749,38
2018	1.378.888,29	1.673.274,85	285.125,15	1.281.419,37	4.618.707,66
2019	1.436.034,66	1.704.956,63	291.397,90	1.256.693,21	4.689.082,40
2020	1.495.793,62	1.737.238,28	297.808,65	1.230.072,72	4.760.913,28
2021	1.558.287,01	1.770.131,15	304.360,44	1.201.462,75	4.834.241,36
2022	1.623.642,35	1.803.646,82	311.056,37	1.170.763,65	4.909.109,19
2023	1.691.993,17	1.837.797,06	317.899,61	1.137.871,00	4.985.560,85
2024	1.763.479,22	1.872.593,91	324.893,40	1.102.675,48	5.063.642,02
2025	1.838.246,80	1.908.049,61	332.041,06	1.065.062,55	5.143.400,02
2026	1.916.449,07	1.944.176,62	339.345,96	1.024.912,28	5.224.883,94
2027	1.998.246,35	1.980.987,66	346.811,57	982.099,04	5.308.144,62
2028	2.083.806,46	2.018.495,68	354.441,43	936.491,27	5.393.234,83
2029	2.173.305,09	2.056.713,88	362.239,14	887.951,16	5.480.209,27

Table 5. Prediction of the total income obtained from the waste management scheme proposed for CUA (Euro)

For the incineration process to be economically efficient and to qualify for additional green certificates from non-conventional energy recovery, it's necessary to combine at least two forms of energy recovery, which usually means combining electricity generation from a steam turbine coupled with at least a household heating system or providing technological steam. Electricity is traded separately from green certificates.

4. Research results

The analysis of the efficiency for an investment project is performed using an instrument of investigation which is represented by the investment efficiency indicators. Using these indicators for measuring the efficiency of investments focuses on a very important aspect: the investments' efficiency can be expressed using a single indicator or multiple indicators. Both in our country's economic practice and the experience of other countries show the usage of a wider or a smaller range of indicators, but never a single indicator. (Binţinţan, 2005)

The time factor has a major influence on the project's performance. The operation through which a certain amount, regardless of its nature, be it income or expenditure, shall be recalculated in monetary units of a reference moment in time, is called an update.

In fact it is a recalculation of that amount and not a simple translation of it from year to year. The update's base is the annual rate of fruition of a monetary unit, which is symbolized by "a" and it is called the discount rate.

This scheme's efficiency was proven by using two efficiency determination criteria, which are very often used in the area:

a. Dynamic criteria or basic on updating, in which we determined the following investment efficiency indicators: (Binţinţan, 2004)
   - the benefit-cost ratio BCR:
     By its content, the benefit-costs ratio indicates the gain, the return or the reward for invested capital, in the form of an absolute value. This indicator summarizes the investment efforts and effects for the entire lifetime of the project; it reflects the difference between the total updated income and the total updated expenditures (including investment costs).
   - the net present value NPV:
     This indicator also refers to the effort and the effects of the investment project for the whole duration of the project; it reflects the difference between the present value of benefits and the present value of costs (capital expenditure and operating costs).
   - the internal rate of return IRR represents the fundamental indicator for accepting a project. It refers to that updating rate for which the present value of the benefits equals the present value of the costs, so the benefit-cost ratio equals 1 and NPV equals 0. The calculation of the internal rate of return IRR is carried out in a succession of approximations, i.e. the net present value is determined for an appropriate discount rate, which is considered to be the minimum rate, for which the NPV must be positive. Then, the NPV is calculated for a rate superior to the minimum rate that is high enough to obtain a negative NPV.

These indicators have been explained in detail in the first project proposed for the investigated area, presented in volume no. 27E/Octombrie/2009. (Scorțar et al., 2009)

b. Double rate criteria or integrated criteria: wherein the following efficiency indicators have been determined: (Todea, 2008)
   - the integrated internal rate of return (IIRR)
   - the integrated net present value (INPV).

The integrated internal rate of return (IIRR) involves the reinvestment of net treasury flows at a reinvestment rate different from the internal profitability index.

Compared to the internal rate of return (IRR), a first advantage of this indicator would be the fact that it allows for a more accurate comparison of two competing investment project against a unique net treasury flows reinvestment ratio.

A second advantage consists in the fact that that IIRR enables comparisons between investment projects of different durations (lifetimes).

$$IIRR = \sqrt[n_{max}]{\frac{\left[\sum_{t=1}^{n} F_t(1+R)^{n-t}\right](1+RR)^{n_{max}-n}}{I}} - 1 \tag{2}$$

where:

I - capital expenditures updated on a specific updating rate;
n - the project duration for which IRR is determined;
F_t - the net flow of treasury for the year t, that may be positive or negative;
RR - net treasury flows reinvestment ratio;
R - updating rate;
n_{max} - the longest lifetime of the projects to be compared
IIRR = 8,03%

The computed value of the integrated internal rate of return justifies a capital investment in this waste management scheme.

This value has been achieved by reinvesting the treasury flows at a 10% reinvestment rate that exceeds the 5% updating rate.

The implicit hypothesis of net treasury flows reinvestment during the implementation period of the intended investment target at a higher reinvestment rate that the updating rate R will be highlighted by starting form the classical net present value (NPV) calculation formula that generates a general integrated net present value (INPV) calculation relation:

$$INPV = \frac{\left[\sum_{t=1}^{n} F_t(1+R)^{n-t}\right](1+RR)^{n_{max}-n}}{(1+k)^{n_{max}}} - I \tag{3}$$

where:

I - updated amount based on a k updating rate of the capital expenditures sustained during the entire project lifetime;
n - the project duration for which IRR is determined;
F_t - the net flow of treasury for the year t, that may be positive or negative;
RR - net treasury flows reinvestment rate;
R - rate that may be a reinvestment rate if the net treasury flows are positive or negative;
k - financing costs, if the net treasury flows are negative;
n_{max} - the longest lifetime of the projects to be compared.
INPV = 12.998.737 euro

This amount represents the income gathered by the end of the 21 years - lifetime of the proposed waste management scheme that is achieved by reinvesting the treasury flows at a 10% - reinvestment rate that exceeds the 5% - capital expenditure updating rate.

The income realized at the end of the 21 years - period is significant, considering the investment start-up expenditures amounting to a total of 14.629.636 Euro, demonstrating the viability of the proposed project.

Year	Up-dating factor (5%)	Total income	Total updated income	Total operation expenditure	Updated operation expenditure	Invest-ment value	Updated investment value
2008	0,952	-	-	-	-	8.000.000	7.619.047,6
2009	0,907	4.061.895,4	3.684.258,88	3.017.940,81	2.737.361,2	6.629.636	6.013.275,3
2010	0,864	4.103.642,8	3.544.880,98	3.065.955,84	2.648.487,9	-	-
2011	0,823	4.163.688,8	3.425.477,13	3.115.073,92	2.562.779,0	-	-
2012	0,784	4.224.908,8	3.310.326,65	3.167.066,69	2.481.479,6	-	-
2013	0,746	4.287.334,0	3.199.274,69	3.333.591,22	2.487.577,0	-	-
2014	0,711	4.350.996,7	3.092.172,16	3.053.404,73	2.169.997,7	-	-
2015	0,677	4.415.930,3	2.988.875,45	3.115.398,10	2.108.624,0	-	-
2016	0,645	4.482.169,2	2.889.246,27	3.181.988,53	2.051.138,1	-	-
2017	0,614	4.549.749,3	2.793.151,44	3.253.424,90	1.997.320,6	-	-
2018	0,585	4.618.707,6	2.700.462,71	3.330.134,60	1.947.060,7	-	-
2019	0,557	4.689.082,4	2.611.056,54	3.412.653,49	1.900.293,1	-	-
2020	0,530	4.760.913,2	2.524.813,96	3.501.570,53	1.856.957,6	-	-
2021	0,505	4.834.241,3	2.441.620,39	3.597.532,80	1.816.998,5	-	-
2022	0,481	4.909.109,1	2.361.365,46	3.701.251,01	1.780.365,0	-	-
2023	0,458	4.985.560,8	2.283.942,87	3.813.505,59	1.747.010,8	-	-
2024	0,436	5.063.642,0	2.209.250,24	3.935.153,24	1.716.894,3	-	-
2025	0,416	5.143.400,0	2.137.188,95	4.067.134,10	1.689.978,2	-	-
2026	0,396	5.224.883,9	2.067.663,99	4.210.479,68	1.666.229,7	-	-
2027	0,377	5.308.144,6	2.000.583,88	4.366.321,41	1.645.620,6	-	-
2028	0,359	5.393.234,8	1.935.860,46	4.535.900,03	1.628.126,6	-	-
2029	0,342	5.480.209,2	1.873.408,83	4.720.575,85	1.613.728,2	-	-
Total			**56.074.881,9**		**42.254.029,3**		**13.632.322,9**
Updating factor: 0,05					NPV (euro):	**188.529,65**	

Table 6. The structure of the proposed investment project (euro)

In the presented scheme, the year 2008 was considered the start of the construction for the investment objective, the capital investment is phased as follows: in the first, when the construction started, the initial investment was 8 million euro, and in the next year the difference of 6.629.636 euro was invested. This project was designed to come into operation in 2009, bringing income as a result of operating the facility. As shown in Table 6, total income and total expenditure have been updated to a 5% updating rate (the rate recommended for environmental projects) and were thus obtained updating income and updating expenditure, required to determine the efficiency indicators of this project.

In the waste management scheme proposed for Comunitatea Urbană Arieş we took into account the pre-existing endowments of the local sanitation systems and this is why we also considered the fact that part of the population living in private houses and part of the economic agents already own adequate containers for waste collection.

In this respect, we included investment cots just for the difference in the containers inventory, costs which would be directly recovered from the direct beneficiaries based on a conscription tariff.

Year	Invest-ment value (in euro)	Operation expenditures	Annual income (euro)	Cashflow (euro)	Updating factor (a = 5%)	Updated cash flow
2008	8.000.000	0,00	0,00	-8.000.000,00	0,951	-7.607.006,81
2009	6.629.636	3.017.940,81	4.061.895,41	-5.585.681,40	0,904	-5.050.376,98
2010	0,00	3.065.955,84	4.103.642,85	1.037.687,01	0,860	892.149,91
2011	0,00	3.115.073,92	4.163.688,86	1.048.614,94	0,818	857.257,54
2012	0,00	3.167.066,69	4.224.908,86	1.057.842,17	0,777	822.318,33
2013	0,00	3.333.591,22	4.287.334,07	953.742,85	0,739	704.975,79
2014	0,00	3.053.404,73	4.350.996,75	1.297.592,02	0,703	912.021,12
2015	0,00	3.115.398,10	4.415.930,30	1.300.532,20	0,668	869.183,86
2016	0,00	3.181.988,53	4.482.169,26	1.300.180,73	0,635	826.262,59
2017	0,00	3.253.424,90	4.549.749,38	1.296.324,48	0,604	783.342,89
2018	0,00	3.330.134,60	4.618.707,66	1.288.573,06	0,575	740.407,91
2019	0,00	3.412.653,49	4.689.082,40	1.276.428,91	0,546	697.400,82
2020	0,00	3.501.570,53	4.760.913,28	1.259.342,74	0,520	654.264,84
2021	0,00	3.597.532,80	4.834.241,36	1.236.708,56	0,494	610.943,18
2022	0,00	3.701.251,01	4.909.109,19	1.207.858,18	0,470	567.378,95
2023	0,00	3.813.505,59	4.985.560,85	1.172.055,26	0,447	523.515,06
2024	0,00	3.935.153,24	5.063.642,02	1.128.488,78	0,425	479.294,18
2025	0,00	4.067.134,10	5.143.400,02	1.076.265,93	0,404	434.658,64
2026	0,00	4.210.479,68	5.224.883,94	1.014.404,26	0,384	389.550,36
2027	0,00	4.366.321,41	5.308.144,62	941.823,22	0,365	343.910,75
2028	0,00	4.535.900,03	5.393.234,83	857.334,80	0,347	297.680,67
2029	0,00	4.720.575,85	5.480.209,27	759.633,42	0,330	250.800,31
Total						-66

Table 7. Determination of the internal rate of return (euro)

The internal rate of return IRR calculation is done through successive approximations, where I determine the net present value at a corresponding updating rate, considered to be the minimum rate and for which it should be positive. Then, I calculate the net present value at a higher updating rate than the minimum rate, high enough to obtain a negative net present value. In the end, the exact determination of the internal financial rate of return is done through interpolation, according to the relation:

$$IRR = R_{min} + (R_{max} - R_{min}) \times \frac{NPV(+)}{NPV(+) + |NPV(-)|} \quad (4)$$

where:

IRR - the internal financial rate of return;
R_{min}- the minimum present rate;
R_{max} - the maximum present rate;
NPV (+) - the positive net present value, obtained at the minimum rate;
NPV (-) - the negative net updated income, obtained at maximum rate.
IRR = 5,17%

The internal rate of return is higher than the minimum rate (5%), which proves the viability of the proposed project.

Graphically, the internal financial rate of return is at the intersection of the abscissa with the line that joins the points of the coordinates minimum rate and positive net present value, respectively maximum rate and negative net present value.

The higher the internal rate of return, the more viable the project is, its efficiency being increased. The efficiency condition of a project from the point of view of this indicator is that IRR should be superior to the corresponding updating rate.

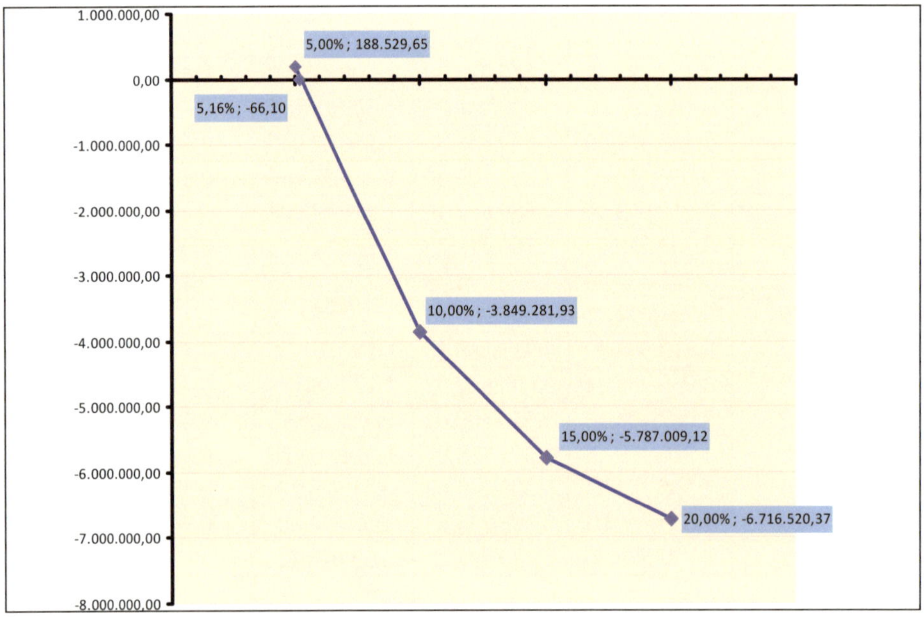

Fig. 1. Graphical illustration of the internal rate of return

The graph illustrates the net present value, determined upon several updating rates (5%; 5,16%; 10%; 15%; 20%) by means of Microsoft Excel.

The net present values calculated on updating rates smaller than 5,15%; the profitability of this project is therefore close to the minimum (5% is considered to be the lowest rate of profitability for the proposed project).

By applying the same calculation method of the financial efficiency indicators determined for the first waste management scheme, the results are:

NPV = 188.529,65 euro
BCR = 1,01
IRR = 5,17%
IIRR = 8,03%
INPV = 12.998.737 euro

By applying the Microsoft Excel calculation software, the resulting integrated internal rate of return for the proposed project is 8,03% in case of financial analysis, which justify the profitability of the project for the area under scrutiny.

The feasibility of the project also results from the determined updated integrated net present values (12.998.737 euro).

The analysis of determined results demonstrates the overall feasibility of the project, though not without a few reservations:

- the initial project start-up investment is extremely high (14.629.636 euro), but the income amounts cover the expenditures required to keep the scheme in operation;
- from the point of view of the dynamic criteria used in financial analysis (net present value, benefit-cost ratio, internal rate of return), this project is very close to the bottom line of profitability, with the values of these indicators being close to 5%;
- the income gained after 21 years of operation, as determined by financial analysis, is not significant (188.529,65 euro), i.e. a financial profitability of 5,17%;
- on the other hand, the values calculated for double rate indicators (integrated internal rate of return, integrated net present value), that are considered to be relevant in the economic practice, prove the profitability of this project against a unique net treasury flow reinvestment ratio (10%).

In order to finish the financial efficiency of the project proposed to Comunitatea Urbană Arieş Cluj county a detailed analysis of internal rate of return sensitivity, by studying the financial indicators' trends according to the evolution of the variables with the greatest impact on the project's profitability.

The sensitivity of this project to the changing of the same relevant variables considered for the first project has been determined by performing the project's internal rate of return sensitivity analysis. To this purpose, the relevant variables have been modified within +/-20% and the internal rate of return has been recalculated under the new circumstances. (Bințințan, 2004).

The selected variables are those whose effects on economic feasibility and financial durability are considered significant.

Critical parameters: **SERVICE PERFORMANCE FEE**

5,17%	0,80	0,85	0,90	0,95	1,00	1,05	1,10	1,15	1,20
	-2,04%	0,62%	2,5%	4,02%	**5,17%**	6,09%	6,86%	7,50%	8,04%

Critical parameters: **SALE PRICE OF RECYCLABLE MATERIALS**

5,17%	0,80	0,85	0,90	0,95	1,00	1,05	1,10	1,15	1,20
	-7,9%	-2,2%	1,24%	3,53%	**5,17%**	6,39%	7,35%	8,11%	8,73%

Critical parameters: **OPERATIONAL EXPENDITURES**

5,17%	0,80	0,85	0,90	0,95	1,00	1,05	1,10	1,15	1,20
	9,3%	8,65%	7,81%	6,7%	**5,17%**	2,9%	-0,78%	-7,81%	-8,5%

Critical parameters: **COMBINED IMPACT (OF THE THREE CRITICAL PARAMETERS)**

5,17%	0,80	0,85	0,90	0,95	1,00	1,05	1,10	1,15	1,20
	2,04%	2,99%	3,81%	4,53%	**5,17%**	5,73%	6,24%	6,70%	7,11%

Table 8. The impact of critical parameters on the internal rate of return

The variables with the greatest impact on the feasibility on the project's feasibility are: (Scorțar, 2010)

a. The variation of sanitation fees for waste materials;
b. The sale price variation of compost and recyclable materials sold to recycling facilities;
c. The variation of operation costs.

The change of the variables within +/- 20% has yielded the following results:

- the change triggers the increase of both incomes collected from sanitation services, as well as incomes collected from recovered recyclable materials, which results in the overall increase of the project's profitability;
- the change of service fees and sale price of recoverable materials within (-20%, - 5%) renders the project inefficient;
- the increase of project-related operational expenditures causes the internal project profitability to decrease as a result of the increasing total expenditures required by the operation of the proposed scheme; a change within (+5%, +20%) renders the project inefficient;
- the combined impact of the three critical analyzed parameters on the project's internal rate of return confirms the fact that the project is sensitive to the change of the variables within (-20%, -5%), which puts the feasibility of the project at risk.

5. Conclusions

Due to the particular features of the analyzed area (distances, population density and behavioral habits of the population) and to the fact that most of the waste materials are biodegradable and processed in the composting facility, this project does not involve a sufficient amount of incinerated waste to make it preferable over the first scheme proposed in volume no. 27 E/October/2009; as the project proposed in this paper has a 8,03% profitability.

The sensitivity of this project has been proven by means of sensitivity analysis, showing that a change of several critical parameters within a certain range makes the efficiency of the project vulnerable.

The benefits that the scheme with incineration may yield are in this case overweighted by both investment expenditures almost twice as high as those required by the scheme without incineration, as well as higher operational expenditures required for providing a temperature that enables complete cremation.

The scheme that includes incineration may have a higher profitability if the service fees would increase, thus quantifying and reflecting the financing effort and also the benefits related to the longer operation of the final deposit.

Furthermore, the analysis has revealed that the efficiency of the project proposed for this area is inconceivable outside a functional recovery system for useful materials extracted from household waste (paper/cardboard, glass, metal, plastics, including organic waste and incineration products).

It is our conviction that, in addition to the responsibilities of the public authorities related to the implementation of the integrated waste management system, the general population also bears the responsibility for the success or failure of such projects. The encouragement of cooperation and communication between citizens and authorities is a crucial part of an effective sustainable waste management program.

In this context, public authorities are responsible for organizing the collection, transport and waste disposal, they are obliged to take measures such as:

- provision of waste collection system, transport, neutralization, recovery, incineration and final disposal;
- waste-collection points with a sufficient number of containers for selective collection;
- Separate collection and timely transportation of all quantities of waste produced in localities;
- Arranging for final storage of waste (landfill), appropriately sized and equipped to protect the health and the environment;
- Prohibition of waste disposal in places other than those for fixed deposits as specified by local authorities;
- Development of guidelines for businesses, institutions and individuals on how to organize waste management;
- Association with other local authorities and contracted cooperation with businesses, in order to achieve public waste management projects;
- Organization of actions on environmental restoration.

In Romania, the current research focuses on waste management basis for strategies, action plans, strategic planning of urban waste management systems. Romanian research in the field of waste makes progress gradually, supporting those actions of the authorities to adopt waste management systems applied in the EU.

Thus, the Ministry of Environment operates two national research institutes in Bucharest (Institutul Naţional de Cercetare-Dezvoltare pentru Protecţia Mediului - ICIM Bucuresti and Institutul Naţional de Cercetare-Dezvoltare pentru Ecologie Industrială - ECOIND) including laboratories and workshops specifically dedicated to research into waste management technologies.

The role of these research institutions is a scientific foundation based on national statistics and national reference databases, adopted local government strategies, and monitoring the implementation of commitments assumed with the signing of the Accession Treaty of Romania to the European Union (through monitoring plans).

But local research themes have been launched for other market players as well, which are quite numerous; these are represented by small business of consultancy and research, by academic structures, local institutions etc. These research topics' role was to either base regulatory decisions for local waste management facilities (environmental impact studies) and local strategies for sustainable development, or to show what happens after the closure of landfills (post-closure monitoring).

Developing this paper and the research results can be a starting point for the official structures of government, but also for some structures with an informational role (institutions, associations, EU bodies).

We believe that the first step to successful implementation of a sustainable waste management project is represented by a change in education, culture and mentality, so that waste should be viewed primarily as a resource and only ultimately as something to be removed.

6. Acknowledgment

I would like to express my gratitude to all those who gave me the possibility to complete this paper.

I would like to express my thanks to Comunitatea Urbană Arieş , Consiliul Judeţean Cluj, Agenţia pentru Protecţia Mediului Cluj, RADP Turda, S.C. SLCIAS S.A. Câmpia Turzii for allowing me to access their statistical data regarding their specific activities, programmes and actions taken during the household waste management activities.

7. References

Binţinţan, P. (2005). *Strategii şi politici de investiţii.* Casa Cărţii de Ştiinţă, ISBN 973-686-798-6, Cluj-Napoca, Romania.

Binţinţan, P. (2004). *Managementul investiţiilor. Aplicaţii. Evaluarea proiectelor.* Presa Universitară Clujeană, ISBN 973-610-334-X, Cluj-Napoca, Romania.

Comunitatea Urbană Arieş, Statutul Asociaţiei, Turda Town Hall, 2005

McDougall, F.; White, P.; Franke, M. & Hindle, P. (2001). *Integrated Solid Waste Management: a Life Cycle Inventory (2nd Edition).* Blackwell Publishing, ISBN 0-632-05889-7, Great Britain

Planul Judeţean privind Gestionarea Deşeurilor, Consiliul Judeţean Cluj, Available from http://www.cjcluj.ro/UserUploadedFiles/File/programe/pjgd%20Cluj%20alternativa%20finala.pdf

Scorţar, L. M.; Lazăr, I.; Popa, M.; Zagan Zelter, D. & Pop, I. (2009). The financial analysis of a modern scheme for managing waste proposed for the Urban Community Aries, Cluj County. *Transylvanian Review of Administrative Sciences,* 27E/2009, pp.220-234, ISSN 1842-2845

Scorţar, L., M. (2010). *Managementul deşeurilor menajere.* Editura Alma Mater, ISBN, Cluj-Napoca, Romania

Todea, A. (2008). *Investiţii.* Editura Casa Cărţii de Ştiinţă, Cluj-Napoca, Romania.

2

The Role of Informal Collectors of Recyclable Waste and Used Goods in Indonesia

Enri Damanhuri and Tri Padmi
Institut Teknologi Bandung (ITB)
Indonesia

1. Introduction

Municipal Solid Waste (MSW) is a problem in Indonesia, particularly in big cities, and is one of the most challenging urban issues for city administrators. Population growth and ever-increasing activities in major cities mean the increase of waste generation and all the inherent consequences. It was estimated that of the entire urban wastes generated in 2006, at most only around 60% to 70% could be transported to final disposal by the institutions responsible for handling wastes. The community were expected to handle the rest themselves.

According to the latest census in 2000, Indonesia's population was 205.8 million. An inter-censal population survey in 2007 revealed a total population of 224.9 million with a population density of 117.6 people per sq. km. Provinces in Java Island had a higher population density than provinces outside. The province with the highest population density was Jakarta (11,968.8 people per sq. km). Provinces in Papua had the lowest density of 5.9 people per sq. km. Although the absolute number of the population was increasing from census to census, its growth was decreasing, from 2.32% in the period from 1971 to 1980 to 1.49% in the period from 1990 to 2000 (Ministry of Health, 2007).

Until the early twentieth century, the urban population in Indonesia was well distributed. According to the 1920 census, 6.6% of Java's population lived in cities. This pattern changed when the urban population of Java started to grow more quickly than the total population of the other islands. It could be said that the larger the city, the quicker the growth. During the 1960s and 1970s inter-provincial migration also began to increase. The main targets of the movements were the Javanese towns, and in particular Jakarta, to which 40% moved immediately, whereas others reached the capital city only after first settling in smaller towns (BPS-Statistics Indonesia, 2004).

It is evident that none of the Indonesian cities was prepared properly to absorb the newly arrived rural migrants as far as employment and housing, including public utilities, were concerned. Once the natural absorption capacity was exhausted, the newcomers established themselves in slums located along railway tracks, waterways, etc. The primary motive behind the rural-urban migration was the lack of employment in the villages (push factor) and not so much the attraction of city life (pull factor) (Donner, 1987).

Though such migrants are not contributing more than their poverty to the cities, they feel that they can make a living in the towns rather than in their villages, although their quality of life will be worse. The number of formal working places is not growing at the same speed as the number of job seekers. In addition, most of them cannot offer any qualifications whatsoever. It is the informal sector that absorbs the bulk of unemployed migrants without capital or professional qualifications. These informal employments are normally not recorded and therefore never appear in any official statistics.

With the increasing growth of population and economic activities, the volumes of waste to be handled increase accordingly. Rapid population growth in urban areas, socio-cultural classes' heterogeneity and community participation that is generally not well directed and well organized have caused complex MSW problems for municipalities. On the other hand, the funding situation and relatively low priority in waste handling among local governments are general trends, along with the limitations in proper human resources, adding to the low performance of municipalities in handling the sanitation and waste in urban areas. Many factors are involved in inadequate MSW management, some of which are lack of support for municipalities to address waste problems systematically, integrally and comprehensively, lack of standard policies that are comprehensive and consistent in matters of waste handling, and lack of discipline among waste managers in terms of applying proper technical procedures (Damanhuri, 2005).

The problems of waste management in developing countries, such as Indonesia, have a number of aspects associated with them, such as technical, institutional, financial, environmental and social aspects. To overcome this problem effectively and efficiently a holistic approach to developing solutions is required. The impact of these aspects varies markedly depending upon the income levels and socio-economic factors of individual countries or cities. Higher per capita income levels in developed countries provide the financial means to maintain appropriate collection systems, treatment and disposal management. The generally higher education levels of the population in developed countries also provide support for the implementation of the mantra 'reduce, reuse and recycle' (3Rs) for waste programmes, public education and strict environmental regulations. Eventually it has become clear that sustainable improvement can be reached only by the integration of socio-economic and socio-cultural elements into the whole scheme. In Indonesia, proper waste management has been a major challenge, but concerns about gradual waste reduction through recycling have been raised in recent years. During the last few years, because the problems of solid waste disposal are all too obvious, public pressure and growing environmental awareness have also caused a change in the policy concerning waste management.

This report will discuss used goods and waste recycling activities, especially among informal sectors, emphasizing discussions on actors' roles and the linkage of one informal sector stakeholder with another in the performance of their activities as waste and used goods collectors, and how material flows and qualities are applied in economic transactions by informal sectors in Indonesia.

2. Municipal Solid Waste (MSW) issues

Indonesia is a country located in South East Asia, which comprises more than 13,000 large and small islands. Administratively, the country is divided into 33 provinces and more than

465 municipalities which consist of 14 metropolitans (one million population or more), 15 big cities (500,000 up to a million population), 56 medium-sized cities (100,000 up to 500,000 population), and 380 small cities (20,000 up to 100,000 population). Indonesia is located at the equator, and it has two seasons every year (dry and rainy seasons) and because of the monsoon rains each season lasts six months. The rainy season is also the fruit season and fruits produce an enormous amount of waste in the city.

Municipal solid waste (MSW) management in Indonesia is the responsibility of municipalities (local government). There is a city/district cleanliness division within the municipality organization. Some big cities contract out part of the services to third parties. In fact, most of the municipalities still give low priority to solid waste services.

The general method currently observed in waste management is collect-transport-dispose. The authorities in urban municipalities transport the waste from designated collection points to a location for its final dumping. Most of the local authorities practise crude open dumping, creating a desperate situation at the landfill sites. The potentials for reuse and recycling have not been fully realized because of a multitude of problems.

The principal source of MSW in Indonesia is households. They generate about 50 to 60 % (wet weight) of the total quantity of MSW per day. Some cities provide their generation data by conducting surveys and sampling but many other cities usually estimate their waste volume by using the estimated generation rate of 2.5-3.0 L/capita/day based on standard national MSW figures. According to a survey of the Bandung area in 2005, the estimated MSW generated in this area was 0.59 kg/capita-day (Damanhuri et al., 2009). In a questionnaire survey conducted in 2007, it was estimated that MSW generation of all municipalities in Indonesia in 2006 was 38.5 million tonnes, as presented in Table 1.

The amount of MSW is normally dominated by organic matter (more than 55% by weight) that mainly comes from food waste. This amount contributes to about 65% of the water content of MSW. This waste consists mainly of food scraps, yard waste, and wrapping materials. It is a mixture of all kinds of waste, organic and non-organic, recyclable and non-recyclable, even hazardous and non-hazardous materials. The other sources are traditional markets, commercial activities/areas, industries (non-hazardous categories), public gardens and streets. Plastic and paper are the two next commonest items. They mainly comprise packaging/wrapping materials and food, beverages, etc. Wood and textiles are the next two important components. According to a survey in 2007 of Bandung metropolitan areas, the average amount of organic MSW taken at transfer stations was around 60% (by weight), as presented in Table 2. The amount of inorganic waste was around 40% (by weight), and about 6% (by weight) was classified as recyclable inorganic components.

Waste collection is the first subsystem of the technical system of MSW management which is part of the municipalities' services. Some households can afford a waste bin made of concrete, plastic or steel built in front of their houses. Others simply store their waste in plastic bags or in used cartons or boxes in front of their houses. These wastes are then picked up by a community collector cart, a small truck, etc. depending on the arrangement in the neighbourhood community. Some large and medium-sized cities have been contracting out part of the collection and transportation to private firms. In the year 2006, collection of MSW in Indonesia covered about 130 million inhabitants or 56% of the total population (Table 3).

ISLANDS	MSW Generation (thousands tonnes)
Sumatera	8.7
Java	21.2
Bali and Nusa Tenggara Islands	1.3
Kalimantan	2.3
Sulawesi, Maluku and Papua	5.0
TOTAL	38.5

Source: Solid Waste Statistics year 2008

Table 1. MSW generated in Indonesia, 2006.

Item		% wet-weight	
		TS-1	TS-2
Recyclable components	Hard - papers	0.92	0.95
	Archives (white) Papers	0.14	0.34
	Bottle - glass	1.77	0.50
	Drinking bottle-plastic	0.29	0.19
	Drinking glass - plastic	0.17	0.34
	Can	0.22	0.32
	PE - plastic	0.03	0.42
	Divers - plastic	1.63	0.47
	Aluminum	0.06	0.05
	Cartoon/cardboard	0.33	0.31
	Newspapers	0.13	0.16
	Metals		0.03
Total of Recyclable Components		5.69	4.08
Organic Component	Food waste	33.90	58.04
	Leaves etc.	12.32	2.21
	Tissue-papers	11.02	1.78
	Textile	0.89	0.90
	Wood	1.98	0.70
Total of Organic Components		60.10	63.62
Others: an-organic non-recyclable		34.21	32.30

Source: Municipal Solid Waste Management in Indonesia, 2010

Table 2. Composition of MSW at transfer station (TS) at Bandung (2007).

ISLANDS	Population (million inhabit)	Population served (million inhabit)	% population served
Sumatera	49.3	23.5	48
Java	137.2	80.8	59
Bali and Nusa Tenggara Islands	12.6	6.0	47
Kalimantan	12.9	6.0	46
Sulawesi, Maluku and Papua	20.8	14.2	68
TOTAL	232.7	130.3	56

Source: Solid Waste Statistics year 2008

Table 3. Population served by municipalities in Indonesia in 2006.

Like any other collection system in developing countries, where the municipal waste from household sources is commonly collected through labour-intensive services (Cointreau, 1982; Joseph, 2010), in urban areas in Indonesia waste is collected by handcarts drawn by one or two crew members. Typically, waste generated by households is accumulated in small containers and placed on the ground to be shovelled manually or left in plastic bags, open cartons or baskets to be picked up by hand. These waste crew collection workers have significant direct contact with solid waste, so they are more likely to encounter potentially toxic and hazardous materials. Containers used for household storage of solid wastes are of many shapes and sizes, fabricated from a variety of materials depending on the economic status of the waste generator. The wide variety of types and shapes commonly encountered within a community creates difficulty in establishing and operating an efficient solid waste collection system (Joseph, 2010).

The waste collection from households to transfer stations is normally organized by the respective neighbourhoods. The system is handled and funded by the communities, who can afford the expenses associated with the activities. This community employs a person who usually lives near the settlement. Generally these are people who do not have a permanent job. They are not scavengers and this activity is considered legal, but the waste crew is categorized as an informal sector. Almost all these waste crews perform valuable waste segregation. In many cases, the waste collected is already sorted by the generator and given to these waste crews. All wastes in the transfer depot, along with non-residential wastes, will be subsequently transported by trucks to landfills by city cleansing division crews.

Collection of waste is conducted by several methods, namely:

- Communal collection: where the community bring their own waste to a temporary collection point whence transportation is carried out by trucks. This method is usually applied in the very thickly populated areas. In certain places, the community bring their waste to vehicles which move along their routes while playing a traditional song, rather like an ice-cream cart.
- Individual indirect collection: where collection of waste is conducted by small vehicles or carts from door-to-door, and brought to the transfer depot where the waste is transferred onto trucks and transported to disposal/treatment sites.

- Individual direct collection: where garbage is collected door-to-door by trucks and directly transported to disposal/treatment sites. This type of collection is conducted in the high-income areas or commercial areas where a large amount of waste is generated.
- Temporary disposal sites are provided and managed by municipalities. From these facilities, collected MSW is transported to final disposal sites.
- Street sweeping is carried out to collect waste from public places and main streets by manual sweepers as well as sweeper vehicles.

So far, most of the existing MSW management systems in Indonesian municipalities have relied on the existence of landfills. The excess has been handled by the community in regular ways, such as burning, burying, composting, and other less regular ways such as recycling or disposal at inappropriate sites, including ducts or drainage channels. Another method of treating MSW in some cities is incineration. There are several small-scale incinerators in operation in different cities, each with a capacity of about 100-200 kg/hour and operating eight hours per-day. Therefore, the system is only able to handle a small percentage of the total MSW generated. Composting of organic waste has also been introduced as part of waste treatment. Compostors are located, normally, in final disposal sites. In principle, the composting system comprises a centralized sorting and shredding system, and thereafter composting of the organic matter is done by means of a simple composting method. According to a questionnaire survey conducted in 2007, the mode of handling of MSW in Indonesia in 2006 is as shown in Figure 1.

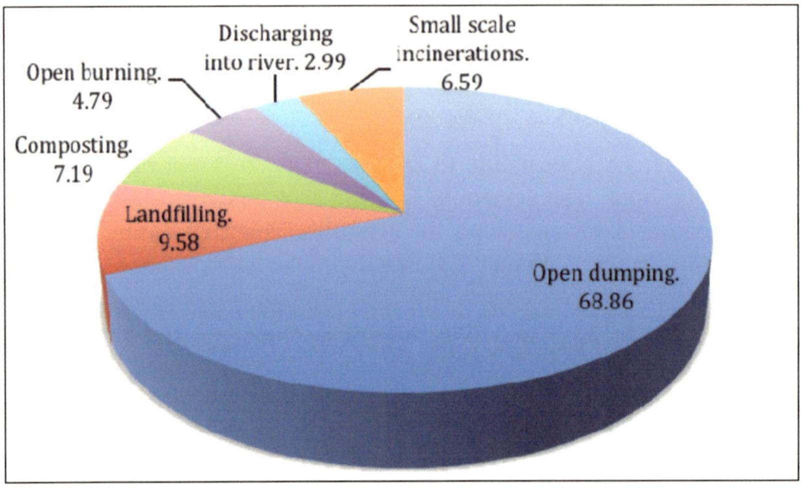

Source of data: Source: Solid Waste Statistics year 2008

Fig. 1. Percentage of MSW handling in Indonesia in 2006.

There are three groups of waste materials that serve as the main objects of recycling activity (Damanhuri et al., 2009):

- Wet waste, especially organic waste, to be converted into compost.
- Dry wastes, especially those with the potentials to be recycled, such as papers, plastics, aluminium, etc.
- Used goods resold by traders.

In 2008 Indonesia introduced the Law of Solid Waste Management (Law No. 18/2008). Some of the central issues of Law No. 18/2008 are as follows:

- Extended producers' responsibility (EPR); every producer should provide a label on their product packaging and/or their final products about reducing and proper handling of waste; and they should also responsible for the packaging of products that are difficult to decompose by means of natural processes (biodegradable).
- Application of waste reuse and recycling through the entire chain of waste transport, from origin to final disposal.
- Selection of waste processing and dumping technologies that are safe and healthy, and conform to the Indonesian situation. Open dumping and open burning are forbidden and five years after the passing of the law open dumping will be completely banned.
- Prohibition on importing waste into Indonesia territories or mixing waste with other hazardous wastes.

The basic aim of this new law is waste reduction through the 3Rs (reduce, reuse and recycle) as the first priority, and the next priority is waste handling. This concept is considered as a new paradigm to replace the collect-transport-dispose concept which is usually adopted in most Indonesian cities. All of the involved parties agreed that the new concept is the best available measure to reduce wastes, and active involvement on the part of the community and other waste generators to reduce waste volume is the key to the success of any waste management system. One of the important mandates in Law No. 18/2008 is the implementation of waste recycling. Any recyclable waste is collected from its respective sources, such as residential areas, commercial zones, temporary collection facilities and the final processing facilities. Wastes are recycled into useful raw materials for production processes (i.e. reprocessing and remanufacturing activities). Activities related to community involvement consist of separation, composting (at source and communal), and recycling. The municipality is then responsible for the transportation of the residue to disposal sites.

A national policy has been established with an initial target of 'reduction of waste as much as possible starting from its sources'. The government has set a target of waste reduction of up to 20% by the year 2014. Some strategies besides 3R education and campaigns have also been formulated to promote the reduction of waste from its sources, starting with households. The Ministry of Environment and the Ministry of Public Works facilitate recycling activities performed in several regions in the country. The top priority in the implementation of these activities is the recycling of organic wastes into either individual or communal level composting facilities. In order to accelerate the multiplication of such activities, the Ministry of Public Works has launched recycling activities in more than 200 locations in 150 cities since 2006 by adopting cost-sharing mechanisms with the local community, NGOs, and municipalities. Components of the project are a composting hall, a plastic crusher, a rotary screen and other ancillary items. Some difficulties arise especially in the formation of a local institution to run the project, and the limited capability of the local community to pay the collection fee among others.

3. Informal sector and waste recycling

Waste definitions as well as MSW definitions vary in different countries. Among developed countries, for instance, the definition of an MSW will encompass any goods that would be defined among developing countries as used goods that still have economic value. In developed countries, the elimination of used electronic appliances, furniture and fixture, used newspapers, used magazines, and used clothes, etc. incurs a disposal cost. Thus in developed countries, these goods are defined as waste, tend to generate problems and require further handling. Conversely, in developing countries, these are regarded as valuable goods and could still be used after being repaired or their components recovered in such a way that they could be reusable. Most Indonesian people of all economic levels have different perceptions of the end-of-life of goods, including consumer goods. These materials are perceived as used objects that still have an economic value, to the extent that they are rarely found in municipal waste management chains, for the reason that these items are actually saleable, or could be donated to others of lower income.

A positive impact derived from the current solid waste management systems in developing countries and economies in transition is the high level of recycling of the inorganic components of MSW. Although the methods employed for sorting and separation of MSW in these countries are considered inappropriate for solid waste management systems as defined by developed countries, these existing methods not only provide an income stream to the hundreds of thousands of people involved in this informal sector but also ensure a far greater amount of MSW generated is recycled. In most countries, plastics, glass, paper, and metals are collected by either the informal sector or municipalities, and these materials are recycled. There are two main recycling flows. In the first flow, collectors, including those in the informal sector, collect recyclable materials at sources. In the second flow, these materials are separated and recycled by the municipality after MSW collection. Recycling activities in this context are all activities reusing objects that have previously been called 'waste', either by directly reusing them or by selling them to waste traders.

Like other major cities in developing countries, the informal sector in Indonesia plays an important role in any recovery effort regarding the usable materials of waste. Recycling of dry waste (inorganic waste) is common practice in large cities in Indonesia. It is known that the recyclable material is reduced en route to the transfer points and to the final disposal. Many stakeholders are involved in the reduction process, e.g. sorting at the solid waste sources, scavenging. Many people treat the informal sector engaged in used goods and waste economic transactions or trading in Indonesia as scavengers. Indeed, it is the group that has attracted most attention owing to its association with social issues faced by most urban areas in developing countries. Actually, the latter is just one of the multiple stakeholders in recyclable collections. The recycling sector includes housewives, waste workers (from the cleansing division), vendors of used articles, and waste pickers (scavengers). Middlemen or intermediary traders are found in all corners of Indonesian cities buying used articles directly door-to-door. These waste recovery activities practised in many cities in Indonesia are mostly done by the informal sectors, consisting of handcart crews, mobile scavengers, transfer depot scavengers, final disposal scavengers, waste traders, recycling business people, and composting units at several points over a city, as presented in Figure 2. So far, the informal sectors' waste recovery activities have not been well organized.

Informal waste recycling is a common way to earn income. Studies suggest that, when organized and supported, waste picking can encourage grassroots investment by poor people, create jobs, reduce poverty, save municipalities' money, improve industrial

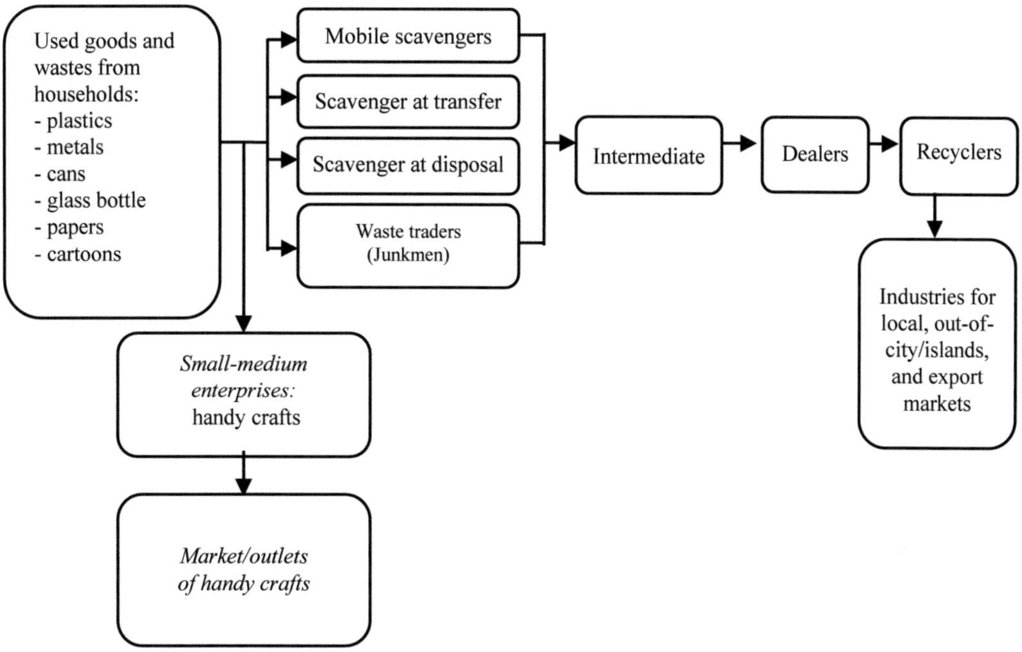

Source: Current Situation of Waste Recycling in Indonesia, 2009

Fig. 2. Interaction between informal sectors in waste recycling.

competitiveness, conserve natural resources, and protect the environment. Three models have been used to organize waste pickers: microenterprises, cooperatives, and public-private partnerships. These can lead to more efficient recycling and more effective poverty reduction (Medina, 2008).

Informal sector activities are not considered illegal in Indonesia. This group performs many economic activities, especially small businesses. All parties in Indonesia, including the government, appreciate its resilience in the face of the global economic crisis which affected the country and other regions during 1997 and 1998. It is documented that this sector has the ability to absorb many independent labours, because the formal sector has failed to provide good and adequate job opportunities. Many city inhabitants in Indonesia who have formal jobs also engage in informal sector business after working hours to increase their income.

Informal sector activities are done openly and have not been deemed as illegal activities so security or law enforcement officers do not need to become involved. These types of activity can be found throughout the cities, either in people's own homes or on the public streets or unoccupied lands. In most cases, however, raids launched by legal officers are because these activities are disturbing civic order, such as using sidewalks for their business, or occupying forbidden areas, such as city parks. Many of these informal sectors go door-to-door, offering goods or services directly to prospective consumers. Usually, these activities are prevalent in cities, such as foods stores, electronic/electric appliance reparation/services, tailors, and other service sectors. The most marked distinction between the informal and formal sectors is that the objects of the former are not taxation objects from their economic activities. In some cases, these informal sector activities have some linkages with the formal sector economic chains, and they are mutually dependent (Figure 3). The informal sectors engage primarily in using wastes generated by a household, especially dry wastes such as plastics, papers, metals, and the like, whereas wastes generated by an industry will certainly be addressed by the formal sectors.

The trading of dry wastes, which are non-compostable, has been the profession of choice of those people generally belonging to informal sectors. Cycles of potentially recyclable and financially valuable wastes start from their sources such as residential areas, industries and so on. These informal sector activities are most attractive for businesses, involving main actors such as scavengers and waste traders, who collect wastes or used goods door-to-door, or their customers/partners. In addition to their contribution to reducing waste handling costs, another benefit is that they serve as a generator of job opportunities.

Stakeholders for the recycled used goods and waste in all sectors (formal and informal) who play an important role in collecting recycled wastes and used goods are:

- Scavengers
- Waste collector crews
- Junkmen (waste traders)
- Intermediates (*lapak*)
- Dealers (*bandar*)
- Brokers

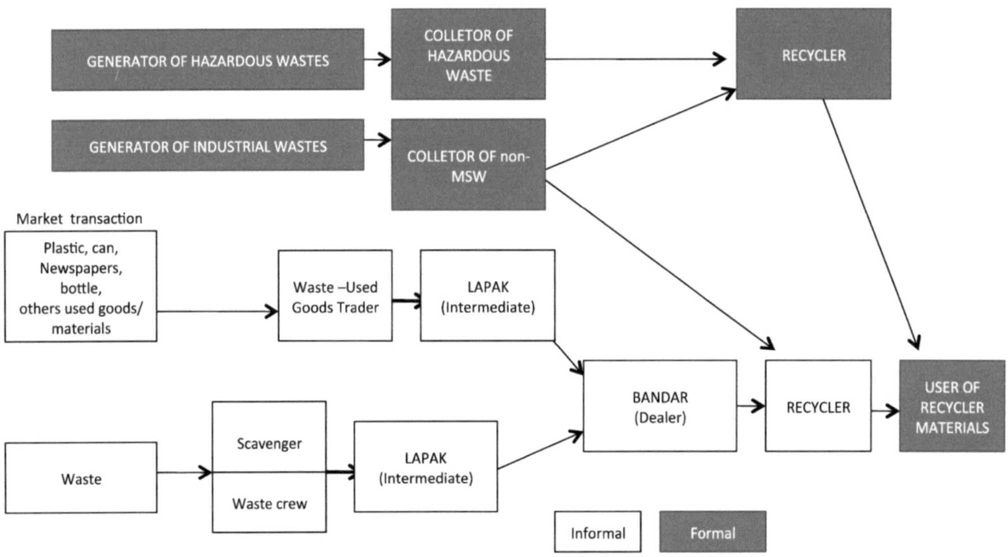

Source: Current Situation of Waste Recycling in Indonesia, 2009

Fig. 3. Formal and informal sectors in used goods/waste recycling.

Waste pickers (scavengers)

Used goods and waste recycling actors usually start by searching for and collecting used goods. The lowest group consists of valuable used goods hunters at the waste sites, and they are known as scavengers (waste pickers). Usually, they are coordinated by the *lapak*, owners or gatherers who accept and buy their used goods. They search for and collecting used goods rigorously. Usually, there are social ties between scavengers and *lapak*, such as similar hometowns. In general, these scavengers are migrants to the cities and not inhabitants. They are assumed to be one of the social urban problems, because they usually have no permanent domicile and live in the unoccupied spaces in the cities. Actually, few want to be permanent scavengers but they migrate to the cities out of economic necessity. Since they are generally unskilled and have no capital, they are forced to be scavengers.

Intermediates (*lapak*) lend money to the scavengers through credit sales for certain used items, similar to future trading, including provision of 'shelters'. Most of them incur heavy debt. Therefore, people who give them loans serve as a significant 'bond' and are very influential; relationships can become those of protectors and protectorates.

The scavengers usually start searching for used goods in waste disposal bins between three and four in the morning and end by four in the afternoon, covering certain areas, usually crowded settlements, transfer waste sites, residential areas, markets, commercial areas, and roadsides. Depending on their locations, scavengers are divided into door-to-door, transfer waste site, and final waste disposal area scavengers.

Door-to-door scavengers are more common than other types of scavengers. The varieties of goods obtained from houses, stores or markets are greater than those obtained by searching the transfer waste sites. The latter scavengers compete with small collectors (waste traders) who stand by the transfer waste sites to obtain used goods or wastes, because these collectors are usually more able to handle waste already separated s thanks to waste transport crews and have established business relationships with *lapak* or dealers.

According to interviews with some of the scavengers in the Bandung area, they sell their goods to the nearest *lapak* or junkmen in dirty condition without any further treatment. Most goods are cup and bottle packaging plastics, newspapers, cardboards, cans and glasses. Scavengers' selling prices are determined by the *lapak*. In general, they prefer collecting plastic packaging wastes and cardboards rather than the other items as they fetch higher prices. Door-to-door scavengers also obtain their items from houses that have separated their wastes into plastic, papers, cans and other non-perishable items. At the end of the day, they deliver their collected items directly to their *lapak* customers under a direct payment scheme. In response to dealers' requests, *lapaks* frequently ask scavengers for one type of item in a certain quantity and provide them with capital. Generally, the scavengers will try to rise to become *lapaks* or small collectors after 8 years engagement as scavengers, such as by borrowing capital from *lapaks* to open warehouses and to buy old cart as collecting vehicle operated by 4-5 local scavengers.

Waste collector crews

Those who also act as recyclable waste collectors act as waste collector crews. In general, city waste handling among urban areas in Indonesia adopts two groups of waste managers:

- Community self-help groups; and
- City waste managers.

The lowest-level community unit in Indonesia is single-neighbourhood, i.e. a group consisting of 30 to 40 households and led by a sub-neighbourhood leader regularly elected from among community members' it is a voluntary group. Ten to fifteen neighbourhoods will form one multiple neighbourhood. As in sub-neighbourhoods, the leader of the neighbourhood is elected regularly. Several neighbourhoods elect a village leader. Village leaders are civil servants and are appointed by formal and official assignment or by the decree of their respective district leader or city mayor.

One of the single-neighbourhood or multiple-neighbourhood tasks is associated with daily waste (garbage) collection of their inhabitants. Subject to the established agreements, waste collecting from households can take the form of either single-neighbourhood or multiple-neighbourhood collection. Usually, these communities hire waste transport crews that transport wastes from households to a waste transfer terminal that has been made available by local government. It is at the waste transfer terminal point that the work of local government (cleanliness department) starts: transporting the collected wastes to waste processors or final disposal sites. These waste collector crews usually serve as recyclable waste collectors as well, the product to be sold to intermediates (*lapak*) at current prices.

In most cases, mid-to-top level income groups will usually give their used goods to these waste crews. Therefore, waste collectors in Indonesia get additional income from their waste collecting services, in addition to revenues from recyclable waste sales that generally are of higher quality than recyclable wastes obtained by the scavengers. Compared with junkmen, these waste collectors gain advantage from acting as recyclable waste collectors. To obtain handcarts they need no money, because these vehicles are made available by the communities they serve and they do not need to buy used goods from their original owners. The only thing that makes them different from waste scavengers is that they reside in the area adjacent to the they serve and they have clear identities and addresses.

In some Indonesian cities, communities, sponsored by their city government, build simple recycling centres independently to manage their wastes. In addition to financing their waste collectors to collect the community wastes, they process wet wastes to be transformed into compost, by simple waste processing and facilities. Many of these centres are subsidized either by local government or central government-. The major tasks of the respective communities are financing and maintaining the continuity of these centres. It is also in these places that wastes are separated into dry wastes available for sale, wet waste available for processing into compost (biodegradable waste) and rejected wastes for transporting to the landfill. Therefore, these waste crews have three income sources, i.e. from compost sales revenue, dry recyclable waste sales revenue and official wages for handling the wastes of their communities. It is this model of waste handling that has been referred to under Law 18/2008 on Waste Management, where any waste handling should be implemented on the basis of the 3Rs principle.

Junkmen (waste traders)

Waste traders are generally found as door-to-door junk people, who buy varieties of used items. They usually work for *lapaks* who lend them money or carts. Generally, they buy items from each house at a price 10% lower than the selling price to the *lapaks*. In middle to high-class residential areas, they usually get their items free of charge from the owners, such as broken irons, blenders, kerosene stoves, shoes and clothes.

Standby junkmen get their items from used goods traders, who come to them directly. They tend to display their goods in fewer quantities than door-to-door junkmen. Interviews with standby junkmen who have been based for five years at a particular location, revealed that this used goods business is quite profitable, especially when they obtain items free of charge that are saleable second-hand such as electric fans, irons, water pumps. These standby junkmen start their business with a certain amount of capital, asking for items door-to-door. After obtaining enough customers, however, they finally decide to stand at a fixed site with the other junkmen (usually at junk markets). Most of them have aspirations to extend their business to include dealers.

Intermediates (*lapak*)

Lapaks' businesses are usually equipped with warehouses, three or four workers, and collector/delivery vehicles such as pick-up cars, trucks or carts. The item types vary from cup and bottle packaging plastics to blank/HVS papers, cardboards, newspapers and cans. They obtain these items from scavengers, small gatherers and door-to-door junkmen from their adjacent areas who serve as their customers on a trust basis.

After receiving items from scavengers or junkmen, they will place them in the warehouses that have been partitioned with plastic tarps, equipped with large plastic drums to store used cans. The usual pre-treatment of these items before delivery them to the dealers involves cleaning the packaging of labels and removing the covers for cup and bottle packaging plastics, milk cans, and glass bottles. In clean conditions, their selling prices at dealer level are higher. Lapaks sell them to the dealers in an incomplete state, but by opening these cans to form sheets to boost their selling prices 50% of their completed states. As regards other items such as papers and cardboard, *lapaks* pack them by sorting according to type and they are subsequently tied in bundles of 1 kg. *Lapaks* deliver their items at three to six-day intervals. The total quantities of these items are not fixed, but depend on the type obtained. The items go to the appropriate dealers, such as plastic dealers, glass bottle dealers, iron dealers and metal or cans dealers, and paper and cardboard dealers.

The relationships between *lapaks* as sellers and dealers as buyers are specified under contracts or agreements between both parties, specifying the details of items, terms of cooperation and item quantities. Frequently, dealers grant *lapaks* loans as capital to search for items, extend their warehouses, buy vehicles or meet other requirements. Dealers usually grant loans to their old customers, such as those who have had business relationships with them for a few years.

Generally, these *lapak* businesses are the extension of a business started by a scavenger or junkman. According to interviews with several *lapaks* in the Bandung area, these used goods businesses started fifteen years ago without any legal status as a formal company. *Lapaks* are usually integral parts of owners' houses or abandoned sites or rented lands for their workers. They are always trying to keep good relationships with workers who help them search for their items and with dealers who accept their items to maintain 'trust' and enhance sustainable cooperation.

Dealers (*bandar*)

The dealers most often to be found in Bandung City are cup packaging plastics and glasses dealers. In addition to iron and metal dealers, and paper and cardboard dealers, there are many plastic dealers in Bandung City, because many plastic recycling factories are located there. There are some *lapaks* or dealers outside Bandung City who deliver their items to Bandung City. The business areas of these dealers are not necessarily similar. In general, dealers do further processing of used goods up to the preparation stage of converting raw materials into materials, such as crushing of cup packaging plastics (polypropylene, PP) and pressing of bottle packaging (polyethylene terephthalate, PET) and pressing and/or grinding food and beverage packaging cans. Papers and cardboards are not usually further processed, because after they have been collected and tied, they are delivered directly to the processing factories outside Bandung City.

Interviews in 2009 of one dealer at Bekasi who is an ex-packaging plastic dealer and the owner and industrial manager of pellet and product recycling businesses, the strength of dealers to survive and grow even during weak economic circumstances lies in having wide networks (inner city, outside the city, or even foreign countries), and the availability of supporting equipment such as press machinery and/or chopping and pelletizing machinery to meet market needs and to secure the quality of products available for sale. In general, these packaging plastic dealers will follow the minimum volume standards to accept items from their peers (dealers) or from *lapaks* inside or outside their cities. Other items such as glass bottles are not valued, because the prices are the lowest in comparison with other items. The glass bottles collected by this dealer have a certain trademark and are subsequently bought back by their original factories.

A used iron and metal dealer at Bandung City revealed that the processing of irons and metals tends to fluctuate. Special treatment applied to irons before delivery to the related industries includes cutting and sorting. Irons and other metals such as copper wires are obtained from intermediates and gatherers of irons or smaller dealers. These irons and metals are delivered to automotive factories in Jakarta according to the respective agreements and ensure good, long-term cooperation between the dealer owners and industrial workers who have special authorization to procure them. According to the dealer's contact owner, irons are the items that have relatively stable selling prices compared with other products such as plastics.

In some cases, dealers in plastic recycling act as recyclers for plastic material processing, which is widely known as the pelletizing industry. It is acknowledged that the pelletizing industry generally needs materials in the form of plastic scraps that should be homogeneous by their respective types of plastics, PP scraps only or PET scraps only. If the industry has grinding machines, it prefers to accept items in pressed form because is a better guarantee of the quality of the resulting scraps.

Brokers

From the interviews conducted in 2009 with one of the plastic recycling players in Bandung City (Damanhuri et al., 2009) it could be concluded that most of the intermediaries or brokers in the recycling business channels serve as the main hubs between small collectors and large collectors, or between collectors and pellet-making industries. These brokers work independently and individually, using territorial bonds or family relationships or phone listings to obtain buyers or sellers. These brokers have authority to determine the quality of any item that would be sold by the sellers, and to be offered later to the buyers or, alternatively, they find the items requested by the buyers. For example, these brokers would search for items in the form of coloured or transparent PET plastic bottles already pressed and ready for the pellet-making industries. Brokers' income is in the form of a commission, i.e. a percentage of the pre-agreed amount.

4. Waste recycling routes

Used goods' trading in urban areas in Indonesia has long been practised as a junk market. The types of junk traded among junkmen, junk stores and junk markets are as follows (Figure 4):

- Home appliances such as irons, blenders, cake mixers, hairdryers, electric fans, TVs, radios and tape recorders;
- Used fabrics and clothes;
- Shoes and bags;
- Used books;
- Used cassettes; and
- Wood from building demolition and furniture.

These used goods are usually kept at special locations, such as stores, street trading sites, and junk markets. The stores accept varieties of usable objects under a revenue-sharing scheme. Unlike junk markets, these stores are actually count as formal business, because they appear as ordinary business stores.

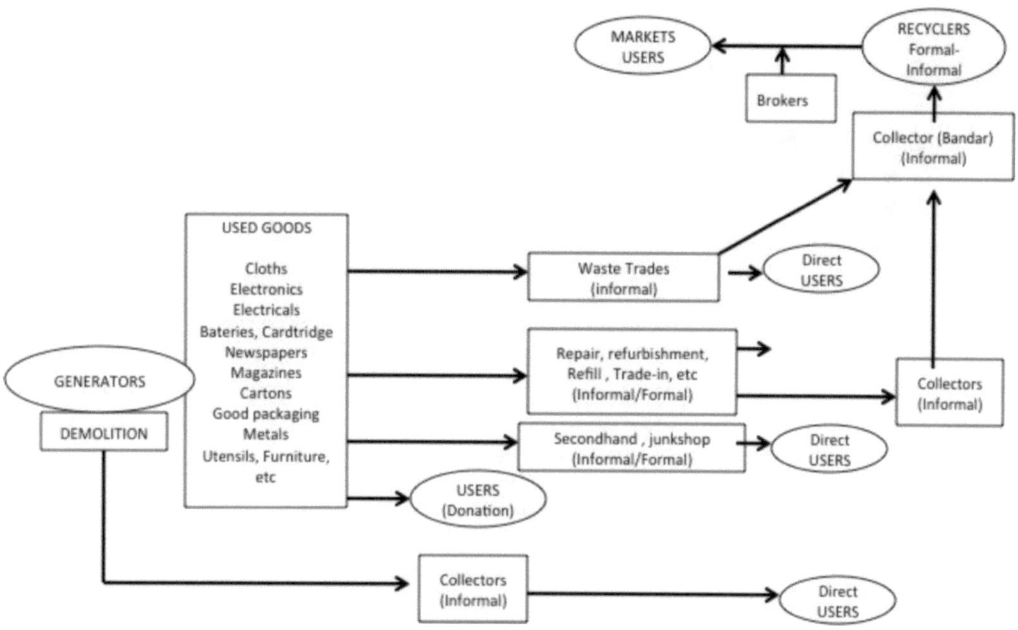

Source: Current Situation of Waste Recycling in Indonesia, 2009

Fig. 4. General destination of recyclable used goods.

In addition to used goods such as electronic/electric appliances, lead batteries and other objects that are considered as non-waste by Indonesians, waste category goods, or any goods that have been disposed of by their owner and are commonly found in public waste bins but still have potential for trading, include (Figure 5):

- Hard plastic packaging (containers and cups/glasses);
- Transparent plastic sheets;
- Papers (blanks, magazines, books, newspapers, writing books);
- Cartoons;
- Metals (nails, irons, coppers); and
- Glass containers

These materials serve as economically valuable business objects among recycling actors from residential level, junkmen, scavengers, intermediates (*lapak*), dealers (*bandar*) to industrial level (recyclers). In general, the traded goods or wastes provided by 'sellers' such as junkmen. Generally, intermediates and dealers will not deal in one single category of goods. They collect hard plastic packaging, glass containers, zincs, metals, cardboards and papers that can subsequently be sold to the major dealers. They trade not only in plastic packaging, but also convert these goods into pellets, in addition to accepting supplies of used metals such as trellis fences and copper wires, used drinking cans and used spoons.

The destinations of potentially recyclable wastes and used goods vary with their respective market circumstances and the availability of a recycler as their end processor. Figures 4 and 5 depict the paths of potentially recyclable wastes, used goods and wastes from their source levels through to the end actors.

Used goods and waste trading start in residential environments. Middle or higher-income residential areas usually donate their used goods free of charge to door-to-door junkmen, waste transport crews in their areas, or scavengers who pass by. In mid- to low-income residential areas, used goods or wastes that still have economic value are additional sources of income as they are sold to door-to-door junkmen or directly to junkmen or trading sites or small craftsmen in the vicinity. The difference between junkmen and scavengers is that the latter get their valuable wastes free of charge while the former have capital to buy these used goods.

Waste pickers often throw out the contents of rubbish bags or bins to take anything of value. In many cases, they take the plastic bag itself, so increasing the difficulty of waste collection by the crew. Waste recovery by waste pickers is often a problem. Unauthorized waste picking can have an adverse impact on neighbourhoods and cities. Municipal authorities do not ban the activity but neither do they support it.

Waste scavengers who sell to the collectors collect some of the recyclable wastes. The collectors separate and classify the wastes into several groups , then sell them to the wholesalers. These wholesalers will then trade these wastes with recycling factories. Some of these wastes are recycled within the cities that produce them, but generally are sold to other cities, or even exported aboard.

The buying prices offered by junkmen or trading sites are affected by the condition of the goods offered by the sellers. Price flux occurs at wholesale level, junkmen locations and trading sites. In general, however, these junkmen or trading sites are situated in the vicinities of the respective residential areas, so interdependent beneficial relationships are

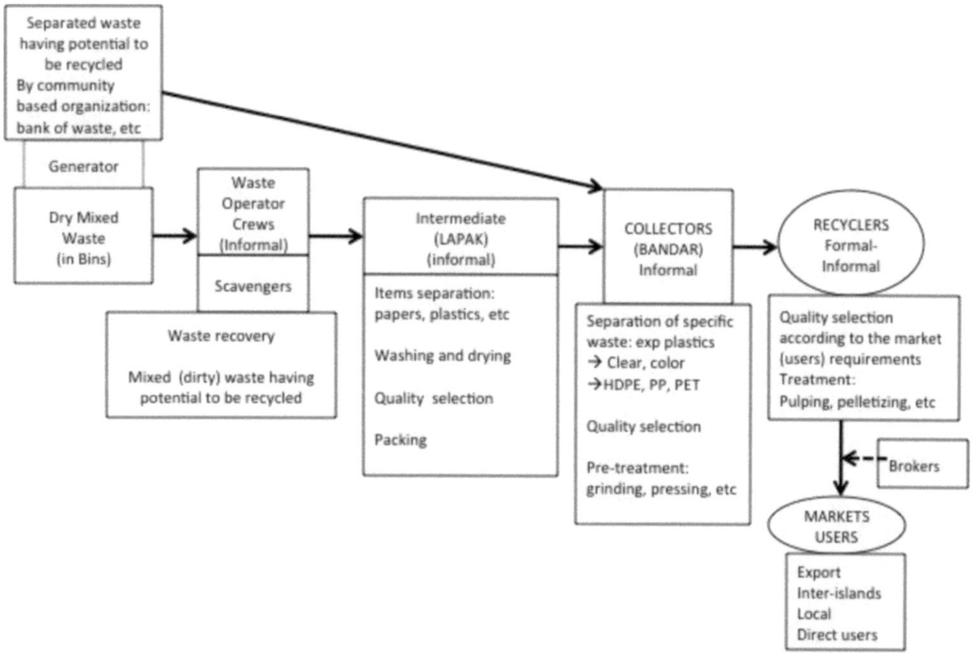

Source: Current Situation of Waste Recycling in Indonesia, 2009

Fig. 5. General pathway of recyclable wastes.

formed. Goods collected at junkmen sites and trading sites require initial processing to boost selling prices at the dealer level. For example, plastic packaging, glass containers and cans should be freed of labels and covers, so that their prices will increase. Cleaned materials increase the selling value from 40% to 50% from base or dirty prices. Figure 4 demonstrates the pathway of second-hand goods that attract the most attention among urban areas.

Dealers as collectors of goods on a mass scale usually arrange for the processing of these goods, especially plastics, to be converted into goods available for recycling. Then they deliver these processed goods to the related industry/recycler within or outside their city or even export collected recyclables to foreign countries such as China. At dealer level, goods categories such as iron, metal or glasses are not specially treated. It is common practice for a dealer to have direct brokerage networks or sell goods to any industry that will use them. The factory that initially produced them will reclaim glass bottles with certain trademarks. The brokers collect good-quality used iron and metals and sell them to automotive assembling companies in Jakarta, for instance. Recycling factories, as end actors, not only produce products, but also process goods such as chopped papers, chopped or granulated plastics and iron scraps.

Recycling industries, more widely known as pelletizing industries, in some cases serve dual roles, either as collectors or as end users of recycled products, depending on their business scale and the completeness of their own production means. A study in 2008 and interviews with one of the collectors (Damanhuri et al., 2009) showed that plastic pellet manufacturing industries generally required materials in the form of homogeneous scrapped plastics in terms of packaging uniformities, such as PP-only scraps or PET-only scraps. As long as these pellet industries have grinding machines, however, they prefer to accept items in their pressed form, because this guarantees the scraps' quality as export products or premium products for local usage. Impurity in the scrap of mixed plastic products from several different sub-collectors is frequently found such as PVC (*polyvinyl chloride*), PS (*polystyrene*), iron rods, broken glass, and aluminium rods.

Figure 4 above demonstrates the trading paths of used goods originating from several sources such as individuals who come directly to the junkmen, junk markets or used goods stores. This is the case for several types of goods such as used electronic appliances like blenders, irons, mixers, tape recorders for automobiles, cassettes, shoes, bags, and books or magazines. The owners come directly to the junkmen or second-hand markets and the buyer usually offers a price commensurate with the goods' condition. The prices for broken but repairable goods are normally 20% to 30% below the market prices of brand-new ones. Irreparable goods are bought by junkmen at a much cheaper price, sometimes at only 5% to 10% of their market price, and there are some sellers who give them free of charge to the junkmen. In the case of irreparable used electronic goods, the junkmen usually disassemble them for the valuable components that can be sold to dealers, such as copper wires, screws, and plastic hard covers, etc. These dealers subsequently sell them to related industries.

Research results in 2007 on waste from mineral water packaging in Bandung City (Damanhuri et al., 2009) showed that the most frequently traded used items were mineral water packaging plastics (bottles and cups), plastic sheets (leaves), newspapers, office supplies papers, packaging cardboard boxes, glass bottles, iron, metals (aluminium, copper) and used electronic debris. This profile is similar to the results of a survey conducted in 2008 among recyclers in the five cities of Batam, Bogor, Magelang, Makasar and Pontianak. The used items most often recycled for local and export purposes comprised plastics, papers, irons and metals.

According to interviews with one of the collectors of used plastic packaging, who is also the owner and manager of a pellet-making and recycled product factory, the power of a broker to survive and expand in weak economic conditions lies in his or her wide networks (inter-city, intra-city, even internationally), the availability of supporting means such as press machines and/or cracking and pelletizing machinery to meet market needs, and the reliability of the products to be sold. Generally, these collectors determine minimum standards for any item to be accepted from their suppliers, e.g. scavengers. These minimum standards are as follows:

- Transparent PP (*polypropylene*) from mineral water cups and transparent or coloured PET from any drinking product bottles are usually subject to an under two-tonnes/week minimum quantities requirement.
- Other types of wastes that can be accepted by some plastic waste collectors such as papers, metals and glasses have no definite criteria.

Waste recycling factories in Indonesia as the ultimate players in the waste recycling business not only manufacture finished goods but also intermediate products (raw materials) such as papers, plastic flakes or granules and scrapped irons. Bekasi City (West Java Province) and its surrounding are famous for finished goods, intermediate products and raw material exports, although their top priority is domestic markets.

The PET bottle recycling factories and other factories manufacture plastic pellet products to be processed further into finished goods (plastic appliances). Some of these plastic flakes and pellets are exported to other countries like Singapore, Taiwan, China, Malaysia and the Philippines, though most products are used for domestic industrial purposes. The capacity of plastic waste-based recycling businesses in Indonesia is evidenced by the existence of plastic recycling associations. They are able to maintain the quality of their products and good business relationships, and thus their PET scrap products can enter international markets.

Generally speaking, other wastes that have specific potential for recycling are papers and metals. The paper- and metal-based waste paths do not significantly differ from those of plastic-based waste. It seems, however, that these paper- and metal-based waste recycling businesses have not yet been accommodated in an association or partnership programme, unlike the plastic-based waste recycling business. It was found that their consumers would discard any waste containing metal elements. Iron- and metal-based scraps from Batam and Pontianak would be delivered to Medan and Jakarta, whereas the same scraps from Makassar and Magelang would be delivered to Surabaya. There are exports of iron- and metal-based recycling products to countries like Taiwan and China, but most of these products are absorbed by domestic steel factories. The areas of Jakarta, Bogor, Bekasi and Surabaya play important roles in the network of the recycling business because the majority of waste recycling activities occurs in these areas.

The same data source from Ministry of Environment (2008) also reported that in paper-based recycling wastes are transported to Jakarta (from Batam, Pontianak, Bogor) and Surabaya (from Batam, Magelang and Makassar). In addition to delivery to Surabaya, paper-based wastes from Magelang are delivered to Magelang's pulp- and paper-based factories. Though here is paper-based waste export, most of these paper-based wastes are absorbed domestically as raw materials of paper recycling.

Another source of used goods, s used clothes, is found mostly in major cities. In Bandung, there are junk markets for used clothes that buy these goods from brokers of imported used clothes from Taiwan, Korea and China at low prices per sack or 100 kg bales of trousers, T-shirts, shirts and jackets. The contents are reopened and mixed with used sweaters and T-shirts, and then sold to junkmen at higher prices. Other types of goods such as automotive electronics and other used electronics are sold individually by their respective owners to junk markets. These junkmen resell them to visiting consumers in a minimally repaired condition.

The other type of waste/used goods recycling category is building demolition by-products in the lumber business. The interviews with several actors in Bandung city revealed that this business has distinct breadth, ranging from sources to end-products. Essential items are wooden frames from old buildings obtained in one of two ways:

- Trading with the owners of old houses/buildings to be demolished in order to obtain lumber from wood frames, roofs, doors, etc.
- Cooperation with construction projects, most of them order wood frames in large quantities.

The buying prices of lumber depend on the buildings to be demolished. Buyers refurnish the collected items by cleaning, repainting and displaying them in front of their stores for being sold.

Other lumber businesses collect wood pellets and used frames from lumber wastes, to be processed into sofa/chair frames and then delivered to sofa production factories and other subscribers in the Bandung area. One trading business explained that they have been operating their business for 10 years with five workers. Low-quality and non-reusable lumber is cut and sold to tofu factories to be used as firewood.

The waste materials, after sorting, cleaning, and processing, are sold to scrap dealers, who in turn sell to industry. The recovery process of the waste component for waste recycling includes various activities, which comprise sorting, collection, washing, crushing, and pelleting. These activities involve scavengers, collectors, agents, suppliers, and recycling industries, which operate in a recycling business chain. Each level in this business chain has a specific activity pattern, which is different from one level to another. At the lower level, the need for equipment, energy, and space is less pressing than at the higher ones. For example, the required facilities on the scavenger level are limited to simple equipment, such as a collecting stick, mask, and gunny bag, whereas at the higher levels more complex equipment, such as a mechanical crusher, extruder, and pelleting machine, may be needed.

Owing to the different treatment of the waste component in each level, the product quality is improved at the higher level. As a consequence the price of plastic material, for example, increases from the bottom to the top chain levels. These prices, however, vary from place to place and are dependent not only on the quality (i.e. degrees of cleanliness and homogeneity), but also on the highly competitive markets (Trihadiningrum et al., 2006). The quality of wastes as used items that have potential to be recycled determines their market selling prices. The recycling business chain players in this activity are collectors (waste traders, and scavengers), intermediaries and recycling industries. They have unique criteria that must be met by their respective business partners.

According to the interview in February 2009 of a big collector/recycler in Bekasi City (Damanhuri et al., 2009), the grade required to produce the end-product in the form of coconut root sweepers, would be met by green, red and blue pellets, depending on the colour of the sweeper frames to be produced. For the end-products in the form of plastic balls, however, they required used grease bottles HDPE (*high-density polyethylene*) pellets. Thus, the essential thing is the homogeneous quality that they accepted from their trusted business partners. This plays an important role in the success of their product and in maintaining good partnerships.

A producer of plastic zips in the west of Java Province expressed similar views (Damanhuri et al., 2009). His company produces zips that require transparent and coloured PET as raw materials. It has an established criterion for items that it will accept from its partner-collectors, i.e. that these items should already be in their pressed form. The main reason for setting this criterion is the limited area of its raw material warehouses and the fact that the quality of the scrap products is generally better if they are produced in-house. Consequently, this industry will only sort pressed materials according to colour for the cleansing, scrapping, washing and drying processes.

The difference between purchasing and selling prices is also affected by the exact status of the respective recycling players. The linchpin of this relationship is that the final price of any metal-based waste is affected by its own quality. Large scavengers receive most of this waste from small waste scavengers, so the items they receive are usually dirtier than the items sold by the used products sellers. This is because the latter purchases his or her used items from households and institutional sources.

5. Conclusion

In Indonesia, waste recycling is an activity strongly supported by all parties concerned. Their efforts are not incorporated into actual and integrated activities, however. The effort that should be performed by the entire parties are how these waste recycling activities in informal sectors could be the integral part of waste handling performed by local government and community's self-help. The main question of municipal waste handling is how ensure waste generated is well-handled so that the overall city environment remains clean and has no negative impact on human health or residential environments. Waste generators and their corresponding local governments would be satisfied if recycling activities helped to decrease their waste problems and at the same time decreased the cost of waste handling that should be provided, owing to the existence of revenues from recyclable waste sales. On the other hand, the main target of waste recycling by people in the recycling business activity is how to manage recovered materials to obtain as high an economic value as possible. Waste problems are not their problems. They will be satisfied if they can get as much waste as possible with a high economic value. They do not care about the remaining non-economic and non-saleable part of these wastes.

There is a different perception in terms of the goals of waste management and recycling activities. The main purpose of city waste management is ensure the city is clean, healthy and environmentally sound. Recycling activity is supported because it is associated with the image of a green city, but very rarely is the waste recovery in MSW management associated with efforts to gain quality recyclable waste in accordance with the needs of the recycling market. Recyclable waste quality is not their concern. The same point of view is found in waste recycling activity. The stakeholders who engage in this activity are only interested in cheap recyclable waste of high quality that meets market demand.

In the case of recyclable and used goods pathways, efforts should be made to enhance the selling process, through selection, separation or reparation and so on. Activities performed in informal sectors in developing countries are pre-treatments such as the melting process, generally performed without consideration for workers' safety and the impact on the residential environment. In the case of large industries, they usually guard their company image and accordingly comply with the appropriate regulations. In the case of small industries, notably home industries, most of them belong to informal sectors; usually they pay less attention to safety aspects, owing to limited capital and knowledge.

The recycling efforts in urban waste management are an integral part of the sustainable waste management concept. The main target is to drive communities to minimize their waste as much as possible, and encourage reusable wastes, safe waste processing and disposal from health and living environment perspectives. On the other hand, recycled goods and waste trading businesses are essentially economic activities, where such factors as prices, product/material qualities, supply continuity based on required quantities, and related profit obey the current market. The two interests, i.e. interest in sound waste handling and recycled product business interests, should be bridged to enhance the continuous growth of recycling efforts along with higher quality of the end-products and an ultimate increase in their economic value.

It is clear that the implementation of waste recycling as alternative materials or energy requires proper regulations and standards to enhance economic value. To process the use of wastes and used goods, which is usually performed by informal sectors, it is important to consider the development of standards, at least the standards and guidance related to market demands, workers' safety and environmental compliance. Cross-sectoral coordination, especially among the parties responsible for regulating the use and trading of materials, is also necessary.

6. Acknowledgements

The authors would like to express their appreciation to ERIA/JETRO Project 2008-2010 and all members of the ERIA Working Group who have enriched the understanding of waste recycling through valuable discussions.

7. References

BPS-Statistics Indonesia. (2004). *Official Statistics and their Development in Indonesia,* BPS–Statistics Indonesia , Jakarta.

Cointreau, S. (1982). *Environmental Management of Urban Solid Waste in Developing Countries,* the World Bank, Urban Development Technical Papers 5, June 1982.

Damanhuri, E. (2005). Some Principal Issues on Municipal Solid Waste Management in Indonesia, *1st Expert Meeting on Waste Management in Asia-Pacific Islands,* Oct 27-29, Tokyo.

Damanhuri, E., Handoko, W. & Padmi, T. (2010). Municipal Solid Waste Management in Indonesia, In: *Municipal Solid Waste Management – in Asia and the Pacific Islands,* Agamuthu, P. & M. Tanaka, (Eds.), ISBN 978-979-1344-78-4, pp. 95-112, ITB Press, Bandung, Indonesia.

Damanhuri, E. & Padmi, T. (2009). Current Situation of Waste Recycling in Indonesia, In: *3R Policies for Southeast and East Asia,* M. Kojima & E. Damanhuri (Eds.), 23-52, ERIA Research Project Report 2008, No. 6-1, March 2009.

Damanhuri, E.,Wahyu, I. M.,Ramang, R. & Padmi, T. (2009). Evaluation of Municipal Solid Waste Flow in the Bandung Metropolitan Area, Indonesia. *Journal of Material Cycles and Waste Management.* Vol.11, No. 3, (September 2009), pp. 270-276, ISSN 14384957, 16118227

Donner, W (1987). *Land Use and Environment in Indonesia,* C. Hurst & Company, London.

Joseph, K. 2010. Municipal Solid Waste Management in India, In: *Municipal Solid Waste Management - in Asia and the Pacific Islands,* Agamuthu, P. & M. Tanaka, (Eds.), ISBN 978-979-1344-78-4, pp. 75-93, ITB Press, Bandung, Indonesia.

Medina, M. (2008). *The Informal Recycling Sector in Developing Countries, Organizing Waste Pickers to Enhance their Impact,* Grid Lines, Note No. 44, October 2008.

Ministry of Environment of Indonesia (2008). *Solid Waste Statistics 2008* (in Indonesian), Jakarta.

Ministry of Health of Indonesia (2007). *Indonesia Health Profile 2005,* Jakarta.

Trihadiningrum, Y., Wigjosoebroto, S., Simatupang, N.D., Tirawaty, S. & Damayanti, O. 2006. Reduction Capacity of Plastic Components in Municipal Solid Waste of Surabaya City, *Environmental Technology and Management Conference,* 7-8 September 2006, Bandung, Indonesia.

3

An Analysis of Policies in Support of Waste Collecting in Rio de Janeiro – Three Case Studies

Maria Scarlet do Carmo
Fundação Getulio Vargas (FGV)
Brazil

1. Introduction

The waste collector has become a locus of public policy activity due to a number of factors: the type of waste produced subsequent to the industrial revolution, the advent of a consumption society, the scarcity of natural resources, the problem of containment, and the development of recycling technologies. The repetitive nature and connectedness of these problems justify the conceptualization of recycling as a discursive practice (FOUCAULT, 1972, 1984; MACHADO, 1981). This repetitive nature encompasses not only the production of new discourses (discursive and non-discursive practices) on waste and the waste collector, but also the emergence of entrepreneurial segments of an economy based on recycling, and a change of habits among consumers, among others.

In this sense, post-consumer waste that was once associated with danger—a threat to health and to the environment—and that once provoked repulsion and aversion (DO CARMO, 2010; EIGENHEER, 2003; RODRIGUES, 1995), is currently viewed as reusable material. It is associated with elements of raw material used by industries and with the income of waste collectors. This association has become possible thanks to an emerging paradigm: practices such as the development of industries devoted to the reuse of waste, the expansion of a market for recyclables (RODRIGUES, 2005), the growing awareness of society and, lastly, the policies that support the work of waste collectors. In this way, what was once repugnant becomes valuable (DO CARMO & PUPPIM DE OLIVEIRA, 2010). To be credited in large part for this change are the growing cadre of authorities, such as engineers and ecologists, who have become involved in environmental education initiatives and different types of recycling projects. It is often the desire to avoid environmental and social harms that is both waste's cause and effect.

This article shows how, just as public sector postures have changed in relation to waste, so too have they changed with regard to waste collectors. In this study, it is defined 'waste collector' as one who subsists exclusively on the sale of reusable material obtained at the source of the generator (DO CARMO, 2009b). Martin Medina (2000, 2007) was the first to discuss the public sector's posture toward his group, and proposed four types of policies used to address waste collectors.

The first approach is a policy of repression. The policy is predicated on the notion that waste collectors belong to an amorphous collection of social marginals —undesirables— and foraging constitutes an inhuman, illegal activity as well as a source of embarrassment and shame. Waste collectors are often kidnapped or expelled, such as in the case of Egypt and Colombia (MEDINA, 2000; RODRIGUEZ, 2003). According to Dias (2002), in Belo Horizonte, Brazil, the municipal government took waste collectors off the streets during the 1980s; a policy that largely reflected the lack of a health-based distinction between refuse and undesirable people, such as beggars, prostitutes, and waste collectors (PORTILHO, 1997). In other countries these policies amounted to coercion, forcing people into other forms of employment. Whether because of low levels of educational attainment, aptitude, or age, these strategies commonly failed to result in enduring work for waste collectors within the formal sector (KASEVA & GUPTA, 1996; MORENO-SANCHEZ & MALDONADO, 2006).

The second approach is a policy of omission or neglect which occurs when public authorities refuse to acknowledge waste collectors and their function. In this way, they are pursued, but also they do not receive support; they are simply ignored. African cities such as Dakar, Senegal, Bamako, Mali, Cotonou, and Benin provide examples of this policy at work (MEDINA, 2000).

The third approach is a policy of conspiracy (collusion) which is characterized by fraud, in which government officials develop exploitative relationships with waste collectors. These relationships often reflect mutual gain and assistance, a sort of political clientilism. Mexico City provides a traditional example of this dynamic, where public authorities and waste collector-leaders, referred to as 'chiefs' or caciques, develop a complex relationship. According to Castillo Berthier (2003) and Medina (2001) these relationship are often illegal, and include not only waste collectors but depot owners, street cleaners, middlemen, businesses, and public authorities. Some of these illegal relationships involve bribing officials to disregard the behaviour of caciques.

Supportive policies [1] (the fourth approach) refers to the most recent and progressive change, whereby governments make way for initiatives to legalize the activities of waste collectors and support their activities through cooperatives. This support typically arises because of the realization that waste collectors are ideally suited to adverse situations. They are particularly important as partners in finding solutions to environmental and economic problems, and they are especially welcome in slums, where local conditions impede the use of American or European technology to manage waste. Some Asiatic countries, such as Indonesia, the Philippines, South Korea, India, and Thailand, encourage foraging through import tariffs on packaging materials and by providing financing for governments to create cooperatives (COINTREAU, 1986; FUREDY, 1984; KASEVA et al., 2002; MEDINA 2000, 2007).

Many of these initiatives have been supported by the World Bank. At the beginning of the 1980s, the Bank produced a master plan for managing waste and promoting local waste-relevant technologies, many of which have been employed in Africa. The Bank's master plan

[1]Medina (2000, 2007) uses the term, "stimulation policies," whereas I use the term, "policies of support" or "stimulus policy".

questioned the wisdom of the mechanized waste disposal systems promoted by developed countries, particularly given the necessities of local environments and waste collectors (COINTREAU, 1985, 1986). In Latin America, the establishment of public-private partnerships and concessions for waste collectors to collect mixed waste provide one example of a supportive policy.

Ojeda-Benitez et al. (1988) insist that it was the public authority's recognition that waste is a source of income that brought about municipal support for policies, increasing employment for low-wage areas. In the city of Rio de Janeiro, strategies to stimulate this sector became a priority at the end of the 1980s and at the beginning of the 1990s. This article suggests that the economic concerns surrounding the issue of waste are the principal reason why support began to emerge, imbuing waste collectors with greater visibility and value.

In this way, the objective of this article is to present the result of policies to support the work of waste collectors in the city of Rio de Janeiro. It does so by analyzing three case studies consisting of an equal number of cooperatives and their strategies. Finally, it seeks to identify the subjective ways in which waste collectors view the processes that guide each one of the three organizations and their respective discourses.

2. Methodology

The purpose of this article is to portray and examine the views of waste collectors interviewed during exploratory case studies. These studies were performed on three government-supported cooperatives in Rio de Janeiro between 1994 and 2003. Three municipal public officials and 66 waste collectors responded to questions during interviews, 17 of which were open-ended and 49, semi-structured. The first case study took place between the months of June and August 2002, the second during April and October 2004, and the third between July and December 2004. Participation in state and national waste collector meetings in 2006 and 2007 provided additional interviews with people linked to the three cooperatives in question. Secondary materials, such as books and journalistic production were also used to carry out research.

This article subscribes to the notion that desires to better the economic and social conditions of waste collectors imply understanding the changing processes of these workers and their work, and how they continue to be constituted. In order to improve conditions, it is first essential to understand the process of constitution: the place of workers with respect to society, the public sector, and buyers (recycling entrepreneurs), among others. In this sense, the article discusses the discursive and non-discursive practices to which collectors were subjected as waste became an object of profit as well as an environmental and social good.

In this way, waste becomes a discursive production (FOUCAULT, 1972, 1980, 1984, 1991ab, 1995) to shed light on the principles, concepts, strategies, and theories that have shaped the issues and industries pertinent to recycling and the environment. It also casts light on what is said—the discursive and non-discursive practices—with regards to the waste collector. The methodology is based on Foucault's discursive framework, which is conducive to understanding history and social construction (HARDING, 2003). It also helps us understand the results of subjective terms and their broader implications. Taking waste to be an object of revelation, the paper proposes to examine the rules and authorities that govern its existence, its historical emergence (its archaeology), and its characterizations (its genealogy). In short, it seeks to understand the transformation in discourse surrounding waste over time.

3. Evolution

Many waste collectors interviewed during the case studies showed that there is an effect of the negative meaning of waste in the process of constructing their professional identity (DO CARMO, 2009ab, 2010). There was a stigma attached to those who work in contact with waste (DO CARMO, 2009ab, 2010; DO CARMO & PUPPIM DE OLIVEIRA, 2010; EIGENHEER, 2003; PORTILHO, 1997; RODRIGUES, 1995). It can be pointed as one of the probable justifications for their manifest lack of economic organization: "*People think we're beggars*" (waste collector, first case). This quote demonstrates the stigma attached to jobs involving the handling of waste—or "dirty work" (AGUNWAMBA, 2003; ASHFORTH. & KREINER, 1999; HUGHES, 1962). At the same time, the other two cases exemplify how the stigma interferes in the process of negotiation (buyers) (second case) or access (the producers) (third case) to the waste: "*They* [buyers and producers] *prefer to negotiate with the leaders* [from the cooperatives] *than with us*". Scrap dealers and intermediaries, people involved in this business in Brazil who merely buy and sell the recycling, are often migrants (Portuguese, Italians, Spaniards, etc.) and seem to not be included in this aspect because today they are only buyers and rarely collect, as in the past (19th century) (ADAMETES, 1998; PORTILHO, 1997).

The cultural aspects within which waste is immersed justify its negative semantic-because the signifier 'waste' (or 'garbage') is pregnant with pejorative connotations. It can refer to concepts such as degeneration, decomposition, leftovers. As an object with no use or value, it may be associated with things of poor quality or things which are out of place, chaos and mess. But it is its unpleasant and troubling characteristics (smell, appearance and formlessness) which lead to its association with things such as death, limit, ending, making it taboo (EIGENHEER, 2003; FREUD, 1976; RODRIGUES, 1992, 1995). These are the aspects which mark what it was termed in previous studies´the negative semantics of garbage` (DO CARMO, 2009ab, 2010; DO CARMO & ARRUDA, 2010). Consideration as a valuable object (commodification) changes its social representation and interests, as if some waste starts to have a positive semantic when associated with recycling (DO CARMO, 2010).

The following sections address discourses surrounding waste, as well as the elements and rules that compose this discursive unity (this association with recycling). Through this analysis, the article identifies the strategies and policies of public support that have affected the conditions of waste collectors.

3.1 Discourses on waste

As a locus of public policy, waste can be associated with three distinct issues: the question of health, the environment, and economics (DO CARMO, 2010, 2008). The issue of health is linked to risks associated with the disposal and storage of waste. During a time when the sun took care of excrement and organic waste in the streets, at some risk to human health, waste was not viewed to be a problem necessitating the attention of city administrators. This picture changed with overpopulation. According to Rodrigues (1995), for example, only at the end of the nineteenth century did the French consider waste something that demanded attention and outlays of public money. Much has changed; each region now has to address this issue to the best of its ability.

As waste became a problem for cities, especially the lack of sewage disposal, diverse efforts were made to address waste. Between 1940 and 1960, the U.S. and Europe chose landfills as a preferred solution. The problem of space, however, encouraged public managers to reconsider the technologies used in public disposal. Incineration emerged as an important strategy, although it proved costly and was restricted chiefly to the more advanced countries.

Unlike developed nations, city governments in many developing nations frequently viewed waste to be an irresolvable problem (GONÇALVES, 2003; PORTILHO, 1997). Many demanded scarce resources for the purchase of expensive technologies (BARTONE, 1990). It was only at the beginning of the 1970s that the management of waste and public health were taken seriously in these countries. Many of the programs launched in the 1970s and 80s were the result of the World Bank, which carried out a mandate to look for solutions to problems of end-disposal (COINTREAU, 1986, 1985; BATOOL, CHAUDHRY, MAJEED, 2008).

Waste as an environmental question garnered attention beginning in the 1950s. This attention responded to the need for new technologies in order to reuse materials and mitigate the harmful effect of overflow (CASTILLO BERTHIER, 2003). It was clear that incineration provided an inadequate solution to these issues and, motivated by environmental and social movements, developed countries became the precursors of efforts to develop new strategies and technologies. In this sense, the relationship between waste and environmental problems was established by environmentalists.

According to Portilho (1997), however, the populational density in Brazil caused by demographic explosions influenced the establishment of the above relationship, because the question of urban space became a central issue for the environmental movement. It was only in the 1960s that the environmental question would become the focus of NGOs and social movements, especially in association with issues of inequality. In the 1970s the movement attracted the attention of politicians, and the business and industrial sectors joined the fray in the 1980s and 90s. It was in the 1990s that concepts such as the '3Rs' began to gain prominence, the concept of product cycles, the waste collector as a recycler or environmental actor and, consequently, strategies reflecting these concepts, including legislation and recycling campaigns. Medina (2007) illustrates this transformation by showing how the denominations used to describe waste collectors began to change:

It so happens that, when transformed into something useful—primary material—waste becomes associated with economic questions, as opposed to merely environmental issues. Economic interest emerges once there is a concomitant connection between the type of waste generated and its suitability for re-use, on the one hand, and on the other hand, the availability of people to collect waste and deliver it for negotiation and rendering. Until this happens, foraging is looked at as degrading, and waste collectors are to be distanced from urban centers by repression or have their activities ignored, thus being omitted from society.

Country or Regions	Traditional terms	New Terms
In Latin America:		
Argentina	*Cirujas, cartoneros*	*Recuperadores Urbanos*
Bolivia	*Buzos*	
Brazil	*Catadores de lixo, Badameiros*	*Catadores*
Colombia	*Basuriegos, Gallinazos*	*Recicladores*
Costa Rica	*Buzos*	
Cuba	*Buzos*	
Chile	*Cachureros*	
Dominican Republic	*Buzos*	
Ecuador	*Minadores, Chamberos*	
Guatemala	*Guajeros*	
Mexico	*Pepenadores*	
Peru	*Cutreros, Moscas, Buceadores*	*Recuperadores*
Uruguay	*Hurgadores*	*Classificadores*
Vanezuela	*Garimpeiros*	
In English-Speaking Areas:		
	Scavengers, Ragpickers, Totters, Rag-and bone men	*Waste pickers*
In French-Speaking Areas:		
	Chiffonniers	*Récupétateurs*
In German-Speaking Areas:		
	Lumpensammler	
In Japan:		
	Gomi-Hiroi (transliteration from Japanese)	
In China:		
	Jian Polan'l (transliteration from Mandarin)	

Table 1. Worlds commonly used for individuals engaged in informal recycling activities (source: MEDINA, 2007)

The general social value placed on waste will continue to grow, as long as new discursive practices keep emerging, particularly among those who previously did not view waste as anything of economic interest. It is worth stressing that new discursive practices surrounding the concept of waste as an economic question, and the search for disposal solutions, emerge concomitantly with the scarcity of disposal options, namely storage space.

Despite the search for solutions, until recently only industrial and commercial waste was recycled; domestic waste was off bounds for the waste collector. But it was the waste collector who was eventually responsible for returning domestic waste to the production cycle. In Colombia, for example, while specialized companies collected industrial trash, waste collectors were the principal gatherers of post-consumer domestic materials (BIRKBECK, 1979a). The waste collector's exclusive interest in domestic, post-consumer waste was due to its mixed organic and inorganic nature. This mixture of waste requires relatively sophisticated technology and a large amount of capital to separate. As such, the waste collector's participation in the recycled material business is only made possible by the difficulty of obtaining the material by other means.

The selective foraging of citizens of Rio de Janeiro, in the mid 1990s, marked a new moment in the way post-consumer domestic waste (now recyclables) was purified, accessed and handled. A social perception of domestic waste's economic value meant that waste collectors began to lose the exclusive access to recyclable materials they previously enjoyed. Many workers in other sectors began to use recycling as a complement to their salaries, collecting waste in their spare time or at work, as did waiters, housecleaners, building managers, and so forth.

Although (when queried) the motivations of people who adhere to a discourse of recycling are ostensibly environmental, they are also economic: environmental awareness campaigns are developed concomitant with strategies to support waste collectors—the traditional salvagers of refuse in Rio de Janeiro—to establish fixed locations for buying and selling recyclable waste. These fixed locations, which aim to facilitate waste collector access, also attract those people who are attentive to the economic value of recyclables.

3.2 The conditions and prospects of waste collectors

According to Castillo Berthier (2003), the first systematic study of waste as a social problem was carried out in Mexico during 1983. This study was undertaken in an era when official statistics and information on the subject was inexistent. In this period, issues relevant to the environment were ignored; neither viewed as a serious problem, nor as an area worthy of study in the social sciences. The difficulty of obtaining data on the work of waste collectors is equally common in Brazil. According to the literature consulted, waste collectors were initially identified in the 1930s. (DIAS, 2002). Many people agree, however, that they only started to gain visibility in the 1970s and 1980s:

> *If I have great production, excellent quality, supply flow, but I don't have a competitive price, then I will buy from whoever is willing to sell… As many are unable to add value, the profiteer establishes a production line of recyclables and fills eight, nine, ten big stores thus adding value from the point of view of economies of scale, supply flow, and price… Nobody sells in the*

industry. Everybody goes through an enormous chain, and that makes the price of the product rise and feeds those profiteers. [municipal public official]

The waste collector has existed since the 70s, the 80s. We've heard the story of his appearance during a national congress. They really belong to the old times. It started like that – people who had lost their jobs, who had no means of survival, collected some tins, some cardboard… [waste collector]

In the course of many talks with waste collectors I have said: 'If you aren't yet, tomorrow you'll be like the coal miners in the beginning of the century. They were extraordinarily valued. During the heavy industry development, the coal miners, although exploited, began to be valued. So, you will be the future coal miners. Today you are worthless, but tomorrow you'll be worth a lot because you'll have recovered raw materials from garbage.' They rested pensive. [municipal public official]

At the beginning was the Portuguese immigrants (like his father), before the cooperative, yes? Right there, with unemployment starting to be the people from the Brazilian Northeast, after that it stars to be a social problem and everybody unemployed, much homelessness and … they take advantage to earn the daily bread. [municipal public official]

Many academic texts affirm that once waste collectors became legitimized and incorporated into the services of waste management, their work began to gain social status. Sicular (1991) cites the example of the City of Mexico, Cairo, and regions of Indonesia, where the status of workers ameliorated as they became integrated into the public service of waste collection. Partnerships of mutual succor emerged in these places. In Brazil, Nogueira (1996) relates the experience of waste collectors in the City of Victoria who, finding work separating recyclable materials in factories, undergo a sort of identity-transformation. Addressing the idea of identity in a different vein, Dias (2002) affirms that the waste collectors of Belo Horizonte gained recognition as lawful citizens from the moment they organized an association.

Defending the notion that waste collectors gain recognition as workers and professionals once they have constructed an identity as such, Bastos (2007) suggests that an identity only emerges once it has been demonstrated that waste collectors are incorporated into the recycling production cycle. The importance of waste collector organizations is corroborated by diverse authors within Brazil and across Latin America (GONÇALVES, 2003; MEDINA, 2000, 2001; RODRIGUEZ, 2003, among others). These observers believe that organizations make the work of waste collectors much more economically viable. Betting on the validity of affirmations such as these, the City of Rio de Janeiro advanced public policies to support the work of waste collectors, principally through the creation of cooperatives.

By the end of 1993, a contract was agreed upon by the municipality and Comlurb [Municipal Cleaning Company], *under Law No. 8,666, to lease spaces for the cooperatives… A cooperative is a mutual action in order to valorize the waste collectors' production. I believe in a promising future, especially to a whole host of people without a chance to study – marginalized, unemployed people with no means of survival. As much as I want to absorb them, I will not succeed. Only garbage can offer these means, under proper safety and hygiene conditions, and with the assurance of their civil rights. The proposition was to freely lease spaces with basic*

> *infrastructure – electricity, toilets, an office, and storage space – and to assist in the management* [of the cooperative]. [municipal public official]

> *A cooperative is an assembly with common ends and goals in order to valorize the product from a four principle point of view: first, to attain an economy of scale in order to acquire advantages in a certain market; second, to improve the production, to qualify it, to obtain a seal of approval to its product; third, to offer the consumer a cycle of frequency in the material supply; and lastly, to compete from the point of view of a competitive market. The production and selling in a small scale are at disadvantage in terms of competitive market when it comes to an economy of scale.* [municipal public official]

In this sense, the City promoted a social change in how waste collectors and their work were portrayed. Interviews confirm that waste collectors sensed this change. In the 1990s they went from being beggars and homeless to being considered recycling 'partners.'

> *Before B* [the municipal public official and coordinator of the program beginning in the 1990s] *we were treated like beggars by the municipal government. There was only the program where the City would round up our belongings off the street* (Cata tralha). *They would take everything and leave it far from the Center.* [waste collector]

> *The foundation of cooperatives aims to reduce garbage volume at the source (the more you collect, the less it goes to the landfill); appreciate the value of the waste collectors' work so that they can be acknowledged in a kind of organization the community can accept, thus ending predatory collectors; and rationalize management practices and costs because the predatory collector litters the streets while he works...* [municipal public official]

This statement helps us think about the transition from policies of repression to policies of support; the transition has not necessarily been linear (FOUCAULT, 1972). This is especially true of the discourse surrounding waste; as indicated by issues previously discussed, in which environmental and economic questions figure prominently. Put differently, the issue's importance to environmental concerns also lends itself to its importance *vis à vis* economic concerns, and vice-versa. In this sense, although environmental and economic issues may be distinct, they are composed of elements that transcend the boundaries of each other's discourses (concomitance).

When new discourses about waste emerge, they inevitably involve new discourses about the environment or the economy. Public authorities provide a good example. They associate recycling as a good for the environment as well as for income-generation. Nevertheless, the norms governing a discourse on waste do not belong exclusively to specialists such as ecologists and engineers, or public managers; other actors must be considered, including psychologists, social workers, and public health workers.

> *There are the plant waste collectors, and the landfill waste collectors, who select the mixed recyclables from domestic refuse... the predatory street waste collectors, who anticipate Comlurb's fleet of trucks, grab what they want and leave a mess behind. There is the street waste collector, who is disciplined and works in an organized way, and who is more socially accepted. He has references and wears an identity card, he goes to condos, to colleges, and to financial centers in order to get the purest material – what the others can't do. The plant waste collector and the landfill waste collector belong to the final destination, and the predatory waste collector*

> *has no access because his presence is culturally aggressive, he lives on the streets, is dirty and has no identity papers. This already has to do with the cooperative… A forth category is the man with a kind of big wheelbarrow who delivers building materials, collects building debris at private properties, moves things (sofas, mattresses, fridges, stoves, beds) and also collects recyclable materials. The street waste collector without a job is one thing, and the person who goes there to sell something but isn't a waste collector, is a servant of a building that has selected something and goes there to sell it during his break—he is something else. He's no waste collector, he does odd jobs. So it's difficult to incorporate everybody in a kind of organization.* [municipal public official]

One thing is for sure, waste collectors are not included among these specialists: they appear as a compositional element in this discourse—as the target of diverse strategies—or as subjects of the discourse (exploited, marginalized). But despite the knowledge they acquire through experience, waste collectors never establish normative discourses themselves. As discussed, Bastos (2007) believes that the construction of a respectable waste collector identity—as a worker and professional—only emerges once s/he is recognized as part of the cycle of production.

Despite this author's affirmation, the objective of supporting policies have been to help the waste collector overcome dependence on the middleman— as long as they provided an economy of scale, added value to the gathered product, helped make workers more applied, and promoted norms of organization through cooperation. Independently of meeting these objectives, in subjective terms it is possible to affirm that, the context permitted those with knowledge to determine a form of work—cooperativism—that was not necessarily suited to waste collectors.

> *The Comlurb set up cooperatives and simply placed the waste collectors within them, but they did not draw up a contract. The cooperatives provided the space and structure, but they wanted to maintain control over the situation and they didn't cede space for the waste collector to negotiate.* [waste collector]

> *Comlurb builds the cooperatives and placed the waste collectors inside, but didn't sign a contract…If there is one, it must be between Comlurb and the space administrator. It (the contract) leased the space and structured it, but wanted to be in charge of the situation and has left no room for the waste collector to negotiate.* [waste collector]

On the other side of the equation, once waste collectors adapted to the system, they did not necessarily meet with economic success—better incomes—as the following section will illustrate.

3.3 The case studies

Breaking with exploitation, ameliorating working conditions and income, and diminishing the build-up of waste in dump sites were the justifications for public policies put into place in the 1990s and 2000s. As well as minimizing health risks by ensuring adequate treatment of waste, these policies also aimed to overcome the stigmas of marginality and informality (BRITO, 2001). The program consisted of installations and spaces—basic infrastructure—as a means of furnishing waste collectors with the ability to accumulate larger quantities of material, providing scale, and adding value through processes such as cleaning (of the recyclables). The Fig 1 below exemplify the recycling chain - the term used to describe the process that starts with the collection of used materials and encompasses all the stages until their final destination, the recycling plants - in Rio de Janeiro city, which is similar to other cases in developing countries.

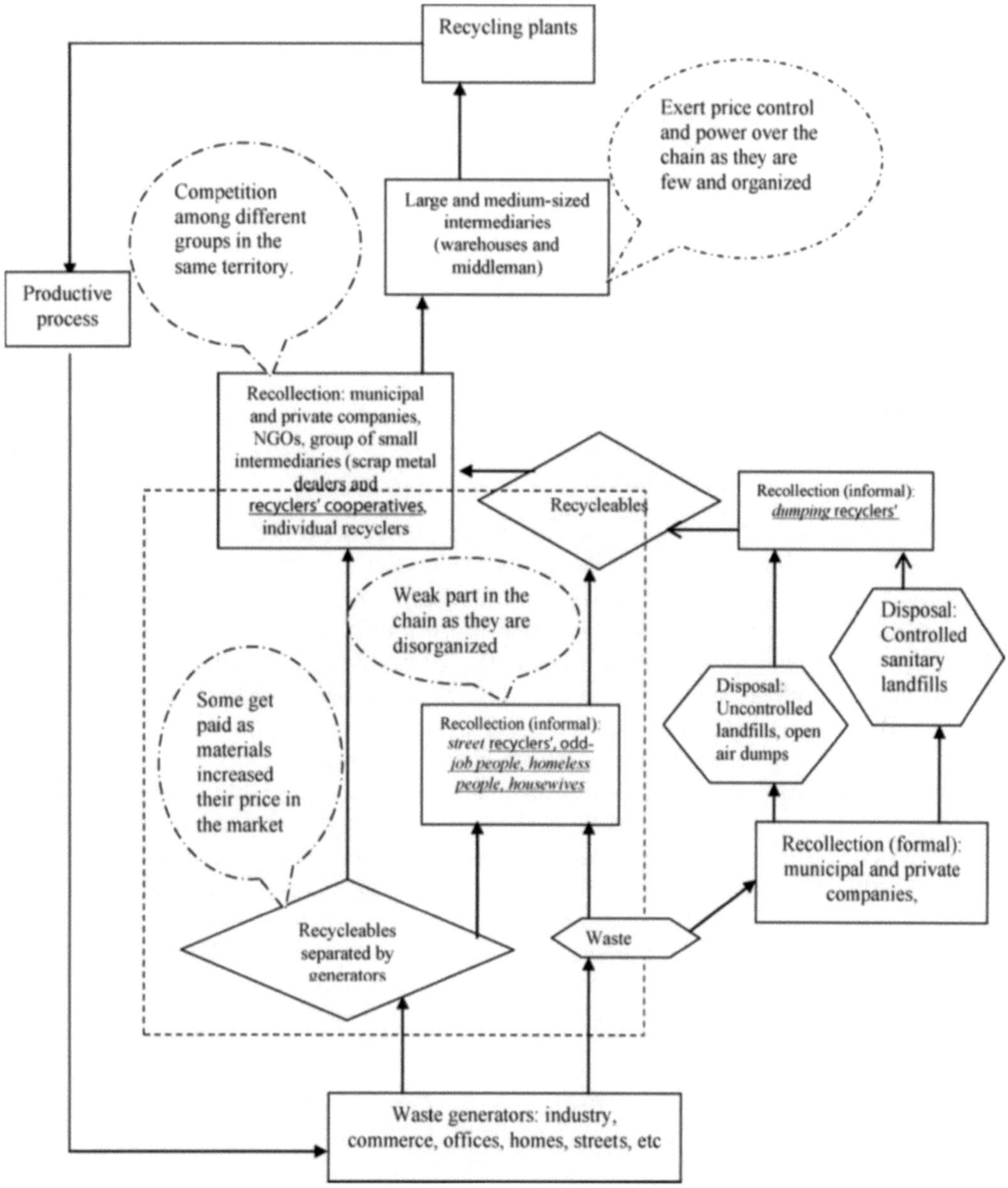

Fig. 1. Waste circulation – the object of this chapter is (first, second and third cases) is surrounded by the dashed rectangle (source: DO CARMO & PUPPIM DE OLIVEIRA, 2010).

The ultimate objective of the public policies, of course, was to help waste collectors negotiate better prices. As well as providing space for materials the Comlurb also offered logistical support to help waste collectors overcome dependence on small middlemen. The earnings that once went to these intermediaries would be shared among members of the group. Among the cooperatives created during the period (18 in 1993/94 and 4 in 2003), three case studies were examined.

> *The hardest part was organizing the cooperative, because the waste collectors used to work alone or in small groups of two or three people, and each dominated a physical space… As they have a low cultural level, it was difficult to establish class consciousness… Our role was to join those groups together encouraging a culture of more solidarity… At first, four cooperative nuclei and a central nucleus were created. Sometimes we tried to help, orienting and assisting them with knowledge so that they could collect more and attain higher prices… The most difficult thing is managing the cooperative… In the process of organizing the waste collectors, it's basic to show them alternatives that will improve their material life… You must always begin with small meetings, talk about the importance of cooperative work, and create commissions.* [municipal public official]

> *In the beginning it was said to be a management course, a technician would be in charge of the administration. Then, this technician would leave and the waste collectors would elect a cooperative member to be the administrator or president. But this never happened, there was only one president (in down town) those cooperatives and he monopolized everything, stole from everyone with the support of Comlurb itself.* [waste collector]

The first case study was undertaken in a cooperative of waste collectors in the Zona Sul or southern part of Rio de Janeiro. The cooperative was administered by a Portuguese immigrant, an ex-waste collector, who bought material obtained at the source (both waste collector and non-waste collector derived) and redistributed it to medium-size middlemen. Unlike what had been proposed by the Comlurb, and despite its scaled operation, this cooperative did not help to ameliorate the earnings of waste collectors. Negotiations were carried out as before, and the proprietor did business with medium-sized middlemen. According to the administrator, the short-term concerns of waste collectors prevented them from building up scale and being able to negotiate better prices: "There is always a segment that doesn't accept organization, which prefers anonymity. They are indeed street people, and it takes time to gain consciousness about the work". In sum, waste collectors did not live or participate within the organization's routines. They restricted their activities to negotiating with it, much the same as people who did not work exclusively as waste collectors and instead used the cooperative as a convenient place to bring recyclables and complement their incomes: "Not so much for me, because I have a job, but for people who don't, this is good. I live nearby. The waiters gather the tins and I bring them. I don't go collecting little tins on the streets, no, not me".

> *There is a lack of jobs, of opportunities; most waste collectors have not got elementary education and have lived on the streets for a long time. Before being collectors they were homeless, or their mothers, their grandmothers, were waste collectors and used to bring* them while they worked. [waste collector]

> *This comes through generations. Today…the waste collector profile is broadening; there are many unemployed people who start to collect waste and don't know how; someone brings a pal who has nothing to do and he works during that weekend but doesn't come back the next, or comes back once in a while. But the waste collector profile is that of someone from Baixada*

> *Fluminense, who hasn't got much job opportunities and discovered scavenging as a means of income.* [waste collector]

The second case study was conducted in a factory for separating select urban waste and recyclables. The factory was established by cooperatives and used competitive public procurement to select cooperatives that would administrate it. It suffered from absenteeism and rotational disputes, and had problems with training, divvying-up chores, auditing goal-attainment, and negotiating the purchase of materials. For its part, the Comlurb provided a storage area for recyclables, took care of the maintenance of the installation, and recruited cooperators among which tasks were divided. The interdependence of the factory—each step depended on achieving another—meant that cooperators had to keep each other accountable. Accountability was particularly important because salaries were distributed equally among cooperators. Widespread dissatisfaction resulted from salary advances (which could constitute up to 70 percent of monthly income), as well as squabbles over the distribution of income and the payment of taxes. As a result, rivalries were common among the cooperators. They blamed each other, the public authorities, and the management, generating dissatisfaction. Supportive policies clearly achieved market success—scale and quality—however, the rules were not to the liking of the waste collectors.

> *If a contract exists, it must be between Comlurb and the space administrator. It loaned the space and structured it, but sought to take charge of the situation and has left no room for the waste collector to negotiate.* [waste collector]

> *Well, if you are leaving a space which offers you rights and a very good profit at the end of the month, it must be because the model isn't good, it doesn't offer you what you want and there is foul play. It's not the cooperative, it's not the waste collector that makes the cooperative bad, it's the people Comlurb places there to assist in the administration... The cooperative generates a lot of money, not for the people who sell, but for the people who buy from us and then re-sell it... I see that Comlurb's intention in organizing the cooperatives is serious, is good... It regarded us as "partners in the cleaning up." But the people placed there to organize it aren't clean; they put a stain on Comlurb's reputation. The first cooperative to end because of that was the Francisco Bicalho (an important avenue at the downtown), it was degraded, its members broke everything. Then the others cooperatives (also ended)."* [waste collector]

The third case study involved an association of waste collectors who worked out of a warehouse provided by the government in the state of Rio de Janeiro. The association first began to receive government support and register waste collectors in 1993, and since 2003 has been administered by an ex-waste collector. This administrator was hired by the NGO responsible for looking after the association, and previously worked with a Portuguese buyer. Incidentally, this buyer owns a recycling depot and buys paper and cardboard from the city in an oligopolistic fashion. The administrator of the warehouse was responsible for maintenance, but did not interfere in the negotiations that took place. Each waste collector operated his or her own work area within the warehouse, and their respective materials: "Collecting is good, because I do it when I want. What is more, I don't have a boss. Nobody tells me what to do".

> *Today, 80% of the street waste collectors have already been through a cooperative and don't want to return... they experienced being fooled, because before joining the cooperative they*

earned, let's say, 150 reais2a week, and afterwards they began to earn 60 reais. So there was no advantage. The cooperative itself should be created to subsidize the waste collectors, to help them to sell the material for a higher price, help them to obtain more material. If there was a true partnership with Comlurb, then it would be possible to double the amount of material because Comlurb has control over all city waste after it leaves your home. So the waste collector would have the opportunity of raising profits and lessening the outflow. But the waste collector's profits diminished and he only paid and paid. [waste collector]

Comlurb created the cooperatives and the waste collector would have to sign up, bring two photos, make an identity card, and be recognized as a member. An internal and an external statute would be created, there would be rules to be followed, discounts on the payment... But everything was done upside-down: first they installed the waste collector in the space, then they created a statute, and afterwards they registered him again. The waste collector had already got used to staying inside the facility to sell his material, not to suffer any discounts on his salary, and to go on with his life. [waste collector]

The biggest problem today is not sending the recyclables directly to the big factories. We have a National Waste Collectors' Commission that meets every two months in Brasilia and we have held a national congress. The questions most posed were about the high tax and the lack of quality (purity) and quantity (scale) to sell to the big factories. They do not come to collect less than twenty *tons...This circles and circles around... It is very difficult to get to the top. [waste collector]*

I would say that 70% of recycled waste comes from people attached to condos, clubs, offices, who make a pre-selection and sell it. But all of them have an activity, they are somehow employed, (...) unlike the waste collectors. They aren't waste collectors, they do it as a supplement. [municipal public official]

The after-use garbage [post-consumer waste] *is very expensive. So, only those who are in charge survive. This feeds a chain of profiteers at the waste collectors' expense. Recycling is a promising market. If you have investment from the government, if you are subsidized, if you strengthen the cooperatives, then it will be a self-supporting activity.* [waste collector]

A waste collector doesn't have to tear the bags; he already knows where the paper is. He has a technique, he feels outside the bag and knows wet garbage from dry (useful garbage). Usually, he who tears the bags is the small waste collector who is neither articulate nor organized. But they're no true waste collectors! [waste collector]

This self-determination had positive effects. Magera (2003) writes that the degree of control exercised by the waste collector over his own material is associated with the degree of manipulation that might occur at the moment of negotiating prices. For the association of waste collectors examined herein, public authorities represented a greater threat (through policies of conspiracy and commission) than they did a source of support (policy of support). Throughout the interviews conducted, the greatest recurring fear of waste collectors was to lose their collection points, especially given the growing interest in waste and recycling among public authorities and society—who now also negotiate recyclables.

---

[2] As of date, US$1 is approximately equal to 1,63 reais.

Discursive Practices	Non Discursive Practices
Sanitary: • Stigma, • Marginality; • Disagreeable; • Symbol of backwardness; • Product of poverty; • Social exclusion; • Unemployment; • Amorphous mass; • Poor working conditions; • Inhuman and illegal activity.	**Sanitary:** • Policies of repression and neglect (kidnapping, expulsion, removal, incentives to change activity; treated as street people/beggars).
Environmental: • Implications for the management of waste and recovery of resources; • From waste collectors (garbage pickers) to Recyclers/environmental agents (see table 1); • Implications in the management of waste and the recovery of resources.	**Environmental:** World Bank initiatives and the questions of the social conditions of the waste collectors (health, work, market).
Economic: • Informality; • Exploitation; • Aptitude for collection in slums; • People who find the means of subsistence in garbage; • Illiterate; • Individualist; • Immediatist; • No class consciousness; • Unorganized/organized (by adhering to the rules of cooperativism).	**Economic:** • Policies of collusion and support (organization of the recyclers into cooperatives, legalization of their activities, labor support laws, concessions for collection contracts, formation of public-private partnerships); • Academic works that approach the topic; • Eco-points; • Selective collection; • Creation of Recyclable Separation Depots[CSRs or Central de Separação de Recicláveis in the city of Rio de Janeiro].

Table 2. The Discourses on Garbage and Elements Attributed to Waste Collectors (source: the author).

4. Discussion

Instead of focusing on the role of the waste collector, the public sector's policies of support and press reports and campaigns have mainly emphasized recycling as an activity of economic value. The establishment of cooperatives has helped contribute to the entry of new actors in the recycling business: waiters, homeless people, domestic employees, and housecleaners, among others. The notion of bettering the work-conditions of waste collectors, as recommended by the public sector, caused workers in the second case study to feel exploited. They were not only upset at colleagues who 'faked' working, but also at managers who maintained unfair information asymmetries—no subjects interviewed knew of the end destination for the material negotiated by managers.

Given that certain skills are required in the identification and separation of material by type—an industry imperative—it is surprising that public authorities perceive waste collectors as knowing little about the industry. This is, after all, an industry and occupation that government has only recently courted.

Contributing to this perception is the evolution of the waste collector as a concept. If it were possible to trace a parallel between the policies of the public sector and the coverage of the media and academic literature, it would be possible to distinguish certain discursive and non-discursive terms, as displayed in Table 2.

From the above table it is possible to identify a few elements that constitute much of what is said (discursive practices) and done (non-discursive practices) in relation to waste collectors. But even if empirical studies were undertaken, what element would they use? Table 2 illustrates how rules conferred upon the waste collector in each case study display different effects.

Case	Rules	Discursive and non-Discursive Practices
First	• Entrepreneurs negotiating with middleman; • High capacity for scale (accumulation) and processing (quality); • Income by individual production (volume collected).	• Neither participation in the negotiation, nor in the advantages of scale and processing; • Temporary work; no participation in collective organization decision-making; • High flexibility and turnover.
Second	• Negotiation with the middleman;	• No participation in the negotiation and, despite sharing the advantages of scale, their income (gain) is lower than in first and third cases;

	• High capacity for scale (accumulation) and processing (quality); • Income by collective production, directly related to group performance (volume collected).	• The structure of the organization creates conditions for the emergence of micropowers due to collective gains; • Temporary work (high turnover); • A feeling of being exploited.
Third	• Negotiation with a medium-sized middleman by themselves (non-intermediaries as in the first and second case); • Capacity for scale (accumulation) and processing (quality) controlled by themselves; • Income by individual production.	• Participation in the negotiation (without intermediaries) with the middleman; • Participation in advantages of scale and processing (controlled individually); • Input on the support they would like to receive from the government; labor affiliation and inherited vocation (in contrast to the turnover in the first and second cases); • Participation in the collective decision-making processes of the organization.

Table 3. Waste Collector Organizations (three case studies) and their Discursive Practices (source: the author).

The above table delimits, specifies, names, and establishes the rules that determine the waste collector's degree of participation in the cooperatives, in accordance with what managers view as being the purpose of their cooperatives. Regarding the disposal of goods, all were successful. With respect to whom the cooperative served—as a source of employment and income—only one case, the third, produced verifiably positive effects due to the agglomeration of groups (cooperators). Success can be attributed to this group not because of the collectivity as a whole, but rather because each cooperator had access to rules governing that organization. By contrast, other cooperatives evinced disinterest and conflicts, and their incomes never bettered what they earned beforehand.

The three case studies were some of the organizational models supported by the municipality of Rio de Janeiro city in response to this new context of waste. The first and the third case studies address the process of how "odd jobbers" and specialized firms respectively attribute value to waste. The second case addresses the consequence of the organization of waste collectors in a factory model of work (according to the municipality's plan).

5. Final considerations

The objective of policies to support cooperatives was to diminish the dependence of waste collectors on buyers. Having observed three case studies, only the third can be called a success. The third case came closest to meeting stipulated public policy goals of scale and negotiating directly with buyers. To what extent did the strategies or policies of support—in the sense of helping waste collectors—attend to the needs of the recyclable market? The flow of goods has improved over the years, as has the response to recycling by society and business. But the same cannot be said about the organization of waste workers.

This idea refers us back to Escobar (1995) who, in conceptualizing poverty, suggests that policies for the development of poor countries are better suited to integrating, managing, and controlling these countries and their populations than in resolving the problems associated with poverty. Waste collectors should be considered authorities on waste, so that they might assume greater participation in the process of solving their own problems. Unfortunately, this ideal was not met in most of the cases examined. Instead, waste collectors gain salience in the media and constitute the target of public policies, but only, however, because specialists in different fields associate them with issues of waste and recycling. Finally, the results suggest that policies did much more to contribute towards improving recycling systems—within a market context— than to improving the economic conditions of waste collectors.

6. Acknowledgment

I thank the Brazilian School of Public and Business Administration of Getulio Vargas Foundation (EBAPE-FGV) for the financial support for this research; the Brazilian Coordination for the Improvement of Higher Education - CAPES (Coordenação de Aperfeiçoamento de Pessoal de Nível Superior) for the scholarship for pursuing y doctoral degree at EBAPE-FGV; to Rita Gabriella Lobo Arruda and Leandro da Mota Damasceno for the invaluable help in part of the data collection; to all people interviewed in this research, particularly the recyclers, and Gregory Michener.

7. References

Adametes, C. M. (1998). O olhar da inclusão: possibilidades de pesquisa com uma catadora de cixo. Unpublished master thesis, Universidade Estadual Paulista Júlio de Mesquita Filho, Araraquara, Brasil

Agunwamba, J.C. (2003). Analysis of Scavenger's Activities Recycling in Some Cyties of Nigeria, Environmental Management, Vol. 32, No. 1, pp. 116-127 ISSN 1432-1009

Ashforth, B., Kreiner, G. (1999). 'How Can you do It?': Dirty Work and the Challenge of Constructing a Positive Identity. Academy of Management Review, Vol. 24, No. 3, pp. 413-434 ISSN 0363-7425

Bartone, C. R. (1990). Economic and policy issues in resource recovery from municipal solid wastes. Resources, Conservation and Recycling, Vol. 4, No. 1/2, (August 1990), pp. 7-23, ISSN 0921-3449

Bastos, V.P. (2007). Na rota do lixo: da casa ao catador o primeiro trajeto da cadeia industrial de reciclagem. Anais do XI Congresso Brasileiro de Assistentes Sociais. Foz do Iguaçu, Rio Grande do Sul, Brasil, October 2007

Batool, S.A.; Chaudhry, N.; Majeed, K. (2008). Economic Potential of Recycling Business in Lahore, Pakistan, Waste Management, No. 28, pp. 294-298 ISSN 0956-053X

Brito, E. (2001). Programa Cooperativas de Catadores (Balanço Sintético - 1993/2000). Rio de Janeiro. Diretoria de Serviços Oeste, Prefeitura Comlurb, (February 2001)

Birkbeck, C. (1979a). Garbage, industry and the 'vultures' of Cali, Colombia. In: Casual Work and the Poverty in Third World Cities, R. Bromley, R.& C. Gerry (Eds.), 161-183, ISBN 9780471997313, NY, John Wiley & Sons

Birkbeck, C. (1979b). Self-Employed Proletarians in an Informal Factory: The Case of Cali's Garbage Dump. World Development, Vol. 6, No. 9/10, pp. 1.173-1.185 ISSN 0305-750X

Castillo Berthier, H. (2003). Garbage, work and society. Resources, Conservation and Recycling, Vol. 39, No. 3, (October 2003), pp. 193-210 ISSN 0921-3449

Cointreau, S. (1986). Environmental Management of urban solid wastes in developing countries—a project guide. The International Bank for Reconstruction and Development/The World Bank. Washington D.C.

Cointreau, S. (1985). Recycling form Municipal Refuse: a state-of-the-art review and annotated bibliography. Integrated Resource Recovery. World Bank Technical Paper Number 30. UNDP Project Management Report Number 1, Washington D.C.

Dias, S.M. (2002). Lixo e cidadania: os impactos da política de resíduos sólidos de Belo Horizonte no mundo do trabalho do catador da Asmare. Encontro Nacional de Estudos Populacionais, Ouro Preto, Minas Gerais, Brasil

Do Carmo, M. S.; Migueles, C. P.; Puppim de Oliveira, J. A. P. (2004). A semântica do lixo, o estímulo à reciclagem e o trabalho dos catadores do Rio de Janeiro: um estudo da relação entre significado e ação econômica. Revista Integração CETS - Centro de Estudos do Terceiro Setor FGV—EAESP ISSN 0011-5258

Do Carmo, M. S. (2008). A problematização do lixo e dos catadores: estudos de caso múltiplo sobre políticas públicas sob uma perspectiva foucauldiana. Unpublished dissertation, Escola Brasileira de Administração Pública e de Empresas da Fundação Getulio Vargas, Rio de Janeiro, Brasil

Do Carmo, M.S. (2009a). A Semâtica "negativa` do lixo como aspecto ´positivo` - um estudo de caso sobre uma associação de recicladores na cidade do Rio de Janeiro, Brasil. Adminitração Pública e Gestão Social. Vol. 1, No. 2, pp. 21-50, abr/jun ISSN 2175-5787

Do Carmo, M. S. (2009b). A semântica negativa do lixo e o desenvolvimento socioeconômico dos catadores de recicláveis - considerações sobre um estudo de caso múltiplo em cooperativas na cidade do Rio de Janeiro. Cadernos EBAPE.BR, Vol. 7, pp.1-16 ISSN 1679-3951

Do Carmo, M.S. (2010). A problematização do lixo e as políticas municipais de apoio ao trabalho dos catadores na cidade do Rio de Janeiro sob a perspectiva analítica de Michel Foucault. In: Novas Idéias em Administração, P.R Motta; R. Pimenta; E. Tavares (Eds.), 179-219, Editora FGV, ISBN 978852250842-6, Rio de Janeiro, Brasil

Do Carmo, M. S.; Arruda, R. G. L. A. (2010). Trabalho com resíduos - considerações sobre reconhecimento social e identidade profissional. Revista de Gestão Social e Ambiental, Vol. 4, pp. 178-194 ISSN 1981-982x

Do Carmo, M. S.; Puppim de Oliveira, J.A. (2010). The Semantics of Garbage and the organization of the recyclers: Implementation. Resources, Conservation and Recycling, Vol. 54, pp.1261-1268 ISSN 0921-3449

Eigenheer, E. (2003). Lixo, vanitas e morte: considerações de um observador de resíduos, EdUFF, ISBN 8522803609, Niterói, Brasil

Escobar, A. (1995). Encountering development: the making and unmaking of the third world, Princeton University Press, ISBN 978-1-4008-2146-4, New Jersey, USA

Foucault, M. (1972). The archaelogy of knowledge, Pantheon Books, ISBN 9780394711065 0394711068, New York, USA

Foucault, M. (1980). Two lectures. In: Power knowledge, C. Gordon, 109-33, Pantheon Books, ISBN 13: 9780394739540, New York, USA

Foucault, M. (1984). História da sexualidade - volume 2: o uso dos prazeres. Graal, ISBN 8570380755, Rio de Janeiro, Brasil

Foucault, M. (1991a). Governmentality. In The Foucault effect: studies in governmentality, G. Burchell, C. Gordon, & P. Miller, 87-104, ISBN 0226080447, University of Chicago Press, USA

Foucault, M. (1991b). Politics and the study of discourse. In The Foucault effect: studies in governmentality, G. Burchell, C. Gordon, & P. Miller, 53-72, ISBN 0226080447, University of Chicago Press, Chicago, USA

Foucault, M. (1995). O sujeito e o poder. In: Michel Foucault, uma trajetória filosófica: para além do estruturalismo e a hermenêutica, P. Rabinow & H. Dreyfus, 231-249, Forense Universitária, ISBN 978-85-225-0833-4, Rio de Janeiro, RJ, Brasil

Freud, S. (1976). Três ensaios sobre a teoria da sexualidade. Obras Completas de Sigmund Freud. Vol. 7, Imago, ISBN 978-85-359-1743-7, Rio de Janeiro, Brasil

Furedy, C. (1984). Socio-political aspects of the recovery and recycling of urban wastes in Asia. Resources, Conservation and Recycling, Vol. 7, No. 2/4, pp.167-173 ISSN 0921-3449

Gonçalves, P. (2003). A reciclagem integradora dos aspectos ambientais, sociais e econômicos, Fase, DP&A, ISBN 8574902659, Rio de Janeiro, R.J., Brasil

Harding, N. (2003). The Social Construction of Management: Texts and Identities, Routledge, ISBN 0415369428, London, United Kingdom

Hughes, E. (1962). Good People and Dirty Work, Social Problems, Vol. 10, pp 3-11 ISSN 00377791

Kaseva, M.E.; Gupta, S.K. (1996). Recycling: an environmentally friendly and income generating activity towards sustainable solid waste management. Case study: Dar es Salaam City, Tanzania. Resources, Conservation and Recycling, Vol. 7, No. 4, (October 1996), pp. 299-309 ISSN 0921-3449

Kaseva, M.E.; Mbuligwe, S.E.; Kassenga, G. (2002). Recycling inorganic domestic solid wastes: results from a pilot study in Dar es Salaam City, Tanzânia. Resources, Conservation and Recycling, No. 35, pp. 243-257, ISSN 0921-3449

Magera, M.C. (2003). Os empresários do lixo: um paradoxo da modernidade: análise interdisciplinar das cooperativas de reciclagem de lixo. Átomo, ISBN 10: 8587585436, Campinas, São Paulo, Brasil

Medina, M. (1998). Border scavenging: a case study of aluminum recycling in Laredo, TX and Nuevo Laredo, Mexico. Resources, Conservation and Recycling, Vol. 23, pp. 107-126 ISSN 0921-3449

Medina, M. (2000). Scavenger cooperatives in Asia and Latin America. Resources, Conservation and Recycling, Vol. 31, No. 1, (December), pp. 51-69 ISSN 0921-3449

Medina, M. (2001). Scavenging in America: back to the future? Resources, Conservation and Recycling, Vol. 31, No. 3, (March 2001), pp. 229-240, ISSN 0921-3449

Medina, M. (2007). The world's scavenger: salvage for sustainable consumption and production, Altamira Press, ISBN 13: 9780759109414, United Kingdon

Mòreno-Sanchez, R.P.; Maldonado, J.H. (2006). Surviving from garbage: the role of informal waste-pickers in a dynamic model of solid-waste management in developing countries. Environment and Development Economics, No. 11, pp. 371-391, ISSN 1355-770X

Munford, L. (1998). A cidade na história: suas origens, transformações e perspectivas, Martins Fontes, ISBN 8533608470, São Paulo, S.P., Brazil

Nogueira, J. (1996). Viver sobre o lixo e sobreviver do lixo: um estudo com catadores de lixo em Vitória Espírito Santo, Dissertação (Mestrado em Psicologia) - Instituto de Psicologia, Universidade Federal do Espírito Santo, Vitoria, E.S., Brasil

Ojeda-Benitez, S.; Armijo-de-Veja, C.; Ramirez-Szelinsk, B. (1988). The New Waste Avoidance and Waste Management Act (WMA). Resources, Conservation and Recycling, Vol. 2, No. 1, (December 1988), pp. 3-11 ISSN 0921-3449

Portilho, M.F.F. (1997). Profissionais do lixo: um estudo sobre as representações sociais de engenheiros, garis e catadores. Dissertação (Mestrado em Psicologia) - Instituto de Psicologia, Universidade Federal do Rio de Janeiro, Rio de Janeiro, RJ, Brasil.

Rodrigues, J. C. (1992). A Cultura do Lixo e sua Angústia. In: Falas em torno do lixo, Rodrigues, F.L., 7-11, Nova/Iser/Polis, ISBN 85-86624-Il-X, Rio de Janeiro, Brasil

Rodrigues, J.C. (1995). Higiene e ilusão: o lixo como invento social, NAU, ISBN 858593624, Rio de Janeiro, R.J., Brasil

Rodrigues, L. A globalização no lixo. O Globo, Economia, pp. 31-32, (November 13th. 2005), Rio de Janeiro, R.J., Brasil

Rodriguez, C. (2003). À procura de alternativas econômicas em tempos de globalização: o caso das cooperativas de recicladores de lixo na Colômbia. In: Produzir para viver:

os caminhos da produção não capitalista, B.S. Santos (Ed.), 281-314, Afrontamento, ISBN 85-200-0605-1 Porto, Portugal

Sicular, D. (1991). Pockets of Peasants in Indonesian Cities: The Case of Scavengers. World Development, Vol. 19, No. 2/3, pp. 137-161 ISSN 0305-750X

4

Group Collection of Recyclables in Japan

Shigeru Matsumoto
Aoyama Gakuin University
Japan

1. Introduction

To promote a sustainable environment, various recycling programs have been implemented worldwide that have collected a wide variety of materials. From 1992 to 2005, the recycling rate of paper and cardboard increased from 33% to 50% in the United States. Over the same period, the recycling rate of glass increased from 51% to 65% in the European Union (OECD 2008). The recycling industry handles more than 600 million tons of recyclables every year and employs approximately 1.6 million people worldwide (Bureau of International Recycling 2011). These data show that recycling has been successfully integrated into modern economies.

One of the distinguishing features of recycling programs is that support from households is required for their success. Household waste consists of a wide variety of waste products such as paper, glass, organic waste, plastics, textiles, and small chemical waste. In order to recycle waste, households are required to separate part of their heterogeneous waste into homogeneous streams (Aalbers and Vollebergh 2008). Additionally, they have to wash and store recyclables at home before taking them to collection sites on designated collection days. These recycling duties require a lot of effort, and the ability and willingness to engage in such activities vary substantially across households. Bontoux et al. (1996) stated that the most important factor that determines the overall profitability of recycling programs is the cost of collection and sorting of recyclables.

Previous studies based on household-level data have examined the relationship between sociodemographic variables and recycling intensity, with the most commonly examined sociodemographic variables being gender, age, education, and income (Saphores et al. 2006). These authors report that female, senior, well-educated, and wealthy people are more active recyclers. However, they also report that sociodemographic characteristics can only partially explain household recycling behavior (Matsumoto 2011). Thus, recent studies have investigated the determinants of recycling behavior other than sociodemographic variables.

Many recycling programs are powered by volunteers and concerned citizens. Therefore, active community participation is considered to be a key factor for the success of recycling programs. The influence of the local community on household recycling behavior has been extensively studied in the past decade. Bruvoll and Nyborg (2004) and Halvorsen (2008) reported that peer influence could be an important factor in household recycling behavior. These authors further demonstrated that people carry out their recycling duties only when they believe their peers carry out theirs.

Although the role of the community in encouraging recycling activities has been examined at an individual case level, it has not yet been investigated at a municipality level. Thus, the aim of this study is to examine the extent to which the recycling performance of the municipality is improved by community support.

In this study, I focus on group collection programs. In a group collection program, local residents form a recycling group and agree a contract with a waste management company. Residents separate recyclables at home and take them to the collection site on a designated day. The contracted company then collects and transports these recyclables to the centralized recycling station. It is expected that recyclables are collected more efficiently in group collection programs and thereby the overall cost of recycling is reduced. Because of these reasons, many municipalities encourage group collection programs and subsidize recycling groups.

Using annual survey data on solid waste in Japan, I conduct panel data analyses to answer the two research questions: (1) to what extent do group collection programs improve the efficiency of recyclables collection and (2) how much cost is saved if a group collection program is implemented?

The remainder of the paper is organized as follows. Section 2 describes the waste management flow in Japan and explains how group collection programs work. In Section 3, I report the descriptive statistics. In Section 4, I compare recycling efficiency across waste collection methods to show that group collection is an effective approach. In Section 4, I also compare the waste management costs of different approaches. Section 5 concludes the paper.

2. Collection of recyclables

2.1 Waste management flow

Figure 1 shows the waste management flow in Japan. In 2003, the total volume of municipal solid waste generated in Japan was 54,435,686 metric tons. Of this, 165,163 tons of waste were treated (either incinerated or buried) at individual homes.

The first stage of the separation of recyclables is conducted at individual homes. Recyclables are extracted, washed, and collected through the group collection method. The volume of recyclables collected through this method in 2003 was 2,829,003 tons.

The collection of solid waste is the municipality's responsibility under Japanese law. All waste is collected by the municipality or the waste management company contracted by the municipality. The total volume of waste collected in this way in 2003 was 51,441,520 tons.

The second stage of the separation of recyclables is carried out at waste treatment facilities. Here, recyclables are extracted from the delivery by the municipality. The total volume of recyclables extracted at waste treatment facilities in 2003 was 8,600,214 tons. The volume of recyclables extracted directly from the waste was 2,271,871 tons, while 6,328,343 tons of recyclables were extracted after treatment. The remaining waste was incinerated for reduction. The volume of the final disposal was 8,451,882 tons.

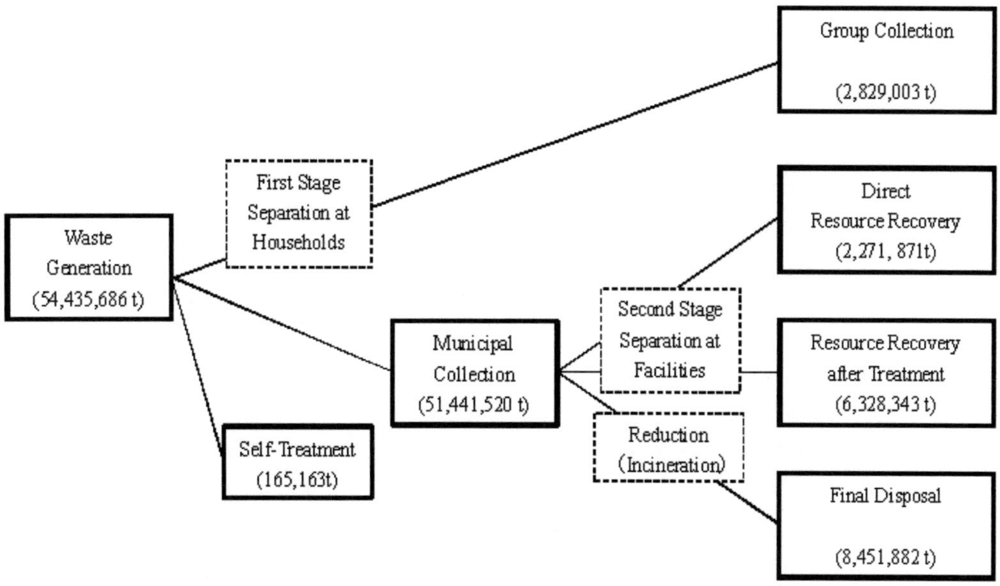

Fig. 1. Waste Management Flowchart for Japan (2003)

2.2 Group collection of recyclables

Table 1 presents the relative share of recyclables extraction in the two stages described above. The share of total recyclables extraction at waste treatment facilities is 69.1%, while the share at individual homes is 30.9%. Paper, metal, and glass are the top three recyclables. Newspapers, magazines, cardboard, and milk cartons are also collected by group collection. A large amount of paper is collected by group collection. Compared with the degree of paper collection, that of metal and glass is relatively minor. (4.1% and 5.8%, respectively).

	Treatment Facility (Municipal Collection)		Individual Homes (Group Collection)	
	Metric tons	(Percentage)	Metric tons	(Percentage)
Total	6,328,343	(69.1%)	2,829,003	(30.9%)
Paper	2,217,419	(45.7%)	2,638,169	(54.3%)
Metal	1,303,260	(95.9%)	55,372	(4.1%)
Glass	865,195	(94.2%)	53,125	(5.8%)
Other	1,942,469	(95.9%)	82,337	(4.1%)

Table 1. Collection Methods and Resource Recovery (2003)

When local residents decide to start a group collection scheme, they form a recycling group. It is common that a preexisting community group such as a parent-teacher association or residents' association adds the role of the recycling group. Once a recycling group is formed, a representative of the group contacts the municipality. Many municipalities encourage their residents to formalize group collection programs, and they provide subsidies according to the volume of recyclables collected.[1]

Picture 1. The Recyclables Collection Process

Once a group collection program starts, residents can bring recyclables to the collection point on a designated day. However, participation in the group collection is voluntary. If residents do not want to participate in the program, they can still dispose of their recyclables. Picture 1 shows how a typical group collection scheme in Japan operates. Households separate recyclables at home and take them to the recycling station. The recyclables are then collected by the waste treatment company contracted by the recycling group. The frequency of collection varies by municipality and type of recyclables.

3. Data

This study used the annual survey of municipal solid waste in Japan as its primary data source. All municipalities record the conditions of waste treatments and recycling activities and report this to the central government. The central government then summarizes the data and publishes it as the annual survey of municipal solid waste. This survey reports the volume of waste collected through each collection method as well as the volume of recyclables extracted from this waste.

These data are downloadable from the website of the Ministry of the Environment (2011). I set the sampling period as 1998 to 2002 for the following reasons. To improve the efficiency of the administrative work of local governments, the central government encouraged the merger of local governments in mid-2000. As a result, the number of municipalities decreased from 3232 in 1999 to 1724 in 2011. This merger of municipalities accelerated after 2003.

---

[1] For example, Yokohama City (2011) and Kawasaki City (2011) showed that these cities subsidize recycling groups at the rate of 3 yen/kg.

Recyclables collection (tons) [a]	Average	Standard Deviation
Total	431.16	4354.37
Paper	271.94	3846.15
Metal	−0.67	723.12
Glass	5.49	612.45
Group Collection (tons) [a]	Average	Standard Deviation
Total	68.55	1499.03
Paper	73.04	1375.54
Metal	−1.58	97.69
Glass	−6.24	139.78
Waste Collection (tons) [a]	Average	Standard Deviation
Total	162.65	6549.93
Mixed Garbage	−25.90	2721.71
Burnable Garbage	393.55	11740.62
Unburnable Garbage	−178.98	2258.98
Recyclable Waste	335.29	5107.55
Other Garbage	5.89	477.27
Bulky Garbage	−54.88	1089.98
Direct Delivery	−312.33	17467.91
Cost (1,000 yen) [b]	Average	Standard Deviation
Collection Cost	−900.22	101544.57
Treatment Cost	−1260.23	350340.74
Final Disposal Cost	951.69	106910.88
Outsourcing Cost	20782.36	120189.24
Kumiai Cost	17679.12	515534.36

Note: a: Sample extent: all 3,212 municipalities, all years from 1998 to 2002. b: Sample extent: all 3,212 municipalities, 1998 vs. 2002.

Table 2. Descriptive Statistics of Waste Collection Data (1998-2002)

The descriptive statistics of the data are reported in Table 2. The total amount of recyclables collected in the average municipality increased by 431.16 tons from 1998 to 2002. Group collection increased by 68.55 tons. The group collection of paper increased, while that of metal and glass decreased. In addition to the recyclables collected though group collection, recyclables were also extracted from seven other types of waste: mixed garbage, burnable garbage, unburnable garbage, recyclable waste, other garbage, bulky garbage, and direct delivery. From 1998 to 2002, total waste collection increased by 162.65 tons. The collections of burnable garbage and recyclable waste increased during the sampling period. By contrast, the collections of mixed garbage, unburnable garbage, and bulky garbage decreased.

In the bottom part of Table 2, I compare the costs of waste management between 1998 and 2002. I include five categories of waste management costs: the cost of waste collection, the treatment cost at waste treatment facilities, the final disposal cost, the cost of outsourcing, and the *kumiai* cost (see below). Municipalities often outsource waste management tasks to private companies. The outsourcing cost is the money paid to these private companies. A municipality sometimes forms a *kumiai* (a waste management organization) with other municipalities in order to jointly collect and treat waste. The *kumiai* cost is the cost paid to the *kumiai*. The data show that the collection and treatment costs decreased from 1998 to 2002. By contrast, the remaining three costs increased.

4. Empirical findings

4.1 Impact on recyclables collection

I first examine how much recyclable material is extracted from waste. The composition of waste can vary substantially across municipalities. For example, the share of used paper is large in one municipality, while the share of used glass is large in another. To account for municipality-specific factors, I employ a difference-in-difference model in this paper. Specifically, I estimate the following equation:

$$\Delta R_{it} = \beta_0 + T_t + \beta_g \Delta G_{it} + \beta_w \Delta W_{it} + u_{it} \qquad (1)$$

where ΔR_{it} is the change in total recyclables collection, ΔW_{it} is the change in total waste collection and ΔG_{it} is the change in group collection in municipality i at time t. The number of municipalities included in the analysis is 3212 and the sampling period is 1998–2002.

I examine how the change in waste collection is associated with the change in recyclables collection. T_t is the time trend variable, while u_{it} is the error term.[2] As mentioned before, all materials collected through group collection programs are defined as recyclables in the Japanese system. Hence, I impose the condition of $\beta_g = 1$ and estimate equation 1.

The estimation result of equation 1 is reported in the first column of Table 3. As the table shows, the coefficient of the time trend variable is positive and significant. On average, the amount of recyclables collected increased during the sampling period. Based on the estimation result, 21.6% of waste is recycled using municipal collection methods.

Group collection is the most efficient collection method because all collected materials are utilized as recyclable inputs. To compare the efficiency of the other seven collection methods, I estimate the following equation:

$$\Delta R_{it} = \beta_0 + T_t + \beta_g \Delta G_{it} + \boldsymbol{\beta}_{wc} \Delta \boldsymbol{WC}_{it} + u_{it} \qquad (2)$$

In equation 2, $\Delta \boldsymbol{WC}_{it}$ is the vector of the changes in the waste collected through these seven collection methods. All generated waste is collected through one of these seven methods.

[2] To account for unobserved heterogeneity, a robust covariance matrix is used in the estimation.

	Total	Collection Methods
Constant	35.321	−0.280
	(79.724)	(31.819)
Time Trend	144.951**	62.304**
	(29.111)	(11.664)
Group Collection (Fixed Parameter)	1.000	1.000
Municipal Collection		
Total	0.216**	
	(0.005)	
Mixed Garbage		0.025**
		(0.005)
Burnable Garbage		−0.066**
		(0.004)
Unburnable Garbage		−0.043**
		(0.007)
Recyclable Waste		0.633**
		(0.006)
Other Garbage		0.105**
		(0.028)
Bulky Garbage		0.129**
		(0.014)
Direct Delivery		−0.065**
		(0.003)
Adjusted R^2	0.282	0.886

Note: Standard errors in parentheses. Superscript * and ** denote significance at 5% and 1%, respectively. Sample extent: all 3,212 municipalities, all years from 1998 to 2002.

Table 3. Efficiency comparison across collection methods: aggregated recyclables (OLS model, N = 12,848)

The amount of recyclables always increases as the volume of collected waste increases. However, if recyclables are collected through an inefficient collection method (e.g., if paper is collected as mixed garbage rather than as recyclable waste), then the overall recycling efficiency decreases. Since the recycling efficiency of the mixed garbage method is much lower than that of the recyclable waste method, the total amount of paper recycled decreases. If the indirect effect of the collection method change dominates the direct effect of the waste increase, then I obtain $\beta_{wc} < 0$.

The estimation result of equation 2 is presented in the second column of Table 3. The most efficient municipal collection method is recyclable waste as expected. Approximately 63% of the waste collected through this method is recycled.

The second most efficient method is bulky garbage collection. Approximately 13% of this category of waste is recycled. These recyclables can be mixed with burnable and unburnable garbage, but their extraction is difficult. Therefore, as the volume of this garbage increases, the overall recycling efficiency is reduced.

	Paper	Metal	Glass
Constant	−26.012	56.310**	2.410
	(18.771)	(12.702)	(8.209)
Time Trend	20.497**	−8.657	−12.090**
	(6.881)	(4.656)	(3.009)
Group Collection (Fixed Parameter)	1.000	1.000	1.000
Municipal Collection			
Mixed Garbage	−0.026**	0.025**	−0.006**
	(0.003)	(0.002)	(0.001)
Burnable Garbage	0.004	0.007**	−0.034**
	(0.002)	(0.001)	(0.001)
Unburnable Garbage	0.076**	−0.070**	0.036**
	(0.004)	(0.003)	(0.002)
Recyclable Waste	0.532**	−0.174**	0.193**
	(0.004)	(0.002)	(0.002)
Other Garbage	−0.043*	−0.014	0.007
	(0.017)	(0.011)	(0.007)
Bulky Garbage	0.152**	0.068**	0.010**
	(0.008)	(0.006)	(0.004)
Direct Delivery	−0.048**	−0.044**	0.016**
	(0.002)	(0.001)	(0.001)
Adjusted R^2	0.947	0.309	0.615

Note: Standard errors in parentheses. Superscript * and ** denote significance at 5% and 1%, respectively. Sample extent: all 3,212 municipalities, all years from 1998 to 2002.

Table 4. Efficiency comparison across collection methods: paper, metal, and glass (SURE model, N = 12,848)

As presented in Section 2.2, paper, metal, and glass are the top three categories of recyclables. I now evaluate the extraction efficiencies of these three categories based on the following seemingly unrelated linear regression equations (SURE) model:

$$\Delta R_p = \boldsymbol{\beta}_p \boldsymbol{X}_p + u_p \tag{3-a}$$

$$\Delta R_m = \boldsymbol{\beta}_m \boldsymbol{X}_m + u_m \tag{3-b}$$

$$\Delta R_g = \boldsymbol{\beta}_g \boldsymbol{X}_g + u_g \tag{3-c}$$

$$\mathrm{E}[u_j | \boldsymbol{X}_p, \boldsymbol{X}_m, \boldsymbol{X}_g] = 0 \tag{3-d}$$

$$\mathrm{E}[u_j \cdot u'_k | \boldsymbol{X}_p, \boldsymbol{X}_m, \boldsymbol{X}_g] = \sigma_{jk} \cdot \boldsymbol{I} \tag{3-e}$$

where $\boldsymbol{\beta}_j \boldsymbol{X}_j = \beta_{0j} + T_j + \beta_{Gj} \Delta G_j + \boldsymbol{\beta}_{wcj} \Delta \boldsymbol{WC}$. It is assumed that the marginal impact of waste increase on recyclables collection varies across recyclables. The disturbances across equations are allowed to be correlated.

The estimation results of equation 3 are presented in Table 4. I imposed the condition of $\beta_{gj} = 1$ as before. The first column shows that paper is extracted from the recyclable waste (53.2%) and bulky garbage (15.2%) methods. By contrast, the coefficients of mixed garbage, other garbage, and direct delivery are negative. This result implies that the extraction of paper becomes difficult once it is mixed with other waste. Metal is extracted from mixed garbage (2.5%), burnable garbage (0.7%), and bulky garbage (0.7%). The coefficient of recyclable waste is negative and statistically significant. Together with the fact that most metal is collected through municipal collection, the separation of metal is not important for recycling efficiency. Metal is extracted at treatment facilities anyway. Glass is extracted from unburnable garbage (3.6%), recyclable waste (19.3%), bulky garbage (1.0%), and direct delivery (1.6%). Unlike metal, this result shows that waste separation is important for glass recycling.

4.2 Impact on waste management costs

To compare the costs of waste management, I include five cost categories: the cost of waste collection, the treatment cost at waste treatment facilities, the final disposal cost, the cost of outsourcing, and the *kumiai* cost. I compare the waste management costs in 1998 with those in 2001 to examine how increases in waste collections are associated with increases in waste management costs.

For this empirical estimation, I use the following seemingly unrelated regression model:

$$\Delta C_{CL} = \boldsymbol{\gamma}_{CL} \boldsymbol{X}_{CL} + u_{CL} \tag{4-a}$$

$$\Delta C_{TR} = \boldsymbol{\gamma}_{RL} \boldsymbol{X}_{TR} + u_{TR} \tag{4-b}$$

$$\Delta C_{FD} = \boldsymbol{\gamma}_{FD} \boldsymbol{X}_{FD} + u_{FD} \tag{4-c}$$

$$\Delta C_{OS} = \boldsymbol{\gamma}_{OS} \boldsymbol{X}_{OS} + u_{OS} \tag{4-d}$$

$$\Delta C_{KU} = \boldsymbol{\gamma}_{KU} \boldsymbol{X}_{KU} + u_{KU} \tag{4-e}$$

$$\mathrm{E}[u_j | \boldsymbol{X}_{CL}, \boldsymbol{X}_{TR}, \boldsymbol{X}_{FD}, \boldsymbol{X}_{OS}, \boldsymbol{X}_{KU}] = 0 \tag{4-f}$$

$$\mathrm{E}[u_j \cdot u'_k | \boldsymbol{X}_{CL}, \boldsymbol{X}_{TR}, \boldsymbol{X}_{FD}, \boldsymbol{X}_{OS}, \boldsymbol{X}_{KU}] = \sigma_{jk} \cdot \boldsymbol{I} \tag{4-g}$$

where $\gamma_j \boldsymbol{X}_j = \gamma_{0j} + \gamma_{Gj} \Delta G_j + \boldsymbol{\gamma}_{wcj} \Delta \boldsymbol{WC}$.

The estimation result of equation 4 is reported in Table 5. This table shows that the impact of waste collection method on waste management costs varies. The cost of waste collection increases as the volume of mixed garbage increases. Based on this result, the cost of mixed garbage collection is 988 yen/ton. By contrast, I find that cost decreases as the volume of burnable garbage increases.

Treatment cost increases as the volume of mixed garbage increases. Based on this result, the average treatment cost of mixed garbage is 6304 yen/ton. By contrast, treatment cost decreases as the volume of recyclable waste and bulky garbage increases. The reductions in treatment cost are 24,830 yen/ton for recyclable garbage and 36,424 yen/ton for bulky garbage.

The final disposal cost increases as the volumes of mixed garbage, recyclable waste, direct delivery, and group collection increase. This result suggests that a certain proportion of the materials collected as recyclables end up in landfill. Based on these results, the final disposal cost of recyclable waste is 3464 yen/ton, while that of group collection is 11,385 yen/ton.

As discussed, municipalities outsource such waste management tasks to private companies. Table 5 shows that the outsourcing cost increases as the volumes of recyclable waste and group collection increase. The costs are 16,725 yen/ton for recyclable waste and 13,257 yen/ton for group collection.

	Collection	Treatment	Disposal	Outsource	*Kumiai*
Constant	4030.4**	11509.3**	452.9	13437.5**	−2620.2
	(1198.3)	(1573.4)	(1495.2)	(1585.6)	(1830.4)
Mixed	0.988**	6.304**	4.097**	1.281**	−1.833**
Garbage	(0.362)	(0.475)	(0.452)	(0.479)	(0.553)
Burnable	−2.826**	−7.882**	−2.394**	−5.957**	12.833**
Garbage	(0.308)	(0.404)	(0.384)	(0.407)	(0.470)
Unburnable	0.170	−4.755**	0.468	−2.029*	5.605**
Garbage	(0.606)	(0.796)	(0.756)	(0.802)	(0.926)
Recyclable	−6.770**	−24.830**	3.464**	16.725**	31.869**
Waste	(0.570)	(0.748)	(0.711)	(0.754)	(0.870)
Other	−5.978	−13.873**	−6.743	−1.017	1.453
Garbage	(3.093)	(4.062)	(3.860)	(4.093)	(4.725)
Bulky	1.710	−36.424**	−0.867	−10.182**	32.376**
Garbage	(1.102)	(1.447)	(1.375)	(1.459)	(1.684)
Direct	0.550*	7.590**	3.147**	4.605**	−12.120**
Delivery	(0.271)	(0.356)	(0.338)	(0.359)	(0.414)
Group	0.008	−2.698**	11.385**	13.257**	8.681**
Collection	(0.788)	(1.034)	(0.983)	(1.042)	(1.203)
Adjusted R^2	0.578	0.939	0.407	0.473	0.962

Note: Standard errors in parentheses. Superscript * and ** denote significance at 5% and 1%, respectively.

Table 5. Cost Comparison across Collection Methods. (SURE model, N=3212)

I also estimated the impact of waste collection on the *kumiai* cost. Increases in recyclable waste, bulky garbage, and group collection lead to an increase in the *kumiai* cost. The costs are 31,869 yen/ton for recyclable waste and 8681 yen/ton for group collection.

Most recyclables are collected through either recyclable waste or group collection. The total cost of the recyclable waste method is 20,458 yen/ton, while that of the group collection method is 30,633 yen/ton. Therefore, the total waste management cost of the group collection method is higher than is that of the recyclable waste method. The benefits of the recyclable waste method, however, heavily rely on cost reductions at the treatment facilities. If the advantage of the treatment cost is removed, then the costs of the recyclable waste method jump to 52,058 yen/ton.

5. Conclusion

The vast majority of municipalities in Japan encourage their residents to use group collection programs, and some municipalities even provide financial support. In this paper, I evaluated such group collection programs based on two efficiency criteria: recycling and cost.

I examined how a change in waste collection method is associated with a change in recyclables collection and found that mixing recyclables with unsorted waste lowers the efficiency of recyclables extraction. In group collection programs, recyclables are separated at home and the mixing of recyclables with unsorted waste can be avoided. Therefore, group collection is the most efficient collection method. Based on the presented empirical findings, the second most efficient collection method is the recyclable waste method. Approximately 63% of waste collected in this way is transformed into recycling inputs. The extraction rate of recyclables becomes much lower if recyclables are collected using other methods. By utilizing group collection programs, more recyclables are obtained.

To evaluate cost efficiency across collection methods, I estimated the impact of waste collection on waste management costs. I included five varieties of waste management costs in the analysis: the cost of waste collection, the treatment cost at waste treatment facilities, the final disposal cost, the outsourcing cost, and the *kumiai* cost. The empirical results showed that the impact of waste collection varies across waste types as well as collection methods. Based on the estimation results, the cost of recycling using the municipal collection method is 20,458 yen/ton. By contrast, the cost using the group collection method is 33,323 yen/ton.

However, because the extraction rate of recyclable waste is 0.63, the net cost of municipal collection is actually 32,473 yen/ton (20,458/0.63). This implies that the cost of recyclables extraction is about the same for group and municipal collections. Contrary to the common assumption that the cost of recycling is reduced by implementing group collection programs, the findings of this paper suggest that the costs are similar.

6. Acknowledgments

This study was supported by the Ministry of Education, Science, Sports, and Culture of Japan, Grant-in-Aid for Scientific Research B (21330056).

7. References

Aalbers, R. F. T. and Vollebergh, H. R. J. (2008) An economic analysis of mixing wastes. *Environmental and Resource Economics* Vol. 39, pp. 311–30.

Bontoux, L., Leone, F., Nicolai, M. and Papameletiou, D. (1996) The recycling industry in the European Union: Impediments and Prospects. Institute for prospective technological studies. A report prepared by IPTS for the committee for environment, public health and consumer protection of the European parliament.

Bruvoll, A. and K. Nyborg, K. (2004) The cold shiver of not giving enough: On the social cost of recycling campaigns. *Land Economics* Vol. 80, pp. 539–49.

Bureau of International Recycling (2011) Recycled Materials Supply 40% of the Global Raw Material Needs. August 6, 2011. Available from: <http://www.bir.org/industry/>

Halvorsen, B. (2008) Effects of norms and opportunity cost of time on household recycling. *Land Economics* Vol. 84, pp. 501–16.

Kawasaki City. (2011) Kawasaki city's subsidy guideline to the registered recyclable collection group, August 5, 2011, Available from: <http://www.city.kawasaki.jp/outline/info1191/file10654.pdf>.

Matsumoto, S. (2011) Waste separation at home: Are Japanese municipal curbside recycling policies efficient? *Resources, Conservation and Recycling* Vol. 55, pp. 325–34.

Ministry of the Environment of Japan. (2011) *Annual Survey of Municipal Solid Waste.*

OECD. (2008) OECD environmental data 2006-2008: waste. Environmental performance and information division. OECD environment directorate.

Saphores J. D. M., Nixon H, Ogunseitan O. A., Shapiro A. A. (2006) Household willingness to recycle electronic waste: an application to California. *Environment and Behavior* Vol. 38, pp. 183–208.

Yokohama City. (2011) Recyclables Group Collection. August 5, 2011, Available from: <http://www.city.yokohama.lg.jp/shigen/sub-shimin/syudan/data/subsidye.html>.

Section 2

Analysis Tools for Recycling System

5

Concept of Dual Traceable Ownership System (DTOS) as a Sustainable Design for Product Recycling

Jun Fujimoto and Dean Poland
AIST/Tokyo University of Foreign Studies
Japan

1. Introduction

According to statistics by Angus Madison at the Groningen Growth and Development Centre, world population has increased four-fold, and the level of global GDP in 2000 was 19 times the level it was in 1900 (Maddison, 2008). This incredible expansion has exposed several kinds of environmental issues such as climate change, natural resource depletion, environmental pollution, and disposal of waste. In the 21st century, there has been further population growth and economic expansion centred on developing countries. It has been estimated that the global population will reach 9,150 million in 2050 (6,115 million in 2000), and the Global GDP will be about four times larger in the following 50 years, if the GDP growth rate will be an average of 2.8 percent annually. The above trend will greatly increase the pressure of resource consumption. Will resources on our planet continue to support this population and economic expansion?

One essential solution is to promote the "3Rs" (reduce, reuse and recycle) globally so as to build a sound-material-cycle society through the effective use of resources and materials (Ministry of the Environment, 2005). 'Reduce' is achieved by using material and design with care to reduce resource consumption and the amount of waste generated. 'Reuse' involves the repeated use of products or parts of products which still have usable aspects. 'Recycle' is the use of waste itself as resources. The minimizing resource consumption and waste can be achieved in an efficient way by focusing primarily on the first of the 3Rs, "reduce," followed by "reuse" and then "recycle." "Reduce" is related to design products in companies. On the contrary, "reuse" and "recycle" are based on complex systems involving several kinds of elements such as institution, business, and people's behaviour. Therefore, it is difficult to achieve "reuse" and "recycle".

This chapter presents potential solutions that can help the progression of 'reuse' and 'recycling' in society. First, some key issues lying behind reuse and recycling systems will be discussed: the lack of management during usage stage, invisible flow caused by discrepancy of personnel expenses between countries, the harmful impact of long-time product life-time, and the purchasing behaviour of people as agents of over-consumption. After such issues have been addressed, a possible solution will be presented which can potentially improve the current situation of "reuse" and "recycle". This solution is linked to a new concept of product ownership: Dual Traceable Ownership System (DTOS). Furthermore, Service Oriented Products (SOP) will be proposed as an example of products suitable for DTOS. Finally, the feasibility of implementing a system such as DTOS will be examined using survey results about consumer behaviour in Japan.

2. Some key issues lying behind recycling system

2.1 Recycling QCD

Business processes are based on a free economy system. In this system, supply and demand are adjusted to the market. If there is a particular demand, manufacturers will meet the demand immediately. On the contrary, in recycling, discarded products are supplied to society for the convenience of users, with or without demand. We cannot clearly recognize the size and stability of a "market" in recycling. Therefore, recycling is operated on like a socialist economy rather than a free economy. Thus, there exists an economic "Berlin Wall" between business processes and recycling (Figure 1).

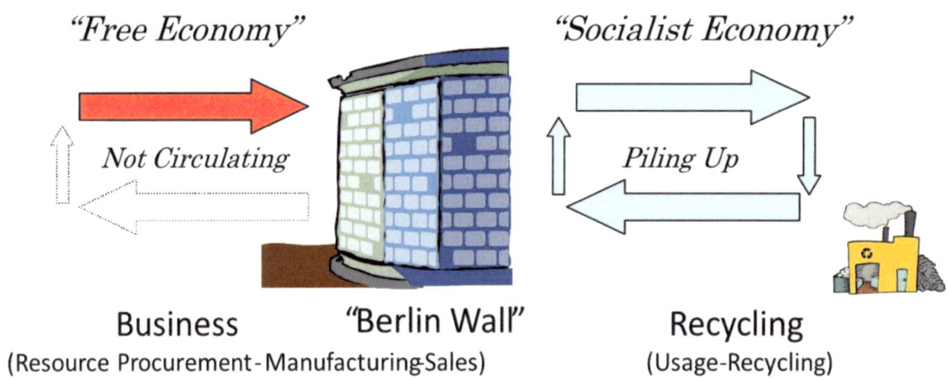

Fig. 1. Economic "Berlin Wall" between business and recycling

While, from the viewpoint of economics, this problem might be a form of economic externalities, from the viewpoint of manufacturing science one of the main causes is derived from the fact that the so-called quality, cost, and delivery (QCD) of recycling systems is not adequate to what manufacturing systems require (Fujimoto et al., 2003). Therefore, to break down this barrier and, as a result, to integrate recycling systems into manufacturing systems, it is essential to control quality, cost, and delivery of the recycling systems. Quality and cost of the recycling systems include the quality and cost of recycled materials and reused components. These need to be competitive relative to virgin materials or to newly manufactured components in order to be consistently used in the manufacturing systems. However, current recycling systems do not satisfy this condition. For example, material from the steel bodies of automobiles cannot be recycled into new automobile bodies because of quality problems, and clothes made of fibres recycled from PET bottles are often more expensive than those made from virgin fibres. Another problem is that the demands of the manufacturing process necessitate a quick response delivery of the recycling systems. The operation of recycling systems depends on the amounts of disposed and collected products (i.e., the push mechanism), while manufacturing systems depend on market needs (i.e., the market pull mechanism). Thus, there is a mismatch in terms of delivery between the push mechanism and the pull mechanism. Table 1 summarizes the relationships between the ease of controlling QCD of recycled materials and types of recycling systems. It is difficult to control QCD of recycled materials within the conventional situation in which there are no recycling regulations and only economically profitable materials are recycled. By adding strengthened recycling regulations, the cost and delivery factors of recycled materials can be improved. By constructing information technology (IT) systems for managing the recycling systems, the quality factor can be improved. Finally, conducting business on a lease basis would make it easier to control QCD in recycling since a company or an alliance of companies can manage a whole product life cycle. To achieve this "recycling QCD", it will be necessary to modify entire life cycle systems by introducing innovative product design, new sales and service patterns, and innovative reuse and upgrade strategies.

	Products owned by customers			Products leased to customers
	Conventional situation	+recycling regulations	+recycling regulations +ICT management system	
Quality	poor	poor	acceptable	good
Cost	poor	acceptable	acceptable	good
Delivery	poor	acceptable	acceptable	good

Table 1. Relation between QCD of recycled materials and Types of recycling systems

2.1.1 The dropped sausage

The difficulties of recycling in relation to quality management of discarded products can be illustrated in the following imaginary situation. If a "new" sausage drops outside a supermarket, can you pick it up and eat it? You may hesitate to eat it, because you doubt the "quality" of the sausage. It might contain some poison, or might be rotten. If you must eat it, you want to investigate the "quality" of the sausage further, perhaps by chemical analysis. However, this kind of analysis would be more expensive than the actual price of the sausage. You don't doubt the quality of the sausage in a supermarket refrigerator because you believe the sausage is under the shop's management. The difference in distance between a refrigerator in the supermarket and the area just beyond the entrance is only several meters. However, there is a huge difference with regard to "quality" management. Discarded products or recycled materials are under a similar condition in "quality" with the dropped sausage, i.e. out of management. All business processes from manufacturing parts, materials and assemblies to selling products are managed by individual companies. Information related to quality management is passed to the next step of businesses like a relay race. However, the relay of information is cut by consumers, i.e. product usage stage. Discarded products have not been "quality" managed during the usage stage, and nobody has information regarding the quality of discarded products.

2.2 Conflict between recycling and energy saving

The feature of long product lifetime has been recommended for various products as a waste reduction measure. The amount of waste from a product, which works for 10 years without failure, becomes half in comparison with that which only works for 5 years. However, the best solution of one issue is not always good for the other. Some products, such as cars and home electrical appliances, consume a lot of energy at the product usage stage. Some LCA (life cycle assessment) researches revealed that energy consumption at the product usage stage showed the largest part of total power consumption for product life cycle. If energy consumption in the usage stage of products has been improved largely over the years, it may be necessary to replace an old product with a new good energy efficient product by suitable years from the viewpoint of energy saving in society. We proposed a new concept "rapid circulation of product", (RCP) and estimated roughly the impact of RCP using product data of four kinds of products: cars, TVs, air-conditioners, and refrigerators. The following assumptions were used in order to simplify the estimation (Koshibu et al., 2003).

1. Conventional lifetimes of four products are similar, 10 years
2. 5 years lifetime was used in RCP
3. Estimation was achieved from 1991 to 2000 in Japan market.
4. Set one type of representative product for individual products, and it is assumed that almost all people in Japan use in same product which has the same specification and capacity.

Figure 2 shows the image of RCP concept. In the conventional case, the stock of products at 2000 consists of products manufactured in every year from 1991 to 2000. On the other hand, in the RCP, that consists of products manufactured in every year from 1996 to 2000.This means that the number of manufacturing and discarding product per year in the RCP was double in comparison with that of the conventional case.

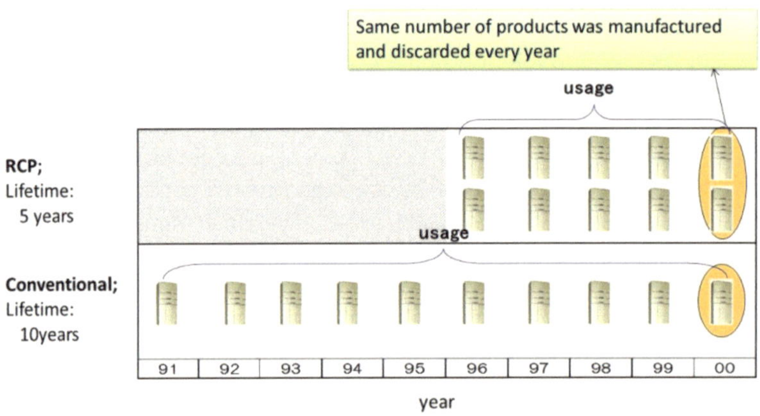

Fig. 2. Configuration of manufacturing year of products at 2000

Figure 3 shows the number of social stocks in the Japanese market and product types used in the estimation in 2000. The number of air-conditioners with both cooling and heating was around 71 million in 2000. The air-conditioner with a cooling capacity of 2.5 kW was assumed as a most popular product. This means the number of 71 million air-conditioners in Japan all occupied by those with 2.5 kW cooling capacity. There were 56 million refrigerators used in Japan 2000. Among these, refrigerators with over 300 liter capacity occupied about 54%, the number of 35 million. We chose the refrigerator with 400 liter capacity as a typical type of refrigerators, and calculated using the number of 35 million. Regarding TVs, 105 million TVs were used in Japan 2000. Among these, TVs under the display size 29 inches were 76 million, and over 29 inches TVs were 29 million. Regarding TVs, we estimated the impact of RCP on the assumption that 76 million TVs were the same display size- 29 inches. There were 52 million cars in Japan 2000. Among these, the cars with the total amount of engine exhaust from 0.66 liter to 2 liter were popular in Japan- 28 million. Therefore, we selected the type of car with 1.5 liter, and 4 doors as an estimation model.

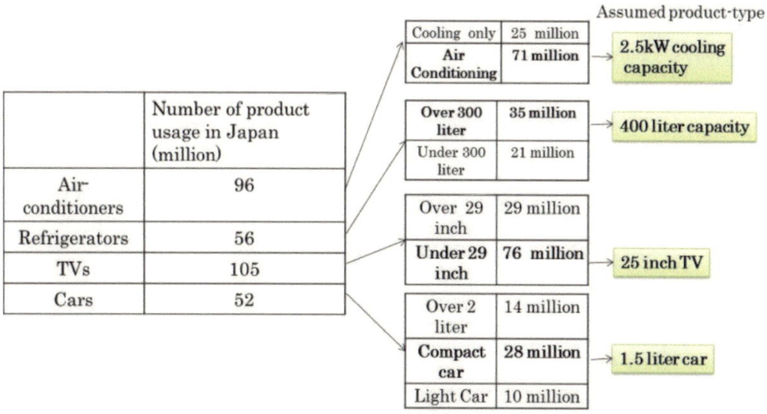

Fig. 3. Number of product usage at 2000 and assumed product-type

The energy consumption in the usage stage was estimated by average usage time per year and energy consumption of the "product model" of individual products, For example, in air-conditioners, the cooling period in summer was 3.6 month and heating period in winter was 5.5 month. In both cases, daily usage time was 18 hours. The energy consumption of the"product model" was obtained according to JIS (Japanese Industrial standards) C 9612 in 13 square meters room of Japanese typical wooden house at Tokyo. The model product was made by Toshiba with 2.5 kW cooling capacity. Figure 4 shows energy saving rate related to 1991 of the model products from 1991 to 2000. The power saving rate of cars do not decrease largely during this period. On the contrary, those of air-conditioners and refrigerators show under the half values of 1991s.

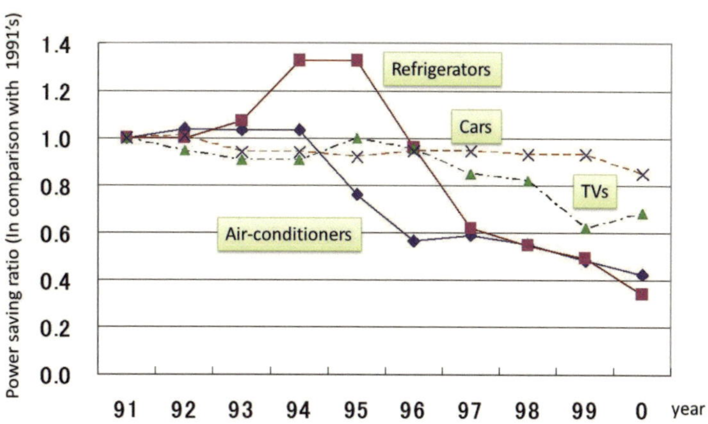

Fig. 4. Energy saving ratio to 1991's assumed product type

If four kinds of products had replaced new products every 5 years in 1990s, what has energy consumption from these products changed at 2000? The differences of the amount of CO2 emission calculated by energy consumption and final disposal amount of waste between conventional 10 years lifetime and RCP (5 years) are shown in Table 2. The minus values present CO2 emission reduction caused by RCP. The CO2 emission from product usage stage decreases in four kinds of products at 2000. However, the CO2 emission from manufacturing and discarding stages increases. Total CO2 emission decreases on air-conditioners and refrigerators, while increase on cars and TVs. The CO2 reduction adding that of air-conditioners and refrigerators is around 13.5 million tons, which is nearly 1% of total emission Japan 2000. On the other hand, RCP is not suitable for cars, because of 8.9 million tons increase. Of cause, final disposal amount of waste, which is a residue after material recycling, increase around 1 million tons, which is nearly 1.5% of total final disposal in Japan. These results suggest that equal 5 years lifetime assumed here is suitable some products, such as air-conditioners, and unsuitable for other products. Do individual products have different optimal replacing years regarding energy saving?

	CO_2 emission at 2000 (million ton per year)			Difference of final disposal amount of waste (thousand ton)
	Difference in usage stage	Difference in manufacturing and discarding stages	Total difference	
Air-conditioners	-11.2	+0.9	-10.3	+110
Refrigerators	-4.2	+0.7	-3.5	+150
TVs	-0.5	+0.6	+0.1	+90
Cars	-1.0	+9.9	+8.9	+600

Table 2. RCP impact on CO2 emission and final disposal waste at 2000

The general solution was derived for obtaining the optimal replacing years of various products. The following assumptions were used in order to simplify the calculation.

1. Improvement of energy-saving rate in product usage stages are constant every year
2. Energy consumption used in manufacturing and disposal are not changed each year
3. Number of shipped products and waste products are similar and does not change each year
4. Product lifetime are constant in an individual products

The average energy consumption per year in one product under the above assumptions can be obtained by the following equation (1).

$$T = \frac{\sum_{i=0}^{n-1}(1-\alpha)^{-i}}{n} + \frac{P}{n} = \frac{1-(1-\alpha)^{-n}}{n(1-(1-\alpha)^{-1})} + \frac{P}{n} \qquad (1)$$

The individual parameters regarding product characteristics are defined as follows:

n: lifetime of products (years):

α : improvement rate of the usage stage energy consumption per year

P: energy consumption required to manufacture and scrap product/ energy consumption required to work product per year

For example, α=0.01 means that the product of power consumption 200 W decreases 200 W*(1-0.01) =198 one year later.

According to the assumption and definition, the average energy consumption per year (T) in one product can be obtained by the equation (1). However, this T value is a relative value when the annual energy consumption of usage stage at the year concerned is "1". The n value to minimize T value can be obtained by differentiating equation (1) with n, and presents optimal replacing years of product regarding power saving during lifecycle of product. The results from the above calculation using various α and P values are shown in Table 3.

P\α	1%	2%	3%	4%	5%	6%	7%	8%	9%	10%	11%	12%	13%	14%	15%	16%
0.0	-	-	-	-	-	-	-	-	-	-	-	-	-	-	-	-
0.5	9.7	6.8	5.5	4.7	4.2	3.8	3.5	3.2	3.0	2.9	2.7	2.6	2.5	2.4	2.3	2.2
1.0	13.5	9.4	7.6	6.5	5.7	5.2	4.8	4.4	4.1	3.9	3.7	3.5	3.4	3.2	3.1	3.0
1.5	16.4	11.3	9.1	7.8	6.9	6.2	5.7	5.3	4.9	4.6	4.4	4.2	4.0	3.8	3.7	3.5
2.0	18.8	12.9	10.4	8.9	7.8	7.0	6.4	6.0	5.6	5.2	5.0	4.7	4.5	4.3	4.1	4.0
2.5	20.8	14.3	11.5	9.8	8.6	7.8	7.1	6.6	6.1	5.8	5.4	5.2	4.9	4.7	4.5	4.3
3.0	22.7	15.6	12.4	10.6	9.3	8.4	7.7	7.1	6.6	6.2	5.9	5.5	5.3	5.0	4.8	4.6
3.5	24.4	16.7	13.3	11.3	9.9	8.9	8.2	7.5	7.0	6.6	6.2	5.9	5.6	5.4	5.1	4.9
4.0	25.9	17.7	14.1	12.0	10.5	9.5	8.6	8.0	7.4	7.0	6.6	6.2	5.9	5.6	5.4	5.2
4.5	27.3	18.7	14.9	12.6	11.1	9.9	9.1	8.4	7.8	7.3	6.9	6.5	6.2	5.9	5.6	5.4
5.0	28.7	19.6	15.5	13.2	11.6	10.4	9.5	8.7	8.1	7.6	7.2	6.8	6.4	6.1	5.9	5.6

α: power saving rate of usage stage compared to previous year
P: power consumption of manufacturing and disposal stage to that of usage stage per year ratio

Table 3. Optimal replacing year of product

Table shows that by decreasing P value and increasing α values, optimal replacing years of product are shortened. When we calculated the impact of RCP of 5 years replacing from 1991 to 2000, we used P=2.39 and α=1.82% of cars, P=1.45 and α=4.16 % of TVs, P=0.44 and α=9.10 % of air-conditioners, P=1.43 and α=11.2 % of refrigerators. These values reveal that optimal replacing years of product was around 15 years of cars, around 7 years of TVs, around 3 years of air-conditioners, and around 4 years of refrigerators in 1990s. The RCP 5 years shown in Table 3 is longer in air-conditioners and refrigerators, and shorter in cars and TVs in comparison with the optimal replacing years. Therefore, CO_2 emission results shown in Table 2 can be improved by setting optimal years in individual products. To be exact, there are various products with different capacities and performances are used in actual society. Furthermore, annual usage time of products and operating condition are quite different depending on users. These suggest that the relation between energy saving and recycling show more complicated situation. However, this paradox between energy saving and recycling actually existed in some products which required high energy consumption at the usage stage. How can we solve this kind of paradox?

2.3 Invisible flow

According to 'home appliance recycling law' in Japan, advanced recycling for home appliances has been conducted from the year 2001. These home appliances are refrigerators, television sets, air conditioners and washing machines. What problems do we have with these practices? The number of reclaimed products using these recycling systems in the above four electric products was 11.2 million in 2004 fiscal year, including about 1.8 million air conditioners, 3.8 million television sets, 2.8 million washing machines and 2.8 million refrigerators (Kuwatani et al., 2004). On the other hand, the total number of disposed products was estimated by using the number of shipped products for previous ten years. This was about 20 million and twice the number of reclaimed products. The number of illegally disposed products were estimated to be about 0.17 million products. What happened to the missing approximately 8 million products (see Figure 5)?

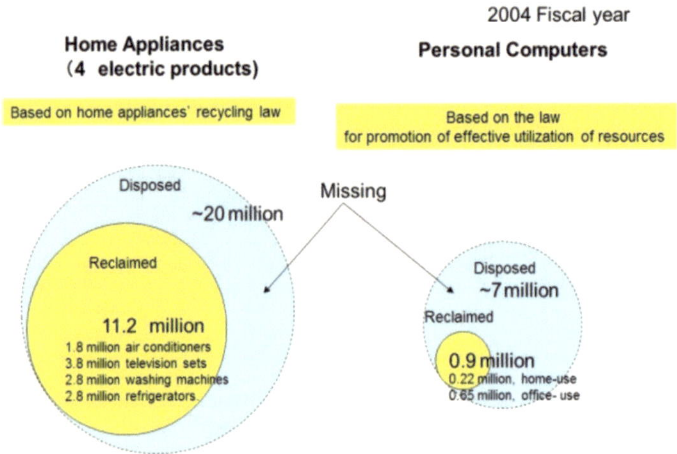

Fig. 5. Number of reclaimed and missing -products at 2004 fiscal year

As for Personal Computer (PC), there are two kinds of usage: office and home. The recycling of discarded PC used in the office has been conducted from April 2001 and that of home-use computer October 2003, based on the law for promotion of effective utilization of resources in Japan. The number of reclaimed products regarding home-use was 0.22 and 0.65 million regarding office-use in 2004 fiscal year (Kuwatani et al., 2004). The total number of disposed products was estimated by using the number of shipped products for the previous seven years. This was approximately 7 million in 2004 fiscal year. In this case, there is a great difference between the number of disposed products and that of reclaimed ones. More than 6 million computers were missing (see Figure 5). Some of the missing products may have been exported to Asian countries as 'used products'; others may have been exported as used parts or materials after they were dismantled in Japan; the remaining ones may have been recycled using different recycling systems from regular recycling ones based on particular laws in Japan. These are chosen according to the economic factors rather than environmental benefits. However, the actual situations of missing products are still invisible. We tried to address these situations using the recycling profit analysis model and the trade statistics of Japan. This will make 'invisible flow' clear. Figure 6 shows a possible image of what happens to missing products.

The disparity in economic conditions among countries has a large effect on the benefits of material recycling. Figure 7 shows the benefit comparison of material recycling in Japan with that of Asia (Kondoh et al., 2009). These values were estimated under the assumption that material recycling would be achieved manually and the disassembly-flow strategy obtained by our analysis model would be chosen. This model presents the different recycling ratios between Japan and Asian countries. The values in Asia included the marine transportation cost from Japan to China. The benefits obtained in Japan indicate all minus values regardless of products. The reason for large variations of benefits in Japan comes from configured disassembly times: from 10 to 20 min. This means that labour costs has a large effect on recycling benefits in Japan. The benefits in Asia are larger than those in Japan, in particular those obtained from PCs and Air conditioners indicate plus values. These show that lower labour costs and higher recycling-material sales in Asia make it possible to cover the extra marine transport cost. The product, which has small volume, consists of several kinds of materials, and takes a great deal of disassembly time, tends to fit the Asia recycling systems.

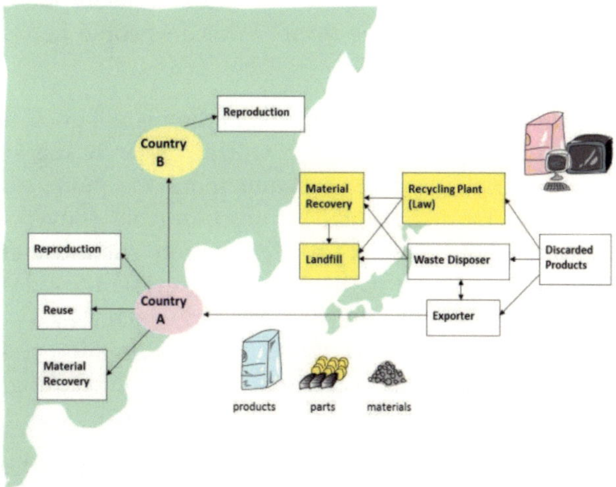

Fig. 6. Situation of missing products

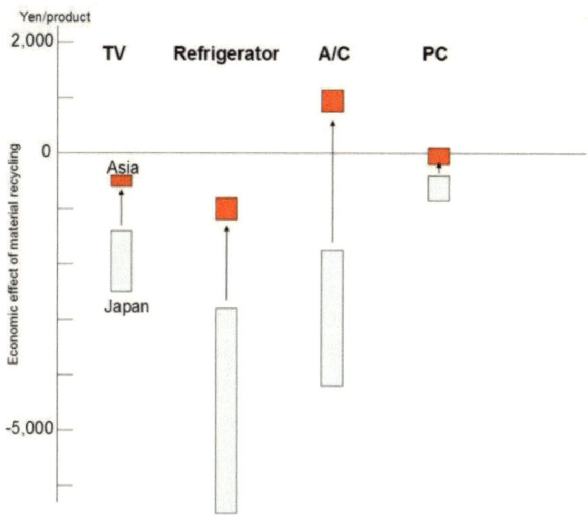

Fig. 7. Economic effect of material recycling

2.4 An unsustainable culture of 'having'

In the societies of 'developed' countries, increasingly materialistic sensibilities and consumerist values have led people to believe that the 'desirable' lifestyle should be founded upon the acquisition of material goods, and the unrelenting search for status and power. Unfortunately, this drive to purchase a 'desirable' lifestyle' is supported by high and unsustainable resource consumption.

Ironically, research also shows that this culture of 'having' does not actually have a positive impact on people's sense of Well-Being. Research has shown that levels of Subjective Well-Being (SWB) do not correlate with high GDP or personal wealth (Myers, 2000; Diener & Seligman, 2004).

If the culture of 'having' manifested in high level material consumption does not increase a person's well-being, then why do people do it? Why do people behave in ways that are unsustainable and seemingly irrational?

Although it is generally understood that overconsumption is not good for the well-being of people, society, the environment or even the world, research suggests that social and psychological mechanisms drive people to consume more and more, encouraging behavior that is both irrational and more importantly unsustainable (Boven & Gilovich, 2003). Furthermore, the institutional structures of modern society such as hierarchy, competition and advertising seem to play a large part in determining the lifestyles people lead. Such forces have been labelled 'weapons of influence', each of which play a part in manipulating peoples' behaviour (Cialdini, 2008). Indeed, the economic foundation of developed countries seems to be based on the principle that to consume, that is, contributing to the continuation of an affluent society, is a social responsibility.

But these ideas are by no means new. Decades ago, Erich Fromm in his seminal work 'To Have or To Be', labelled this kind of lifestyle a mode of 'having': "To acquire, to own and to make a profit are the sacred and the unalienable rights of the individual in the industrial society" (Fromm, 1976). Fromm concludes that ultimately the slogan of industrialized societies is: "I am what I have and what I consume." And it is this attitude of 'Having' in general which has contributed to the current situation of unsustainable consumption. Research by Winter and Koger showed that "Much of the irrationality of our environmentally unsustainable behaviour could be attributed to a 'false self' system" (Winter &Koger, 2004). In this system, external objects are used to express people's identity. The consumption of products function as symbols of some ones 'self' in relation to others. For example, cars not only have the function of transporting people from one location to another, but also make a statement to others about who they are. Winter and Koger conclude that much of our overconsumption may be driven by a 'false self' system. Striving for endless increases in material wealth; placing high value on social status and deriving our identity from what we 'Have' rather than who we 'are' or what we 'do' have all contributed to the creation of a mechanism of unsustainable lifestyle, and "constitutes a vicious circle of consumer-buying" (Fromm,1976).

The condition of "Hyperphagia" i.e. over-eating disorder, the mechanism shown in Fig. 8, may serve as a useful analogy for many resource/energy over-consumption issues. Over-eating disorder requires "Excessive food" consumption. This leads to the problems of disease and obesity, and then creates "Excessive Demands". For example, medical and exercise services. The root cause of Hyperphagia could be said to be mental disorder. In order to suppress resources energy consumption, it would be essential to cure "Hyperphagia"(Fujimoto &Poland, 2009).

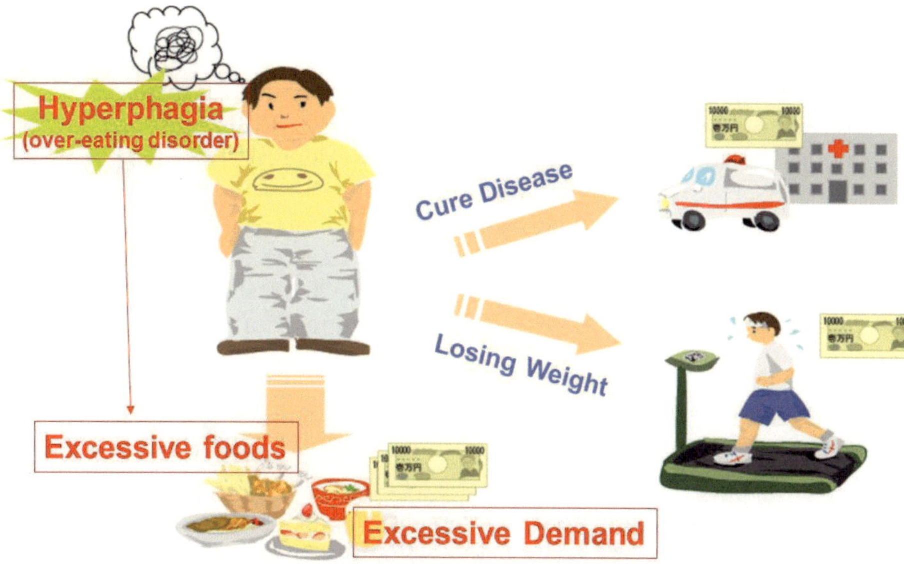

Fig. 8. Mechanism of resources and energy overconsumption

3. Concept of DTOS as a potential solution

3.1 DTOS

As described in this article, the problem of 'invisible flow' is a key issue which must be solved if a multilateral recycling system is to be feasible. To prevent 'invisible flow', the root cause of this phenomenon must be addressed.

Ultimately, the root cause is connected to the concept of 'ownership'. What does 'ownership' mean for Japanese and Asian people? What does it mean to 'own a product' in today's society? In the current economic and cultural situation of Japan, ownership seems to mean freely possessing a product without taking responsibility for the products 'life cycle' especially with regard to the products 'end of life'. This style of ownership must be radically changed, if the problem of 'invisible flow' is to be solved. Against this background, the relationship between 'ownership' and 'responsibility' will be considered and grounded upon this new understanding, we will propose a new holistic approach: Dual Traceable Ownership System :DTOS (Fujimoto, et al., 2009).

DTOS is a system based on a different attitude towards 'ownership' which makes traceability of a product visible. In this system, there are two styles of 'ownership':

1. Individual consumer has complete ownership of the product.
2. Company has complete ownership and the consumer hires the product.

The first type differs in the level of responsibility for a products 'end of life' in comparison with the current practice of 'ownership'. In this style of ownership, the following requirements must be met:

1. An identification number is attached to individual parts of the product and this number corresponds to the owner.

2. When transferring ownership during product use, the owner has to follow a set of procedures laid down by law.
3. When discarding the product, the owner takes responsibility for the recycling process, for example choosing an appropriate recycling trader. In this case, an owner must consider and be responsible for such things as recycling cost, recycling processes and the legitimacy of the recycling company.
4. If the product which was owned by the consumer, i.e. now in a post-used state, is discovered in an unsuitable or illegal situation, such as a 'black market' recycling process, the owner will receive a severe penalty. The ownership can be easily traced by the identification code which was attached to the product during the time of purchase.

On the other hand, the second style of 'ownership', i.e. the company retains complete ownership responsibilities and the consumer hires the product have the following characteristics:

1. The consumer pays money not for the product itself, but for the services or functions which the product provides. Consumers may pay more money in comparison with the 'consumer ownership' style because of paying long-term monthly fees, but there a several advantages to be gained.
2. The consumer can enjoy the product without worrying about its disposal. The company or 'seller' will take responsibility for all the recycling duties that come with 'ownership.' Through the identification code system, the product and its parts can be easily traced back to the company. Therefore, the company will be liable to pay a large penalty if the product is disposed of inappropriately.
3. It may be possible for the consumer to receive new services quickly and at minimal or no extra cost. For example, consumers may update the product for a 'new model' without the inconvenience of disposing of the old model. Moreover, the consumer may be able to receive repairs and general maintenance of the product for no extra charge.

An increase in the number of consumers who choose the second style of ownership, may lead to manufacturers changing their style of manufacturing. Current products are designed based on the premise of consumer ownership. However, due to this new ownership style, manufacturers may consider a product design with easier functional upgrades or part reuse. Companies will have motivation to provide different styles of 'service-oriented products' (Fujimoto et al., 2003). They can provide consumers with selection of 'service courses' from a 'service menu', made on the basis of service contents that several types of products offer, and on a lease basis obtain new products that incorporate the selected service course. Manufacturers will be free to make various products according to their service menu, by combining functional components, the reuse of which enables them to be used up until the very end of their life cycle. Figure 9 presents prototype service-oriented facsimile products (Fujimoto et al., 2003)

Therefore, the DTOS system will not only give choice to the consumer who can choose one of the two styles of 'ownership' according to their personal preference, but also gives companies incentives to develop more innovative service plans and manufacturers to develop environmentally friendly product design.

The notion of 'hiring' a product rather than 'buying' a product is certainly not new.

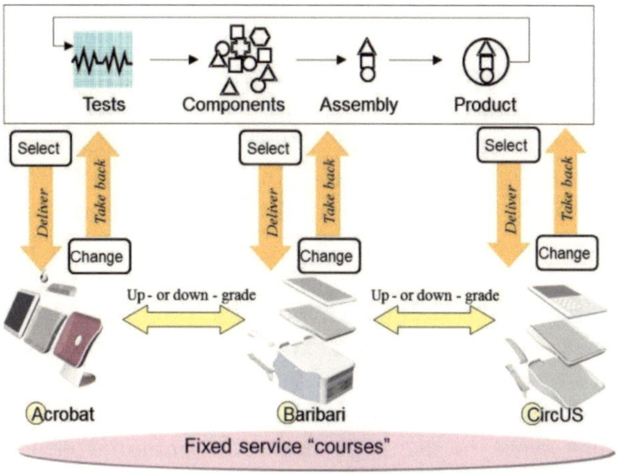

Fig. 9. Example of "Service-oriented products" applied to consumer fax

However, unlike previous systems DTOS emphasizes the notion of 'ownership responsibility'. A key characteristic of DTOS is that the consumer will be confronted with a clear choice of 'ownership' styles on a wide range of products. When faced with the choice of privately owning or hiring a product, the consumer will have to consider the advantages and disadvantages of both styles. This will create an awareness of the impact of consumer behaviour on the environment, society and economy, thus creating social and environmental consciousness which will lead to a more eco-friendly lifestyle. This 'consumer consciousness' will hopefully reduce excessive and frivolous consumption. If DTOS is implemented, it may also have a large impact on social and lifestyle issues. The consumer can choose one of these two styles according to his or her life-style. For example, consumers may want to 'own' a computer and take on the responsibility that this new style of 'ownership' requires, but do consumers really want to own a Plasma TV or refrigerator? For a consumer 'ownership' of such impersonal and large products may not be appealing, and thus would rather take the 'hiring and service' path. The implications of such a system are far reaching and would have an impact on such areas as crime, dwelling styles, social behaviour, ethics and so on. There is no doubt that due to the ease by which products can be traced back to the owner, and the penalties that will accompany violations of the 'ownership contract', DTOS will help to solve the problem of 'invisible flow' in Japan. This means that upon this foundation, an adequate multilateral recycling system will be able to be constructed.

Furthermore, DTOS will increase the levels of social responsibility and the environmental consciousness of consumers. The key philosophy that underlies DTOS and that citizens will come to understand, is that 'ownership' equals 'responsibility'.

3.2 SOP

To control QCD of product life cycles, one promising approach is to servicify products and operate the business on a lease basis by employing the concepts described previously (Brezet et al., 2001). We call such products service-oriented products (SOPs). SOPs can revise traditional thought which holds that only products comprising new materials and components are valuable (Fujimoto et al., 2003). SOPs should be able to provide customers with new benefits not provided by ownership. Enabling customers to enjoy products they lease and to eventually exchange them for other products would be a good means of providing various services to customers. SOPs should also increase manufacturer profitability; the lease system would enable them to own the products they produce until the end of their product life. Purchasing and recycling cost reduction can be achieved by controlling product life cycles. Service of SOPs includes not only the service a product performs but also installation support, operation assistance, expendables delivery, maintenance, upgrading, and collection of post-use products. Required services differ according to usage pattern, the customer's knowledge, awareness of the environmental issues, and so on. For example, one customer may require optimal performance, be able to repair a product by himself, but not care about environmental loads of disposed products. Another customer may want to use a product with minimal effort, have expendables provided, and care strongly about disposal. In anticipation of these differences, we prepare several "service courses" for a product group (e.g., facsimile machines) and call this set of service courses the "service menu" of the product group. A service course consists of a set of services covering all phases of a product life cycle. A service course, in one instance, consists of installation support, an easy-to-use product with basic performance, operation support, expendables delivery, maintenance support, and a post-use collection service. This idea is similar to so-called "mass customization". However, while the mass customization focuses a variety of products, SOP offers a variety of services throughout a product life cycle that includes a variety of products.

In these service-oriented products, a customer selects a specific service course from a service menu and obtains, on a lease basis, a product incorporated into the selected service course. They can obtain many services in accordance with the course they select and also alter the service course, if desired, for an extra charge. Manufacturers create several kinds of products according to their service menu by combining functional modules. Reuse of functional modules enables them to be utilized until the very end of their lifetime. Table 4 summarizes the differences between conventional products and SOPs. With SOPs, a closer relationship, until the end of the service provision period, is maintained between the customer and the business. This closer relationship enables manufacturers to better understand customers' needs, which are often difficult to ascertain because they tend to change dynamically, and to control the QCD of post use products.

		Conventional products	SOPs
Customer	choose	product specifications	"service" course
	purchase	hardware	service(hardware rental)
	replace	buy a new one	change with extra charge
Business	profit	product sales	service provision
	manufacture	assembling parts	combining module
	Post-use	disposal/recycling	reuse/recycling
Relation between business and customers		weak (in post-sales periods)	close (until use period ends)

Table 4. Comparing SOP concept with conventional products

3.3 Is the implementation of DTOS feasible?

In a study by Van Boven and Gilovich it was shown that 'happiness' is increased when people focus more on the acquisition of life experiences than the acquisition of material possessions. They demonstrated that materialistic people tend to report lower levels of subjective well being than non-materialistic people; concluding that experiences make people happier because they play a more meaningful part of their identity and contribute more to social relationships (Boven & Gilovich, 2003). More recent studies have supported their initial findings. For consumers, satisfaction with experiential purchases such as going to a restaurant or taking family vacations starts high and increases over time, whereas spending money on material goods may feel positive at first but after time makes people feel less happy (Carter &Gilovich, 2010).

It is important to note that previous research in this area has pointed out the challenge of delineating between 'material purchases' and 'experiential purchases' (Boven& Gilovich, 2003). A useful way of illustrating this problem is if we look at the purchase of a car. On the one hand a car is a material object, a possession made of metal, plastic and so on, which has an impact on environmental and social issues. On the other hand, a car can facilitate certain experiences such as travelling. Therefore, in this case is a car an example of a 'material' or 'experiential' purchase?

Boven and Gilovich addressed this problem by focusing on a consumer's 'intention'. For example, a person might buy a car because it is an object of material desire: a status symbol, something to be possessed. Such a person would also likely have the desire to buy the latest model of a car, perhaps buying a different car every year in order to satisfy their desire 'to have'.

On the other hand, perhaps a nurse buys a car because she needs it for visiting her patients. She doesn't care about the car as a material object; her only concern is that the vehicle helps her meet the challenges of doing a satisfying job. In this case, she is likely to keep the car for as long as it serves her 'non-material' purpose.

To sum up, it seems a move from a materialistic 'having' to an experiential attitude towards purchasing could affect consumption levels in two ways:

i. If consumers begin to purchase 'experiences' rather than material goods such as cars, TV's and so on, this may have a direct impact on levels of material consumption.
ii. If experiential purchases increase levels of well-being, then this could disrupt the cyclical mechanism of overconsumption suggested by Fromm (Fromm,1976) and expressed in the 'False Self' system (Winter &Koger ,2004)

Finally, we discovered that young Japanese are already moving away from material purchases such as a car, to a more experiential and communication based lifestyle.

3.3.1 Questionnaire survey regarding Japan's people purchasing behavior

We conducted a questionnaire on over 1200 people in Japan on February 2010 via the Internet examining peoples purchasing behavior. Their age demographic was divided into four, 18~29, 30~37, 38~47, and 48~57. Each individual demographic had around 300 participants (Fujimoto &Poland, 2011) .

The survey focused on what goods and services people want to have. The surveyed people chose their top 10 desired goods and services out of a selection of 34. The goods and services which people desired most was allocated "point 10", and want to have 10th was allocated "point 1". The total number of points of individual goods and services was obtained by multiplying the percentage of individual rank of goods with the previous points. Therefore, the higher values of products and services showed they wanted to purchase them strongly. Figure 10 shows the top ten ranking of goods and services which people wanted within each individual age demographic. In general, the trend was that a lot of services were ranked within the top ten, such as "domestic travel", "eating out", "international travel", and "watching". On the contrary, the choice of car, TV, AV, PC, and home appliances, which are often regarded as symbols of "material consumption", were ranked lower. This suggested that Japanese society is maturing away from a materialistic attitude. When analyzing age demographic differences, material goods such as car, TV, AV, PC, and home appliances were desired by the older age range, as shown in Fig. 10. Particularly, the low rank of car in young generation was very impressive. The number of points regarding the desire to purchase a car in the age demographic 18~29 man was 2.6, that in 30~37 man was 2.5, that in 38~47 man was 3.3, and that in 48~57 man was 3.1. The Questionnaire Survey also asked about car possession in relation to age demographics. These results showed 58.6% of 18-29 males possessed their own cars. The same was true for 77.4% of 30~37 males, 82.2% of 38~47 males, and 84.2% of 48~57 males. Though 18~29 man had low car possession rate, their desire for cars was low. On the contrary, older males who already possessed a car, still desired a car.

	-29M	-29W	-37M	-37W	-47M	-47W	-57M	-57W
1	PC 4.4	Fashion 6.4	PC 5.5	Domestic Travel 5.9	Domestic Travel 5.8	Domestic Travel 6.2	PC 5.6	Domestic Travel 5.9
2	Domestic Travel 4.2	Domestic Travel 5.7	Domestic Travel 4.7	Fashion 5.5	PC 5.3	Eating Out 5.7	Domestic Travel 5.5	Eating Out 4.8
3	Game 3.4	Eating Out 5.3	Eating Out 3.8	Eating Out 5.2	Eating Out 4.2	Fashion 4.7	Eating Out 3.5	Fashion 3.6
4	Music(CD, Concert) 3.1	Music(CD, Concert) 3.3	Game 3.1	International Travel 3.0	Car 3.3	Watching Movie 3.5	Car 3.1	Watching Movie 3.5
5	Animation, Manga 2.8	Cosmetics 3.2	TV 2.7	Furniture, Interior 2.7	TV 3.1	International Travel 3.1	International Travel 3.0	International Travel 2.9
6	Fashion 2.7	International Travel 3.1	Investment 2.7	Domestic Appliance 2.7	AV Equipment 2.7	PC 2.8	TV 3.0	PC 2.8
7	Eating Out 2.6	Book 2.6	Watching Movie 2.6	Book 2.6	Watching Movie 2.4	Book 2.8	Watching Movie 2.7	Book 2.7
8	Book 2.6	Watching Movie 2.4	Music(CD, Concert) 2.5	Cosmetics 2.6	International Travel 2.0	Cosmetics 2.5	Book 2.6	Music(CD, Concert) 2.7
9	Car 2.6	PC 2.0	Car 2.5	Watching Movie 2.4	Music(CD, Concert) 2.0	Music(CD, Concert) 2.4	Visiting Famous Spot 2.3	TV 2.5
10	Watching Movie 2.2	Animation, Manga 1.9	AV Equipment 2.5	Music(CD, Concert) 2.4	Book 1.9	Domestic Appliance 2.4	AV Equipment 2.2	Cosmetics 2.4

Fig. 10. Overall Chart of Purchasing Behavior

Our survey results were supported by previous research e.g. "2008 Passenger cars market trend investigation" (Japan Automobile Manufacturers Association, 2008). On the surface it seems young people's consciousness is moving away from an interest in purchasing goods, especially with regard to purchasing a car. This suggests that more service oriented consumerism such as DTOS may be feasible in the future.

4. Conclusion

Resources and energy have to be used effectively when considering future population growth, economy expansion in developing countries, and the preservation of natural resources for posterity. Furthermore, environmental issues, such as environmental pollution caused by the disposal of products across borders, and the paradox between energy-saving and waste are problematic. Against this background, "recycling" has become one of the essential solutions but recycling strategies and methods must become far more sophisticated. The conventional recycling system has serious defects from the viewpoint of "Recycling QCD". Businesses stimulate people's desire, sell a lot of products to people, and then lose their attention of the products after sale. These businesses are also supported by people's overconsumption disorder. To improve these situations, we proposed a new holistic approach: Dual Traceable Ownership System (DTOS): a system based on a different attitude towards 'ownership' which makes traceability of a product visible. Moreover, we found DTOS would help suppress the invisible flow of discarded products. The fundamental style of 'ownership' proposed by DTOS is that companies have complete ownership and the consumer hires the product. In this style, the service oriented product was proposed. DTOS can be realized depending on whether the company creates an attractive service model or not. In addition to this, it would be necessary to reduce people's attachment to the notion of "owning" consumer goods. Regarding consumer behavior, data suggested a departure from the materialistic notion of "owning" consumer goods in Japanese young people, suggesting a system such as DTOS may be feasible in the future.

5. Acknowledgment

We appreciate the financial support of NEDO (New Energy and Industrial Technology Development Organization, Japan) and MSTC (Manufacturing Science and Technology Center, Japan) in helping us to make our prototypes (SOP). This research was conducted as part of Grants-Aid for Scientific Research (KAKENHI) Project of Japan from the fiscal year 2004–2006 and 2008–2012, supported by MEXT, Japan

6. References

Boven, L.V. & Gilovich, T. (2003). To Do or to Have? That Is the Question, *Journal of Personality and Social Psychology*, Vol.85, No.6,(2003), pp.1193-1202

Brezet, H.; Diehl, J.& Silvester, S. (2001). Institute of Electrical and Electronics Engineers: Piscataway, *Proceedings of EcoDesign 2001*, Tokyo, pp.605-612

Carter,T.J. & Gilovich, T. (2010). The Relative Relativity of Material and Experiential Purchases, *Journal of Personality and Social Psychology*, Vol.98, No.1 (Jan 2010), pp.146 -159

Cialdini, R.B.(2008), *Influence: Science and Practice. Pearson A and B* (5 edition), Pearson Education, ISBN 0205663788,

Diener, E. & Seligman, M.(2004). Beyond Money: Toward an economy of well-being, Psychological Science in the public internet, American Psychological Society, Vol.5, No.1, pp.1-31

Fujimoto J.; Poland, D. & Matsumoto,M. (2009). Low-Carbon society Scenario: ICT and Ecodesign, *The Information Society*, Vol. 25, No. 2, (March-April 2009), pp.139-15

Fujimoto, J. ; Umeda, Y. ; Tamura, T. ; Tomiyama, T. & Kimura, F. (2003) Development of service-oriented products based on the inverse manufacturing concept, *Environmental Science and Technology, ACS Publications*, Vol. 37, pp.5398–5406.

Fujimoto, J. ; Kondoh, S. & Poland,D. (2009) Ecodesign of multilateral recycling systems in Asia, *Int. J. Environmental Technology and Management*, Vol. 11, No. 4,pp.276-289

Fujimoto, J.; Poland, D. (2011), Purchasing Behavior- 'Having Material Possessions' or 'Having Communication', *Proceedings of SUSTAINABLE INNOVATION 11, UK, October 2011*

Fromm, E (1976). To Have or To Be, *Continuum International Publishing Group* (May 1996), ISBN 0826409121

Japan Automobile Manufacturers Association (2009), 2008 Passenger cars market trend investigation, Data of access, Available from: <http://www.jama.or.jp/lib/invest_analysis/pdf/2008PassengerCars.pdf> (Japanese)

Kondoh, S. ; Kuwatani, M. ; Fujimoto, J. & Umeda, Y. (2009) Cost-profit analysis of multilateralrecycling system in Asia, *Int. J. Environmental Technology and Management*, Vol. 11, No. 4, pp.257-275

Koshibu, H.; Fujimoto, J.; Umeda, Y.; Suga, T. & Kogure, K. (March 2003), *Eko dezain kakumei (EcoDesign revolution)*, Maruzen, ISBN 4621072161, Japan (Japanese)

Kuwatani,M. ;Fujimoto,J. ;Umeda, Y. ;Kondoh, S. ;Li, Z.& Nakamura K. (2005), Ecodesign of Multilateral Recycling System in Asia -2nd Report; Recycling Profit Analysis Model-, Proceeding of EcoDesign 2005, ISBN 1-4244-0081-3, Tokyo, Dec 2005

Maddison , A.(2008), (Groningen Growth and Development Centre) ; Statistics on World Population, GDP and Per Capita GDP, 1-2008 AD, *Groningen Growth and Development Centre* Data of access, Available from: <http://www.ggdc.net/MADDISON/oriindex.htm>

Ministry of the Environment (2005), The 3R Initiative; Data of access, Available from: <http://www.env.go.jp/recycle/3r/en/outline.html>

Myers, D. (2000). The Funds, Friends and Faith of Happy People, *American Psychologist* , Vol.55, No. 1,(January 2000), pp. 56-67.

Winter, D.N., Koger, S. (2004), *The Psychology of Environmental Problems*, Lawrence Erlbaum Associates, ISBN 0-8058-4630-1, pp.44, USA

6

The Economic Aspects of Recycling

Beatriz Ferreira, Javier Monedero,
Juan Luís Martí, César Aliaga, Mercedes Hortal and Antonio Dobón López
Packaging, Transport and Logistics Research Centre (ITENE)
Spain

1. Introduction

Over the last years, environmental issues have gained remarkable significance as a result of the high level of environmental awareness in society and also due to the widening of the scope of the environmental legislation and regulation. One of the major concerns within this context is the high rate of packaging waste generation. Nowadays there is a variety of technologies to deal with waste treatment among which recycling is one of the most preferable. Recycling allows waste to be converted in resources and thus its economic value is prevented to be buried in landfill. Back in 2000, the World Business Council for Sustainable Development (WBCSD) proposed improving recyclability as a critical aspect for a sustainable development in business since this end-of-life treatment was foreseen as one of the most sustainable. Furthermore, international legislation regarding end-of-life of waste is pushing stakeholders to adopt more sustainable strategies in the long run. For instance, the European Union Directive 2008/98/EC establishes the five-step waste hierarchy, in which the most sustainable step is the waste avoidance, followed by reusing and recycling. Up until now, waste is taken into account as a product since it is possible to obtain economic benefit from it. After those treatments, recovery (including waste-to-energy) and landfill are considered as the last steps. The day when waste can no longer be avoided and we reach the maximum reuse rate possible, recycling will be the most preferable strategy. This approach is influenced not only by the point of view of the potential economic value of waste, but also by its eco-efficiency.

The benefits of recycling could be summed up in energy savings, natural resource conservation and reduction of waste disposal to landfill. Energy savings of recycling have been demonstrated in a wide range of materials (Stromberg, 2004). The fact of avoiding the use of virgin materials allows reducing natural resources consumption and thus preventing problems to future generations. Moreover, the high quantities and extensions of landfills is a serious disadvantage in a lot of countries like in Japan where land is scarce and it is hard to locate new final disposal sites. Moreover, a great amount of landfill sites are currently full and other solutions should be searched.

Thus, the recycling advantages from an environmental and an economical point of view are influenced by different elements. Environmental and economic life cycle analysis in combination with market studies can be used for the definition of the most suitable waste treatment process in a specific situation.

This chapter is focus on analysing the characteristics of the recycling processes of the main packaging materials, i.e. paper and board, plastics, metals, glass and wood, from an economic point of view. Firstly, a brief market analysis for the recycled materials is carried out, followed by a review of the steps to be considered in the recycling supply chain, and culminates in the description of different methods used for developing an economic evaluation of the recycling processes. Finally, some case studies have been used to illustrate the economic analyses and its relation with both, the environmental evaluation and the sustainability concept.

2. Markets for recycled materials

The market is one of the main aspects to be taken into account to decide whether a recovery process is economically viable. Indeed, there is a vast variety of influential indicators in the market and the prices of the material tend to fluctuate strongly over time.

Both market situation and prices ought to play a decisive role when it comes to choosing the appropriate waste treatment process in a certain context. This is due to the fact that the recycling process depends greatly on the efficiency and effectiveness of the previous and posterior processes needed. Example of such processes are collection of waste, transport, separation, conditioning, as well as post-treatments which are necessary for the transformation of the obtained product to a marketable product.

Obviously, another aspect that should be taken into consideration is the recycled material market price. For instance, as example for plastic waste, the selling price of the recycled material depends greatly on the price of virgin polymer, which is linked to the crude oil price as well as the electricity cost. On the other hand, the prices of the plastic waste bought by recyclers are rarely influenced by the oil price and financial profit influences it more than environmental concern (EuPR, 2010). The price of industrial and consumer waste polymers appears to be modified by:

- The export of materials, above all to emerging countries
- High demand for high-quality recovered plastics
- Collection and sorting companies trying to boost their income
- The plastic characteristics regarding cleanliness, colour, etc.

It should be stated that sometimes the value of waste materials involve negative figures. In other words, the user is paid for acquiring the recovered materials. Nakamura and Kondo (2006) explained that Japanese steel mills accept waste plastic against payment and use them, after pretreatments, as a reduction agent in a mixture of pulverized coal in blast furnaces. Thus, they avoid buying more expensive materials as raw material in mills.

The law of supply and demand is an influential factor on the material price setting. For instance, material exportations to other areas of the world like Asia influence the price. When the demand in the Asiatic market is high, the availability of such material in Europe decreases and consequently prices tend to increase. On the contrary, high availability of either virgin or recycled material usually causes a reduction of prices.

The geographical location plays also an important role in the price setting caused by differences in prices of machinery, materials, and production and labour costs. For example, production and labour costs in emerging countries are known to be lower than in Western countries, but inaccessibility to raw materials could revert to a less advantageous position in the global market. According to statistics, developed countries produce higher waste quantities per person (IPCC, 2006), although high-populated emerging countries like China could vary this trend. Anyhow, recycling constitutes not only a challenge in environmental terms, but also a great opportunity to make profit from it.

Mixtures of recycled and virgin materials can be used instead of purely virgin material in manufacturing processes. Indeed, there are numerous studies showing that the addition of certain amounts of recycled material to virgin, does not damage the properties and characteristics of the resulting material (McGregor, 2009).

However, there are cases where recycling might not be the best way to deal with waste (Leiter, E., 2010). For example, when the price of the recycled material is highly priced with regard to the virgin material, it would be more profitable to choose the recycling option. On the contrary, in periods where the recycled material is low-priced would rather be more beneficial to focus on waste-to-energy technologies and obtain profit from the sale of energy.

On the other hand, recycling could have significant economic impacts as it replaces materials commonly obtained, transported and manufactured outside a specific region with materials collected and processed usually within the region.

Therefore, the selection of the recovery process that performs best in a given context requires a multi-perspective approach including issues such as financial cost, environment, market, supply, demand, etc.

2.1 Paper and board

The major paper producers in the world are Asia, the CEPI countries[1] and North America. These figures are closely related to population and level of industrialisation of countries. According to CEPI 2009, the paper production decreased as a result of the world economic crisis in 2008 and still continued in 2009. This fact was partially balanced by the pushing of the Asiatic countries, especially China (RISI, 2010). Nevertheless, pressure on the recovered paper sector set up by governments appears to explain part of the statistics and the use of this material in the paper and board industries at the expense of virgin fibres. In 2010 the negative tendency was reversed and the utilisation of recovered paper was increased. On the other hand, 2009 was the fourth year in which generation and consumption of recovered paper outstripped virgin fibres ones, but it should not be forgotten that virgin fibres are indispensable in producing high quality paper. Furthermore, the use of fresh fibres is required to strengthen the mixture of recovered paper during the papermaking process (METAFORE, 2006). This is due to the degradation that fibres suffer in the process of recycling and the use of recycled fibres usually varies between three and eight times depending on the quality of the recycled paper.

---

[1] CEPI countries are as follows Austria, Belgium, Finland, France, Germany, Italy, The Netherlands, Norway, Portugal, Poland, Spain, Sweden, Switzerland, United Kingdom, Czech Republic, Hungary and Slovak Republic.

Taking a look at the figures for the consumption recovered paper[2] in Table 1, the influence of Asia is stated. The differences between collection volumes and apparent consumption are explained by quantities in stock by collectors and consumers at the end of the year and a certain quantity of paper cannot be taken advantage like the toilet paper.

	Collection (tonnes)	Imports (tonnes)	Exports (tonnes)	Apparent consumption (tonnes)
Asia	83,108,000	35,980,000	7,710,000	111,334,000
Europe	62,980,000	12,950,000	24,650,000	49,250,000
North America	49,900,000	1,650,000	20,880,000	30,670,000
Latin America	10,020,000	2,105,000	670,000	11,455,000
Australasia	3,340,000	160,000	1,470,000	2,096,000
Africa	2,140,000	90,000	60,000	2,172,000
Total	211,488,000	52,935,000	55,440,000	206,977,000

Source: Magnaghi, 2009.

Table 1. Consumption of recovered paper in 2009.

By countries, the Table 2 shows the importance of Asia and particularly China as it comes to imports of recovered paper.

Imports of recovered paper		
Country	2008 (tonnes)	2009 (tonnes)
China	24,200,000	27,500,000
Germany	3,556,000	2,860,000
Netherlands	2,472,000	2,964,000
Indonesia	2,080,000	2,290,000
India	1,755,000	2,135,000
Belgium	1,500,000	1,536,000
Mexico	1,436,000	1,510,000
South Korea	1,307,000	1,120,000
Austria	1,305,000	1,190,000
Thailand	1,218,000	970,000
Spain	1,170,000	902,000

Source: Magnaghi, 2009.

Table 2. Major importers for recovered paper.

Although, most of the recovered paper is consumed in the country of origin, the exports are measured in millions of tonnes and represent a huge international market since is the main material used to make new paper such as newsprint for newspapers and cardboard for

[2] Recovered paper consumption is calculated as the recovered paper collected plus imports minus exports

packaging. Asia remains the most significant importer area due to the constant development of its paper and board industries. Regarding China, a fact that normally is not considered is that over the last years the Chinese government has encouraged to close many traditional mills because of uneconomic or focus of serious water pollution problems. In return, modern factories with highest capacity have been built principally in the east coast (Magnaghi, 2009). As a result, the China´s capacity for processing recovered paper has been increased and estimations assure that this fact will continue in the next years. Thus Chinese producers arbitrate the world market export pricing.

As far as recovered papers prices are concerned, the price of selling them is rather dependant of their quality and desirable uses. For instance, according to EN 643 "European List of Standard Grades of Recovered Paper and Board", recovered paper grades can be divided as mixed grades (1.01), corrugated and kraft (1.04), magazines (1.06) and deinking paper (1.11). As a summary, in the Figure 1 the price of recovered newspapers is shown. One can state the extremely variations from both perspectives, the modifications in a country and the comparisons between countries. The reasons for these figures have been explained previously in section 2.

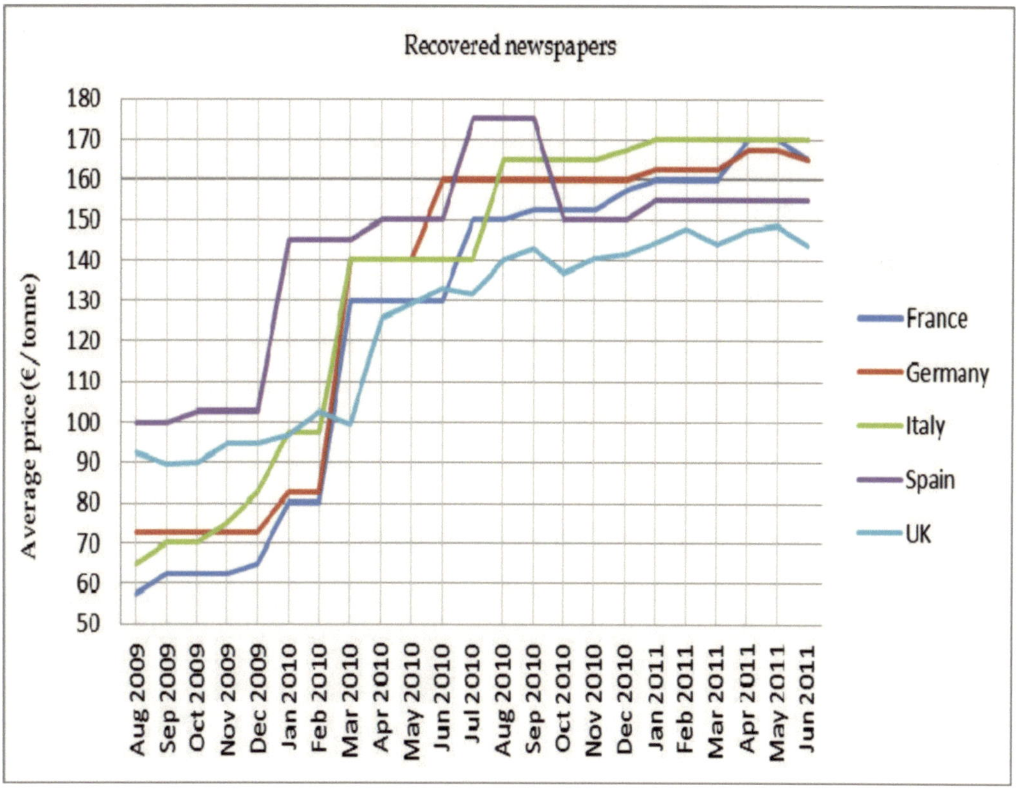

Source: RISI

Fig. 1. Average price of recovered newspaper for selected countries.

2.2 Plastics

Plastic waste can be recycled by mechanical and chemical recycling. Not all plastics can be recycled since there are chemical difficulties that make it impossible or with undesirable results. Mixed plastic packaging is not usually recycled due to problems in the separation stage and it is under research currently as it comes to collection and processing. Nevertheless, there are initiatives in order to adapt it to the industrial scale. Thus, the plastics which can be recycled are PET (polyethylene terephthalate), HDPE (high density polyethylene), LDPE (low density polyethylene), PP (polypropylene), PVC (polyvinyl chloride), PS (polystyrene) and others. The new uses range from jackets, coat and food packaging for PET. HDPE can be recycled in order to make tables, roadside curbs, benches, lorries cargo liners, trash receptacles, etc. LDPE is used as shop bags after the consumer stage. Recycled PP is suitable as small beans and battery boxes. Recycled PVC is appropriated for making weep pipes. PS can be converted in flowerpots. Furthermore, uncommon products from recycled plastics are being studied in order to add value to these materials. Examples of such innovations constitute the intelligent textiles (Greenpac, 2011).

World plastic production had been increased from 1950 to 2008. In 2008 the economic crisis made total tonnes of produced plastic to drop, but it seems like it is slowly recovering in 2010. The higher demanders of plastic are the NAFTA countries[3], Western Europe and Japan. Focussing on the European market, the demand for plastic was in 2009 as shown in Table 3.

Polymer	Demand in percentage (%)
LDPE	17
HDPE	12
PP	19
PVC	11
PS	8
PET	8
PUR (Polyurethane)	7
Others	18

Source: PlasticsEurope

Table 3. Plastics demands by converters in 2009. Breakdown by types.

Plastic recycling has been increasing its tonnage by around 11% per year over the last 10 years in Europe. Nevertheless, as stated above, in 2009 this growth fell to 3.1% as a direct impact of the economic crisis. This number is also explained by stronger activities of some packaging collecting and recycling systems and the increment of the exports outside of Europe, i.e. to the Far East. Derived from these facts, the quantity of plastic that end its life in landfill is reducing.

[3] NAFTA: North American Free Trade Agreement is a trilateral trade bloc established by the governments of USA, Canada and Mexico

As it comes to the Asian market, the same tendency is observed. From 1990 to 2008 waste plastic trade was increased by 100 times as a result of the increasing demand of such material. High prices in oil were related to high prices in virgin plastics which caused an increment in recovered plastic demand. Thus, 80% of waste plastic was sent to Asia in 2007.

The most common plastics which are currently recycled are PET and PE as it comes to prices of recycled plastics. The data figures are usually presented varies with the type of plastic. As an example, the PET prices differ from whether the material is in form of bottles or flakes, if it is colourless or mixed colour, or if is suitable for food contact (food grade) or not. In Figure 2, the prices for recycled plastics are presented for the British market.

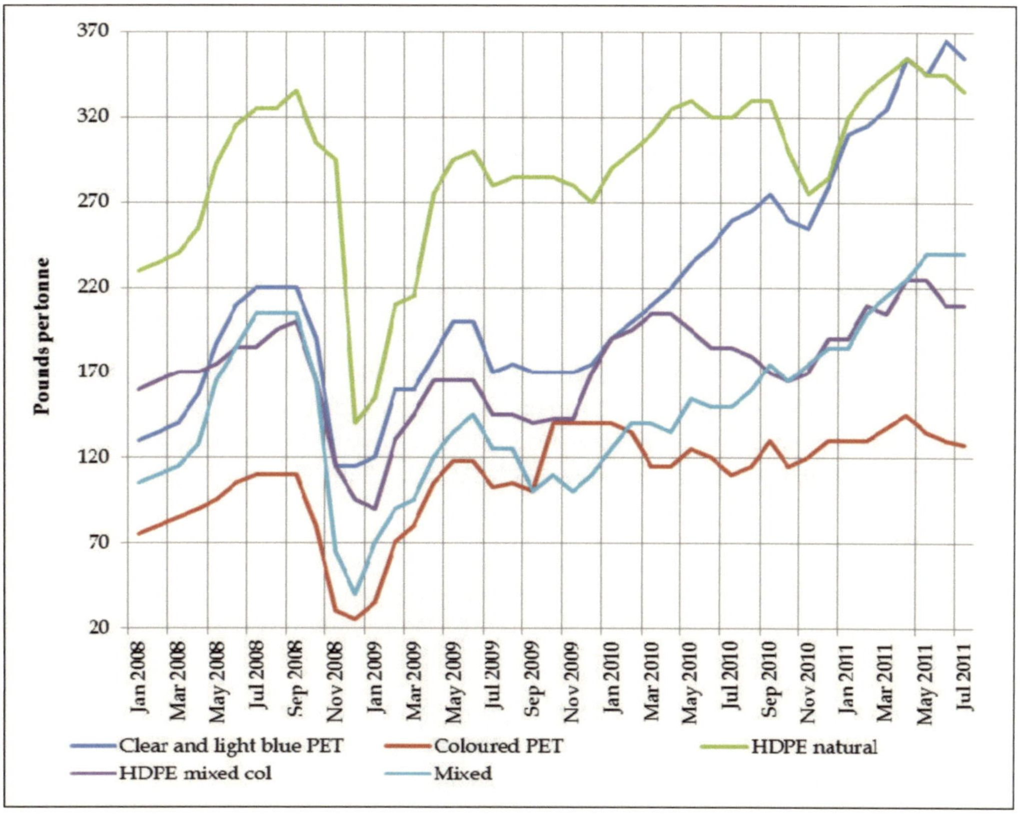

Source: Letsrecycle.

Fig. 2. Average prices for different kinds of plastic bottles in the British market.

A clear drop in 2008 and the subsequent increment are shown for high quality sorted materials such as PET and HDPE bottles. High demand from Chinese processors, depreciation of currencies against the US dollar[4], increasing demand for food grade recovered plastics, tight domestic supply, related to the lower consumption due to crisis and higher virgin plastics prices can explain this behaviour. Plastic demand is strongly sensitive

[4] Prices are typically set in US dollars

to consumer spending habits, and therefore, economic downturns. Nevertheless, a recovery is observed for every type of plastic. The line for mixed plastic bottles remains more horizontally than the other kind of recycled plastics because of uncertainties in their process of recycling and the need to study deeply their recyclability. On the other hand, and focussed on the plastic container sector, disruptions are influenced by mergers and acquisitions of companies especially in the USA due to large players that are fighting in a mature market.

2.3 Metals

The advantages of recycled metal versus virgin one, from an environmental point of view, are the lower use of natural resources and the savings in the energy required for its processing. Recycling saves 95% of energy in aluminium production, 85% in copper production, 74% in steel production, 60% in zinc production and 65% in lead production (BIR, 2009). In economic terms, this is translated to savings in the energy bill and the possibility of taking profits from an interested market. Such importance lays on the metal market has raised their demand in China, India and the Far East due to the development of the commercial, residential and industrial construction. Moreover, industrial machinery, cars and armament demand these materials. Metal scraps help the primary metals industry to achieve these needs not only in the developing countries, but also on the Western countries.

Steel is by far the most-recycled material in the world, followed by paper and aluminium. As it comes to economic figures, the main exporters in 2010 were USA (20.56 million tonnes), the European Union with 18.97 and Japan (6.47), whereas the main importers were Turkey (19.19 million tonnes), South Korea with 8.09 and China (5.85) (BIR, 2010).

The steel scrap consumption in the European Union increased in 2010, whereas USA slightly reduced it in that year. The main reason for that behaviour is that electric arc furnaces (EAFs) only consume steel scrap because of saving in energy consumption, but were operating at lower rates. Moreover, the EAFs in USA utilized around 40% iron alternatives.

Asia had become a net centre of recyclable import with the 80% of the world's copper scrap imports, the 50% of the aluminium scrap and the 30% of iron and steel scrap in 2007. China has converted in the world leader in steel production.

Figure 3 shows the prices paid by medium to large size recycling companies which receive material from local collectors. As shown in Figure 3, the economic crisis has not influence the metal prices so deeply in comparison with the previous materials and prices are recovering slowly. This is due to the constant demand of these materials. For instance, in the aftermath of the recession the steel production in the world reached 1.412 billion tonnes in 2010, which represented an increase of 14.8% over 2009 and a new record (BIR, 2010). Only copper reduced its price in 2008 and roller coaster movements have been done in the last months. This uncertainty in the price is partially due to the crisis of the Europe sovereign debt, the downturn of the America´s economy and the process of restructuring the Chinese economy.

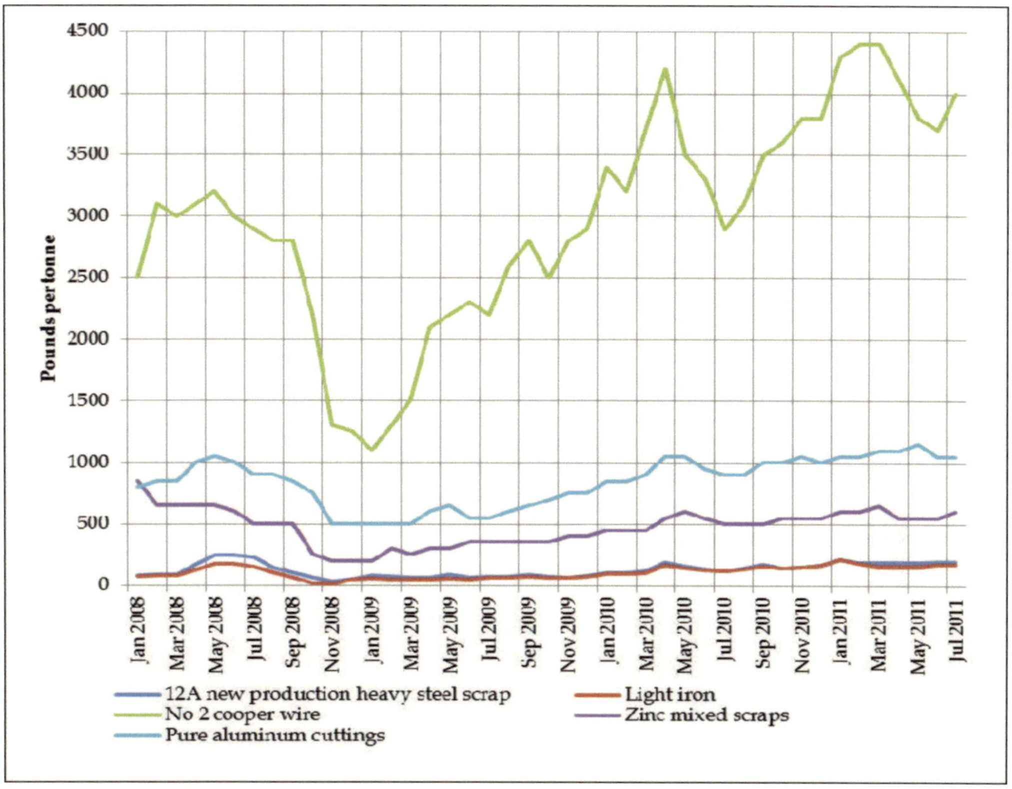

Source: Letsrecycle

Fig. 3. Average values of scrap metal prices in the British market.

Not surprisingly, the price of non-ferrous metals is appreciably incremented in comparison to the one of the ferrous metals. The main reason is the limited availability of those materials, so the unrestricted flow of non-ferrous scrap from country to country is crucial according to the law of supply and demand.

The precious metals constitute a very special case. The precious metals´ price is increasing almost continuously. They act as a safety refuge when global markets are unstable. Furthermore, when the price of oil is moving up and down, investors look for these kinds of metals with relative stabile value on the market.

2.4 Glass

The process of glass recycling is quiet interesting from an industrial and environmental point of view. Glass can be recycled endlessly without losses in quality or purity (GPI, 2010). Energy costs drop around 2.5% for every 10% cullet[5] used in the manufacturing process. In addition, cullet makes the manufacturing mix less corrosive and reduces the melting temperature extending the furnace life. Moreover, glass recycling is a closed-loop system which does not produce additional waste or by-products (GPI, 2010).

[5] Cullet: Glass crushed and ready to be remelted.

Demand for cullet has grown over the last years with new glass processing plants and alternative markets under developing. Thus, the domestic market for recycled glass remains usually at high level and this fact makes the price paid for this material to be quite stable. In most countries this circumstance is added to the non-dependence of the glass industry on the export market. As a result, long-term contracts between local authorities, which are the main providers of glass for recycling, and glass recycling companies come with a good degree of security.

On the other hand, in the last years companies have adopted strategies in order to improve their environmental credentials as a way of making their green marketing better. Derived from this fact, actions like reducing the weight of their glass bottles in important percentages have been doing. 99% of glass containers in the world (in weight) belong to food packaging and it is easy to imagine the global avoided impact of these actions.

Imports and exports in the glass market are rare because of its high weight and the high fuel consumption associated to their transport which increases prices. Moreover, multinational companies have plants in several countries and the raw material recovered glass is mainly provided by the national market.

In the Figure 4 the prices of glass containers in the British market are shown. Completely mixed glass has the lower value due to strange colours obtained after melting. This fact makes

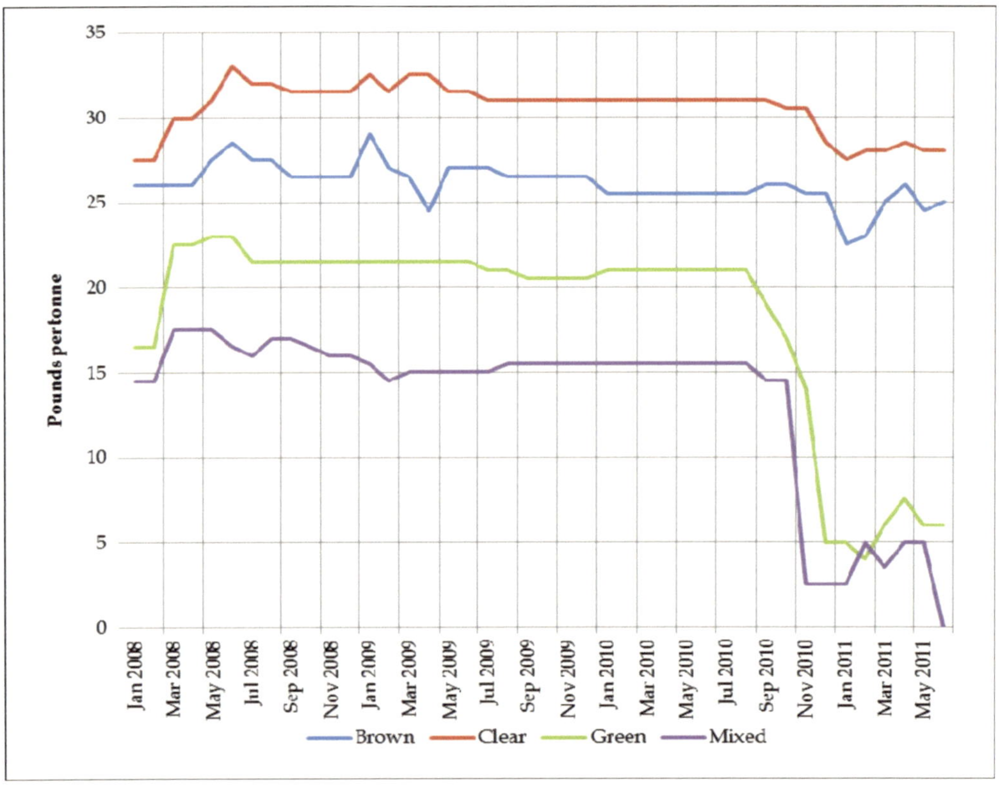

Source: Letsrecycle

Fig. 4. Average values of glass prices of containers delivered to collectors in the British market.

it difficult to sell it since customers demand pure colours. Instead, it should be derived to alternative uses. Green glass has the third lower price because it is the most abundant in the United Kingdom waste stream. The clear glass is the most expensive since it is the most demanded one in this market. Nevertheless, internal issues could affect the glass price. For instance, in the Figure 4 the slump in prices from September to December of 2010 was related to the continual decline in Packaging Recovery Notes[6] revenues (Letsrecycle, 2011).

Apart from glass containers, alternative uses of recycled glass are coming out. Such new approaches include glass as grit blasting, use in road surfaces or water filtration.

2.5 Wood

As it comes to recycling of wood, two types of materials should be distinguished. i) High grade wood comes from clean white softwood, wooden pallets, timber offcuts, packaging crates and joinery waste. They should be free of paint and coverings. ii) Low grade wood includes plywood, doors and window frames, roof timbers, panel products and so on. Both kinds of materials come from construction and demolition, commercial, industrial and household sources.

The main market of recycled woodchip has been the panel board industry due to historical reasons. This material is used in the production of chipboard. Nevertheless, the wood industry has been searching for value-added markets and new uses have been discovered, i.e. animal bedding, equine surfacing, garden mulches markets have steadily developing. This fact has made wood industry to increase in the last years. The demand of woodchip is expected to continue growing since governments are promoting the generation of renewable energy.

In the Figure 5, the price evolution of recovered mixed wood delivered to a wood recycler can be seen. Negative prices indicate that the recycler is paid for these materials. Nevertheless, this fact can change because of increasing demand of clean wood pallets and sawmill round wood towards small payments by suppliers. In addition, wood waste prices will vary considerably depending on cleanliness of the material, volume and location.

3. The recycling supply chain

Recovered waste materials, before being recycled must be collected, transported and separated among other processes and as a result, the general performance of the recycling process depends greatly upon the efficiency and effectiveness of those "minor" processes. Similarly, from the economic perspective, the recycling process must be preceded from also efficient and economic sub-processes. Otherwise, high cost of recycling could overweigh the environmental benefits of recycling. In addition to the recycling process itself, a broader approach including the prior and posterior stages of recycling is needed.

---

[6] Packaging Recovery Notes is a document which provides evidence that waste packaging material has been recycled into a new product in Great Britain.

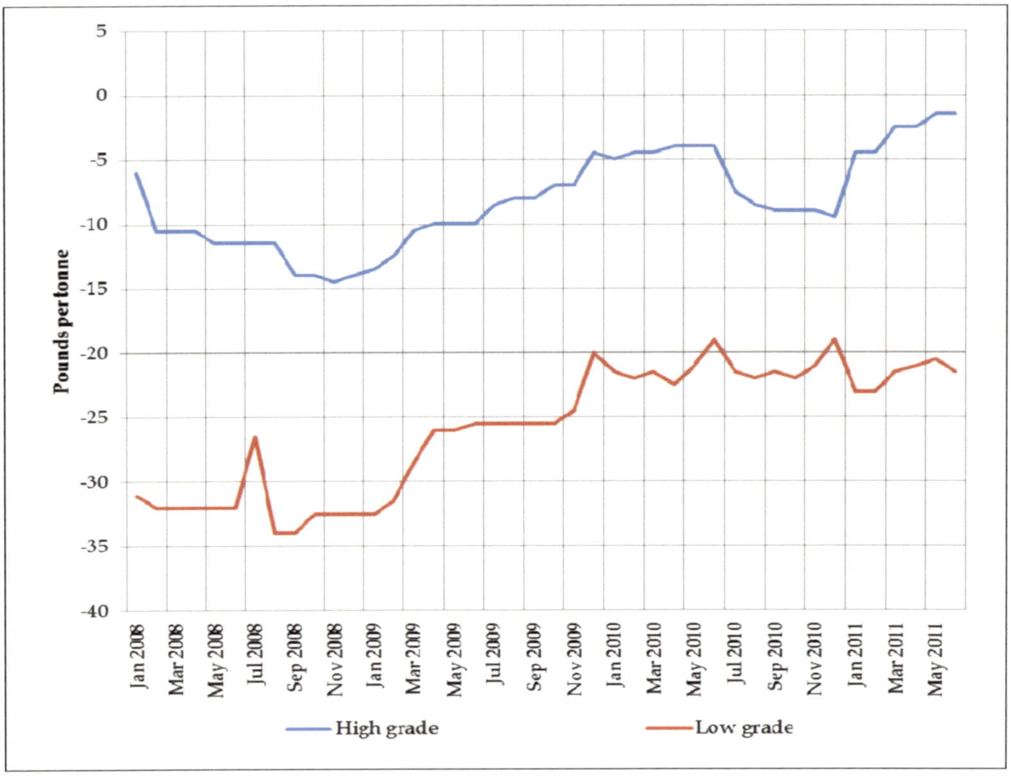

Source: Letsrecycle

Fig. 5. Average values of wood prices of mixed wood delivered to a wood recycler in the British market.

3.1 Collection

The municipal solid waste generated in the European Union was 514 kg per inhabitant (Eurostat, 2010) and 719 kg in the USA (EPA, 2010) in 2009. Such high quantities should be treated under the most convenient processes. The end of life of materials starts wherever is produced, whether it is generated in a house, store or industry. It is necessary to separate out waste in appropriate fractions according to the facilities in which they are going to be treated in order to make further stages more effective. Throughout this action waste is treated easily and economic costs are reduced, allowing making recycling a competitive treatment. Ideally, waste should be set apart as paper and board, plastic, metals, glass, wood, batteries, textile and organic. This can be done at source, having selective collection in the cities or at specific sorting plants, where all recyclable materials are sorted out and sold to recyclers. Nevertheless, the availability of these facilities varies from country to country and is close related to environmental policies.

Transport plays an important role in the collection stage. A wide range of vehicles can be used, but lorries have become as the most used in this issue due to their load capacity and the particularities of towns. Furthermore, these vehicles can sometimes incorporate a pre-waste disposal unit which enables compacting waste and making collection more effectively.

3.2 Sorting

Once waste is collected, sorting is a mandatory activity in order to have separated the different recyclable materials and to avoid undesirable materials which can decrease recycling efficiency and recycled material quality. For example, ceramic is considered a disadvantageous material when glass is recycled. Furthermore, separating a material in specific fractions (for instance different colours, etc.) allows taking advantage of differences in sell prices as has previously been shown in the elements studied. Paper and board´s costs differ on the quality of fibres and, at the same time, on the origin. Plastic should be separated according to the type of resin because some recycling processes are specific of each material. For instance, methanolysis only works with PET polymer (Aschilias, 2004). Glass is required to be sorting according to its colour. Coloured glass is unable to produce clear one and mixed glass presents the lower price. Metal values depend on their kind as we have seen and sorting is a must in order to separate them according to their nature.

The cost of sorting depends on the technique selected. For example, in case of using optical separators the purchasing cost of the equipment could be more expensive than carrying out a density separator, i.e. a pool with water and salt (to increase the density) or alcohol (to reduce it) is enough for separating plastics. However, this last treatment could lead to an increment in the final cost by considering also the price of energy in order to dry wastes as well as the substances used for adapting the water density and the water treatment after the process.

Then we have two flows: the materials which will continue with the recycling process and the sorted materials that can be sold in order to obtain revenues.

3.3 Shredding

Shredding is required before the recycling process and is considered a key pre-treatment since it facilitates further processes. Its main advantage lies on the volume reduction, which minimize the storage space as well as reduce transportation. Moreover, as it comes to glass, it allows reaching the melting point by using less energy and time, and improving furnace feeding. Small pieces increment their surface exposed and the energetic yield. Thus, less fuel is required in order to melt glass and metals. It is desirable that metals arrive at the processing facility in bales or briquettes since this way the efficiency of the remelting process is increased. Nevertheless, the most common is not to receive such material in this form and a shredder constitutes usually a stage in the recycling facility. In addition, it helps to liberate trapped contamination, upgrading the quality of materials like glass, metals or plastics and its value. On the other hand, the shredding of paper and wood is not high recommended since fibres will be damaged. Wood recycling requires shredding when the final use is to make agglomerate boards or smaller pieces.

3.4 Recycling process

The recycling process differs depending on the material to be treated. The recovered paper is put in a pulper with water and is pulped with mechanical and hydraulic agitation in order to disintegrate paper into fibres. For purposes of deinking some chemicals are sometimes used, like deinking agents and NaOH. Contaminants are removed during the operation due to differences in physical properties. The pulp slurry is pumped from the pulper to hydrocyclones. The organic rejects are often burned in order to take advantage of their calorific value. In general, screening at lower consistency in order to separate undesired particles is more effective, but it requires additional machinery installations and its energy consumption of the process is increased (INERIS, 2010). After that, a fractionator is used in order to separate the pulp in two fractions, creating a short-fibre stream and a long-fibre one, to apply different treatments. In the case of long-fibres, dispersion can occur with the finality of achieving better fibre-to-fibre bonding, strength characteristics, and to reduce dirty specks in size. Furthermore, refiners improve optical and strength characteristics, but its main disadvantage is the impressive energy consumption. Continuing with the industrial paper recycling process, the mixture enters in the paper machine after a cleaning substage and a fine screening (INERIS, 2010). Moreover, a flotation deinking stage is recommended when recycling paper so that ink is removed and a better brightness is followed. In addition, bleaching chemicals can be added before entering at storage tower. Finally high quantity of waste water is produced during the process and should be conveniently treated in order to reduce pollution.

In case of plastic recycling, after sorting and shredding, a washing step should take place in which impurities are removed. Next, the melting process is done throughout an extruder which applies heat by friction. It also allows homogenizing and filtering in order to produce high quality recycled material. In the end, the pellet conformation is produced by a pelletizer.

The recycling process of metals consists of a melting process. The temperatures should reach from hundreds to thousands degrees, but alloys can reduce such high values, high energetic costs, to a more reasonable one. There are particularities in the melting process due to the wide range of recyclable metals and stages for overcoming potential problems. For example, as it comes to aluminium alloys, a degassing step is necessary to reduce the amount of hydrogen in the liquid metal. High hydrogen concentration could result in gas porosity that deteriorates mechanical properties. Continuing with the process, the molten metal is poured into molds. Once the metal is solidified, it is removed from its mold. A degating stage is necessary in order to remove head, runners, gates and risers from the casting (AFSSOCAL, 2011). For this purpose, cutting torches, bandsaws or ceramic cutoff blades are used. This metal must be remelted as salvage increasing, thus, the yield and reducing the costs. Surface cleaning is needed for a better presentation. Usually, sand or other molding media might be adhered to the casting and metal is cleaned using a blasting process. In other words, a granular media is propelled against the surface of the casting. The media, propelled by compressed air for instance, strikes the surface at high velocity and tears any impurity of the surface. Finally, grinding, sanding or machining steps are done in order to achieve the desired dimensional accuracies, physical shape and surface finish, as well as painting in case prevention of corrosion was needed and improve visual appeal.

Once again, melting is also the main step in the glass recycling process. Cullets are melted in huge furnaces. Decolorizing and dyeing is the next stage. Firstly, oxidizing of the melted glass cullet is required. For green glass, the colour turns from green to yellow-green and manganese oxide is then mixed until a grey colour appears (All-recycling-facts, 2009). For brown glass, zinc oxide is added to oxidize it to blue or green cullet. If clear recycled glass is required, erbium oxide and manganese oxide are added to help clear all the colours from the glass cullet. In the end, the recycled glass is moulded into the final product.

Taking into account that contaminants have been removed before, the recycling process for wood consists in a two-steps process in which wood is introduced in a tub grinder, horizontal grinder or wood chipper. Wood is grinded into chips which are ready to be sold for use in particle board, chipboard, pulp and paper products, animal bedding, mulch, biomass fuel and compost. In the panel board industry, an adhesive is necessary to stick wood pieces together.

3.5 Distribution of recycled material

After recycling, two additional steps should be also considered: packaging and transport to customer. As it comes to packaging, the type is influenced by the transported material. Paper is carried in coils, and for plastic pellets sacks are usually required. Moreover, metal and cardboard can be transported with strapped sheets. Glass recycling and manufacturing are done in the same facilities and transportation to glass maker is not required, although transport from glass maker to filler is needed. Here, packaging entails of palletizing and shrink wrapping. Initially, the distribution stage does not modify substantially the purchasing cost of the recycled material, but it is greatly influenced by the distance, weight and way of transport.

4. Economic evaluation of recycling

The aim of the economic evaluation in recycling processes is, on the one hand, to assess the economic impact of recycling and on the other hand, to identify weak points, or the less economically efficient stages of the process, in order to be improved. This data is valuable for decision support from a public and private point of view. In regard to public organisms, it allows to select the best waste treatment alternative. As it comes to private companies, it also provides information about the recycling process´ profitability and recovery.

The economic evaluation of recycling processes can be determined through several methods. Some of the most relevant are the Life Cycle Costing (LCC), the Cost-Benefit Analysis (CBA) and Input-Output methods. LCC is a method which follows a life cycle perspective. All the inputs and outputs are determined in economic units and several types of LCCs can be found (see section 4.1). CBA has been developed for major public investment plans and compares the total expected costs against the total expected benefits. This way the difference between costs and benefits is determined and measured. The Input-Output method is a linear model in which the interdependencies between links in a chain are presented on the basis of the output of one industry is the input of another. This quantitative economic technique is used to compare branches of national economy or between competing economies.

Despite the fact that the economic evaluation of recycling can be worked out throughout diverse methods, in this chapter LCC will be the most selected economic method since it is a precise and useful tool suitable for economic evaluation of recycling.

4.1 Life cycle costing

LCC analysis, which is a cost management method, is carried out on existing products and is used to monitoring and managing costs over the product or process life cycle. From a time perspective, it can be used for comparing past alternatives or future ones.

Environmental LCC analysis is a methodology that allows:

- To identify the processes that are most relevant for the overall cost
- To compare life cycle costs of alternatives
- To detect direct and indirect (hidden) cost drivers
- To identify trade-offs in the life cycle of a product
- To use the full costing to identify new products

One of the first steps to carry out a LCC is to determine the functional unit (FU) to which all the costs will be referred. Examples of FU in a recycling process could be one tonne of recycled product. However, the FU should be adapted to the specific case considered in order to reflect the real costs. For instance this means considering the same time frame for all costs and including, if necessary the effect of inflation. The system boundaries should also be defined. System boundary limits which stages, inputs and outputs are taken into account in the LCC. Ideally, considering the whole life cycle is the best, but sometimes it is impossible due to missing data. In this case, one can make a bibliographic search to overshadow the handicap, although in case of not finding the data the system boundaries should be modified. All the data will be expressed in economic terms and converted into present-value costs by using discount rates. Another piece of advice highly recommended is to break LCC in cost elements so that it is easy to use. Not all costs should have a great impact on the final results, but breaking costs constitutes a good starting point aim at considering those that contribute most to the total cost.

There are many ways to classify the costs involved in a LCC. One of them is the approach of (Bovea, 2003) according to which the costs are:

- **Internal costs (IC)** along the life cycle of the products (i.e.: production, use or end-of-life expenses), which are the costs for which the company is responsible over a period of time. These type of costs include:
  - Conventional costs (CC): direct costs borne by the company when manufacturing a product (i.e.: raw materials, electricity, transport, etc.).
  - Hidden costs (HC): general costs related to license expenses, waste management costs, etc.
  - Less tangible costs (LTC) which are often not included in the company accounts due to their probabilistic nature. These costs include expenses on marketing, improving the image of the product, safety measures for workers, etc.
- **External costs (EC)** that are envisioned to include monetized effects of environmental and social impacts not directly billed to the company, consumer or government. These costs are also called "externalities" on life cycle management forums (like costs related to depletion of natural resources, impact on human health). Quantifying of negative effect of external costs is a critical issue in LCC practice. According to Kloepffer (2008) external costs to be expected in the decision-relevant near future (e.g. cost occurring in the future due to legal requirements in order to fight against climate change or special

requirements for radioactive waste) are difficult or even impossible to estimate. Several attempts for the monetisation of external costs have been done, i.e. monetisation of emissions, road congestion and noise negative effects.

(SETAC, 2008) proposed the definition of the internal costs as the costs "*directly borne by an individual or organization in supplying or consuming a product*". And the external costs were defined as "*the market costs, not directly borne by an organization in terms of costs of labour, capital, and taxes, but as costs for purchases from other firms in the system, covering the internal costs of these other firms*".

Consequently the LCC can be calculated as follows:

$$\mathbf{LCC = IC + EC = (\Sigma\ CCi + \Sigma\ HCi + \Sigma\ LTCi) + \Sigma\ ECi}$$

Figure 6 represents the stages of the life cycle of a product. As can be seen, the life cycle of a product can be a close-loop in case recycling is used as a waste management. Costs in LCC correspond to the stages in life cycle. This way, finding the most disadvantage phases is easy.

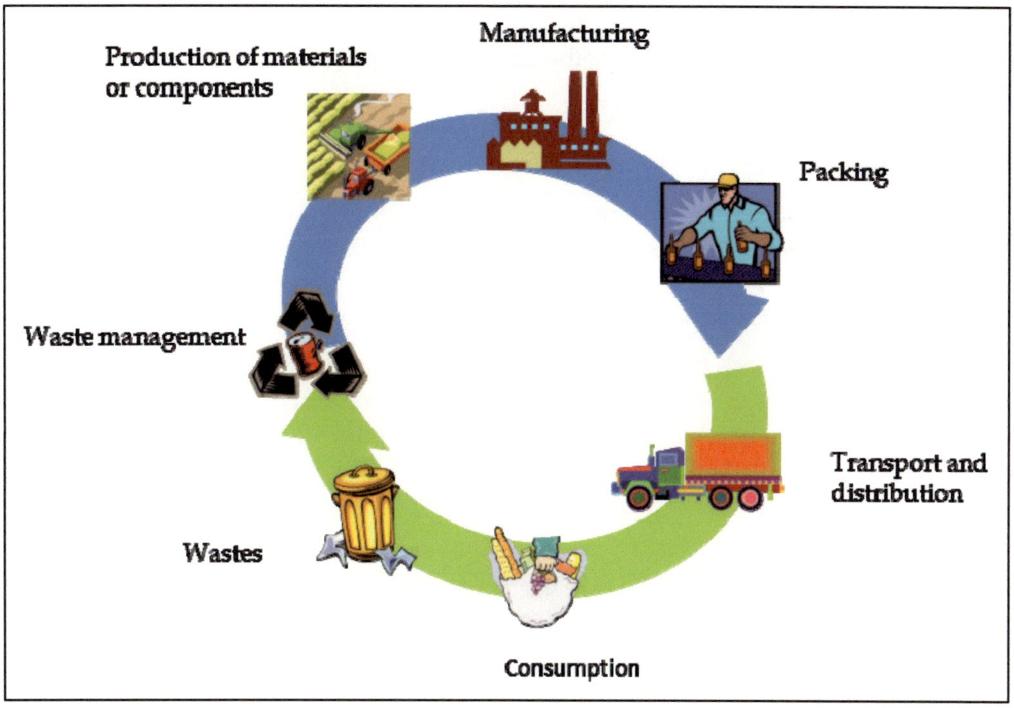

Fig. 6. General life cycle of a product.

The SETAC-Europe Working Group on Life Cycle Costing has defined three types of LCC, depending on the costs considered in the study: conventional life cycle costing, environmental life cycle costing and societal life cycle costing (SETAC, 2008).

4.1.1 Conventional LCC

The assessment includes only internal costs excluding on some occasions the End of Life (EoL). The perspective usually refers to one actor: the manufacturer, the user or the consumer. The conventional LCC is similar to the profit-loss account carried out by companies and does not include social impacts. It is usually not associated with separate LCA results.

4.1.2 Environmental LCC

According to (Rebitzer & Hunkeler, 2003) and (Kloepffer, 2008), environmental LCC may be defined as "*an assessment of all costs associated with the life cycle of a product that are directly covered by any one more of the actors in the product life cycle (supplier, producer, user/consumer, EOL-actor), with complimentary inclusion of externalities that are anticipated to be internalized in the decision-relevant future*".

The assessment includes internal cost plus external costs expected to be internalized. Complete life cycle is taken into consideration and the perspective used may refer to one or more actors. Environmental LCC takes into account the financial costs associated to environmental impacts from LCA. The fact of using the same functional unit allows the comparison between LCA and LCC. This type of LCC allows the assessment of costs along the life cycle of a product maintaining the economic pillar of sustainability separately from the environmental and social pillars.

4.1.3 Societal LCC

The assessment includes both internal and all external costs and complete life cycle is considered. The perspective refers to society including governments and considerers both, the present and the long-term future possible situation. Societal LCC includes all of Environmental LCC and additional external costs (SETAC, 2008).

The main difference among the three types of LCC is that only environmental LCC follows a functional unit as a reference unit for the economic assessment. This aspect is extremely important, since the environmental LCC is the only kind of LCC that follows the same approach like LCA. Although not existing an international specific methodology for carrying out LCC studies, the reference (SETAC, 2008) can be used to carry out this kind of analyses.

Once the types of LCC and costs are explained, a review of some case studies will be carried out in order to know more about this kind of analysis and better understand how to interpret the results.

4.2 Case studies

In section 4.2, the description of the main techniques and problems regarding LCC in recycling is shown by reviewing case studies. It is of the opinion of the authors not to show all the results since they would change depending on the case studied, country and year of the study.

4.2.1 Selection of alternatives based on economical parameters

As has been stated, LCC can be used to select the best economical alternative in the EOL of products. Craighill & Powell (1995) carried out a study on recycling household waste from economic and environmental points of view. They quantified the cost of recycling aluminium as 111.41 £/tonne, glass 67.20 £/tonne, paper 73.79 £/tonne, steel 31.64 £/tonne, HDPE 12.07 £/tonne, PET 21.25 £/tonne and PVC 11.55 £/tonne. When comparing the recycling cost with the waste disposal cost of every material, the net benefits were favourable to recycling except in the cases of plastics.

As an example, in 1999 the more attractive economic end-of-life of municipal solid waste was disposal with an average value of 105.91 $ per tonne in the state of Maine, United States, whereas in the case of recycling it cost 107.37 $ per tonne (Maine, 1999). Although each region faces a very difference set of circumstances, this fact would be extrapolated to other areas. From the actual point of view, it is clear that recycling technology has been constantly upgrading and advances in the state of the art allow reducing cost. Moreover, legislation is pushing stakeholders to adopt the environmentally best alternative in EOLs of materials.

A special case appears when recycling contaminated waste. Chang & Farr (2000) focused on contaminated low-level mixed waste pre-processes. They centred their efforts on a multiobjective evaluation to treat heavy metals, radio-nuclides, PCBs, halogens, pesticides and chlorinated waste. This study constitutes an example of some problems that decision-makers should face when one or more strategies of recycling of waste with hazardous materials. Firstly, it is necessary to determine which contaminant or contaminants are tainted within waste. Secondly, the targets are not only costs and resource recovery, but also volume reduction, simplicity of process, health risks and energy efficiency. Thirdly, a review on the technologies and their costs is necessary. Once these stages are reached, the authors proposed a partially subjective evaluation in order to have a single score with which being able to select the best option depending on the type of impurity.

Once the LCC is carried out, several tools can be used in order to prioritise the components recycling of a product. Park et al. (2006) used four methods to help decision-makers to prioritize washing machine components to be recycled from an economic and environmental perspective. The four methods selected were the two-dimensional diagram, eco-efficiency, the monetary method and the multi-attribute decision making (MADM). The two-dimension diagram is a graphical representation of the environmental impacts of components calculating in a LCA along the X-axis and the economic values, for instance as benefits or selling prices, along the Y-axis. Eco-efficiency was worked out as the ration between the economic benefits in the numerator and the environmental impacts in the denominator. Derived from that, a component that has a lower economic benefit and higher environmental impact is considered to have the highest priority in improving the economic and environmental performance. In addition, this fact is considered a key issue in ecodesign. The monetary method consists of expressed the environmental impact as monetary value. There are several tools in order to make this like the LIME method. This way the environmental and economic results are in the same units and adding or subtracting is easy. MADM is used in selecting one or more alternatives or ranking from a multiple alternatives.

Numerous methods have been proposed for carrying out the MADM. Among others, the simple additive weighting (SAW) was selected by the authors. The SAW method runs as the weighted sum of the performance ratings of each alternative based on the evaluation of all criteria.

The authors reduced variance of prices of recycled materials by selecting average sale prices. Moreover, weighting factors were used since 0.25 corresponded to environmental aspects and 0.75 to economic ones. The study showed similar results for the two-dimensional diagram, the monetary method and the MADM for the recycling of the washing machine components. The results of the eco-efficiency were slightly different. This fact sums up the answer of how important is to considered additional analysis beyond LCC and LCA. By applying several methods the information on the recycling process is more complete and stakeholders are in a better position to make decisions.

4.2.2 Influence of product design or the kind of material in recycling

According to Porter (2002) recycling has two major financial costs: collection and reprocessing. This is true only in the cases in which recycling involves a single recyclate, such as PET or PE. When a product is compounded by more than one material, the influence of the design is highlighted.

Nakamura & Kondo (2006) proved that dissembling is the first and most important stage in the recycling process as it comes to electrical home appliances (TVs, refrigerators, washing machines and air conditioners). Nevertheless, this idea is likely to be extended to different kinds of waste like cars, industrial machinery, etc. all of which are compounded by several materials. The design for disassembling increases the efficiency of disassembling and raises the purity of some recovered materials, in the end the selling price. Improving the design for disassembling means saves from both an economic and environmental point of view (Nakamura & Kondo, 2006). Costs and revenues are determined to a high percentage in the design phase, and thus the importance of carrying out a proper ecodesign before launching the product to market. This will allow reducing disassembling costs.

Wright et al. (2005) also measured the effect of improving the recyclability of a fibre optic cable by redesigning it. The authors modified some of the materials which compounded the cable in order to increasing their recyclability. In the redesign aluminium and organic gel were removed. The costs were joined according to the stages in the life cycle: PBT tube manufacturing, sheathing, stranding and LLDPE recycling, and divided by materials. In addition, LCAs were carried out in order to complete the analyses. Results showed that the largest contribution to costs in the cable system is from the material inputs. The fact of avoiding materials without compromising the function and security of the cable is translated in significant reduction in costs (up to 40%).

Replacing materials was the option selected by Ungureanu et al. (2007) in passenger cars, i.e. heavy material like steel was replaced by lighter aluminium. Different close-loop recycling percentages were assumed. Once the materials reached its end-of-life, it is supposed to be melted and sheet manufacturing ready to be used. Nevertheless, data regarding the kind of fuel used and its price were not shown. The authors composed a table in which the prices for both materials as scrap and recycled ones are presented. Based on that and taking into account only the recycling (post-use) stage, the calculations were easy and showed that, in this case, post-use was not considered significant for computing the total life-cycle cost. This is due to the high contributions of the manufacturing and use phases, especially as a result of the litres and gas price and the kilometres driven in the life of a car. Nevertheless, the premanufacturing costs depend greatly on the percentage of recycled material. The more recycled material, the less costs. Comparing the price of recycled steel, one can state that steel is cheaper than aluminium, but here additional circumstances should be considered. The aluminium scrap is also more expensive than the one for steel and its lightness offset the costs during the car´s lifespan.

An example of the influence of the recovered material is found in Craighill & Powell (1995). When comparing the recycling cost with the waste disposal cost of the selected materials, the net benefits were favourable to recycling except in the cases of plastics. According to the authors, this fact is due to the higher volume to weight ratio. In other words, one tonne of plastics occupies more volume than one tonne of glass, metals, paper, etc. As a result, this point explains a greater number of kilometres travelled per tonne of material, and thus, it results in negative net benefits. Derived from this observation, the importance of plastic collection trucks with integrated compactors is highlighted in order to favour recycling.

4.2.3 Economic evaluation and other techniques

LCC, as a decision tool, provides more information if it is presented with an environmental analysis (LCA), constituting the ecoeficiency. Furthermore, the combination of economic, environmental and societal analyses provides what is considered to be the sustainability.

Norris (2000) indicated the impossibility of separating the environmental study of a product from the economical one. Both should be studied in parallel since a separation of LCA and LCC does not allow characterizing completely the important relationships and trade-offs of alternative product scenarios. Moreover, the sum LCA plus LCC permits to know which modifiable stage provides the greatest combined economic and environmental leverage. Related with the previous point, taking into account the double-approach enables to determine the incremental costs of environmental improvement for diverse options, i.e. when comparing economic behaviour of recycling and landfilling.

In Kim et al. (2009) the recycling potential of electrical appliances components was calculated, i.e. in waste television, washing machines, refrigerators and air conditioners. Basically, the considered formula consisted in multiplying the economic score by a weighting value and the environmental score by another weighting value. The environmental weighting was used in order to reduce the number of decision variables into a manageable amount and to better communicate results from environmental studies. The weighting values derived from an analytical hierarchy process (AHP) method. On the one hand, the AHP reduces decisions to one-on-one comparisons and synthesizes the results. The weighting factors are determined based upon multiple factors obtained by interviews. On the other hand, subjectivity represents the main problem in this method. One way to solve this problem is questioning high-prepared people. Thus, relevant personnel in the industry and academia were interviewed. Results from questionnaires showed that the economic value is more important than the environmental value by three. As a result, the economic value was 0.75 and the environmental was selected as 0.25. The economic value was calculated taking into account the collection fee received from the generators, the disassembly cost, the maintenance cost, the residue disposal cost and the selling price of the recycled materials. After calculating the economic and the environmental value of each material and the weighting factors, the recycling potential was worked out. As it comes to economy, recycled copper showed the highest value due to its selling price, followed by steel, glass and circuit board, aluminium and plastic.

The ecoefficiency of a motor and this motor with a frequency converter were measured in Lyrstedt (2005). The formula of the ecoefficiency proposed was 1-(EDC[7]/LCC). On the one hand, EDC is based on the indicators measured in a LCA: greenhouse gases, acidifying gases, ozone depletion gases, gases contributing to creation of ground level ozone, emissions contributing to oxygen deficiency in water and consumption of non-renewable energy reserves. On the other hand, ECD and LCC functional units and system boundaries, in manufacturing, usage and disposal phase, were consciously selected equal since it allows comparing figures, and thus maintained the formula coherence. Scrap materials prices were negative inputs since the prices of selling the metal components are incomes, recycler pay you in order to obtain raw material.

Similar than in Lyrstedt (2005), in Lightart & Ansems (2007), shadow costs[8] were followed by considering the shadow price per environmental effect category under the CML2 LCA method (see Table 4). There the shadow prices were expressed in price per unit of emission reduction for the most expensive measure to be introduced to achieve an objective. The shadow costs were calculated by multiplying the quantity of equivalents by environmental category found in a previous environmental analysis by the shadow prices.

Effect category	Unit	Shadow price (€/kg eq.)
Abiotic mineral resources depletion potential	Sb eq	0
Acidification potential	SO_2 eq	4
Eutrophication potential	PO_4^{3-} eq	9
Fresh water aquatic eco-toxicity potential	1.4-DCB eq	0.04
Global warming potential	CO_2 eq	0.05
Human toxicity potential	1.4-DCB eq	0.08
Marine aquatic eco-toxicity potential	1.4-DCB eq	0.0001
Ozone depletion potential	CFC11 eq	30
Photochemical ozone creation potential	C_2H_2 eq	2
Terrestrial eco-toxicity potential	1.4-DCB eq	1.3

Table 4. Shadow prices per environmental effect category. Source: Lightart & Ansems (2007).

[7] EDC: Enviromental damage costs

[8] The shadow cost of a resource is the additional profit generated by that resource and it usually has important environmental impacts. It expresses the environmental burden of a product or other system in a monetary unit.

Although 1,000 disposable polystyrene cups recycling showed a shadow cost of -0.29 €, it should be stated the fact that recycling a material do not necessarily involves negative shadow costs. In the same study, 1,000 disposable paper cups recycling were equivalent to 0.09 €. These points are explained by paper cups were composed by paper and coated with polyethylene. Recycling paper cups had environmental impacts when recycling, but recycling polystyrene cups offered a net negative impact in some environmental categories. The negative shadow costs of these categories counteracted the positive ones of the other and, as a results, the net shadow cost appeared negative.

In regards with sustainability, environmental concern has made people to prefer recycling to other EOL alternatives like landfilling or incineration. A clear economic impact due to social behaviour is the choice of a social discount rate. The social discounting rate is related to the ethic of intergenerational equity and is influenced by country, standard of living, purchasing power... It is possible that most people do not willing to pay extra money for recycling, but for obtaining some benefits not necessarily economic ones. These are basically subjective and difficult to measure. The willingness to pay concept is applied to recycling when citizens are asked to pay extra money to use recycled containers. As an example we should not forget that the first stage in the recycling process is to collect recyclable materials. Few studies have been carried out in relation with this issue. One of them (Gillespie & Bennett, 2011) determined the willingness to pay for the recycling collection service. Inhabitants from Brisbane, Australia, were asked about the possibility and frequency of establishing a recycling service collection scheme through questionnaires. Figures showed that people was able to pay 131.49 Australian dollars (A$) per year for a fortnightly recycling service and an extra 18.30 A$ to increase the frequency of this service to weekly. However, on average, respondents declined by 34.18 A$ per year if general waste collection increases from weekly to twice a week. In other words, they did not want to pay more for duplicating the general waste collection. This is a clear indicator on how environmentally concerned people can push to what they consider as more beneficial in environmental, social and economic approaches.

4.2.4 Methodological problems of LCC

The used of technical tools usually implies some points that must be considered. Regarding the main difficulties, three concepts should be explained: Types of LCCs, discounting and uncertainty of results.

4.2.4.1 Types of LCCs

Not all the case studies on LCC correspond to environmental LCC. In some cases, authors have chosen different kinds of economic analysis in order to consider further approaches. For instance, Utme (2009) proposed a non-conventional LCC. Based on literature, she modelled the LCC as a sum of capital costs, costs produced in the operational phase, costs resulted from occupational accidents and fatalities, environmental expenditures and decommissioning costs in fishing fleet. The fact of including costs like the ones derived from accidents and fatalities is an indicator of a societal LCC. It is relatively easy to adapt this approach to the recycling process. Capital costs, costs produced in the operational stage (recycling), costs derived from occupational accidents and fatalities are produced during the process, the last with a small probability. In addition, environmental expenditures could materialise in the near future, for instance with the adoption of a green tax based on the CO2 emissions. Decommissioning costs could be assimilated to transport of recycled material.

A different approach was followed by Dahlbo et al. (2007) in which costs were calculated through a societal life cycle costs (SLCC) and environmental impacts were measured by LCAs. SLCC refers to the costs produced in the entire life cycle of a product or service within the system boundaries defined in the LCA. Similarly as before, this approach considers both the direct costs (labour, energy, etc.) and the shadow costs, but taking into account also social issues. The fact of defining the costs from a social point of view instead of an agent one implies, for instance, that taxes levied on stakeholders and obtained by society are cancelled out when summing the costs. Thus, the difficulties of measuring the social costs are highlighted and rarely find in literature due to subjective perception of impacts and its economisation vary from country to country and from one year to other.

4.2.4.2 Discounting

When carrying out a LCC on a recycling facility another problem appears: to consider changes in costs through time. For overcoming it, the time value money expressed as discounting is used. It considers inflation, cost of capital, investment opportunities and personal consumption preferences. Actually, here we have also uncertainties because of nobody knows exactly what value will have inflation for instance in five years' time.

There are two types of discounting:

- The one used in the global LCC
- The discounted cash flow of the monetary flows which take part in product life cycles

In economic studies which consider medium and long run, the discounting process must be taken into account as discounted cash flow. It means, expressing the value of a future cost item in today´s monetary terms. Logically, products which are consumed in the short time like most of the packaging do not need to consider the discounted cash flow, but recycling facilities works for years and waste flow also vary in time. One can state that higher yield of a recycling facility as a result of increment in waste production or technological advance makes it possible. Because these present values are not unique in the sense that the choice of discount rate affects the outcome, a sensitivity analysis must be conducted to find out how crucially the choice of the discount rate affects the costs (SETAC, 2008).

In addition, discounting in the global LCC varies and should follow different rules depending on the kind of LCC (SETAC, 2008):

- Conventional LCC: It is recommended, although usually not applied
- Environmental LCC: Discounting of the total LCC result is not applied, although use of discounted cash flows for money flows occurring at different times within 1 product life cycle (usually for periods no longer than 5 to 15 years) is commonly applied and does not violate the steady-state assumption. Use of discounted cash flows shall be considered as function of the goal and scope and time duration of a product life cycle

- Societal LCC: Determination of an appropriate discount rate for societal LCC is iterative and requires a sensitivity analysis. Is very complex and depends directly on the impact category and the product/process intended to be analysed

4.2.4.3 Uncertainty of results

Environmental LCC is influenced by uncertainties as a result of the way of ecological and social systems change in the future should be considered in studies, not to mention economic changes like the inflation. Issues not considered as problems today could be in the future. For instance, back in 1950 nobody feared DDT (dichlorodiphenyltrichloroethane), but it was banned for agricultural use in the USA in 1972. Imagine that a LCC on DTT was carried out and it was not considered all the impacts both on human life and nature. How much millions would it represent? How to measure it?

Moreover, a double counting could be produced if considering this approach. Environmental costs could be partially internalised through policy efforts aim at reducing these costs like air pollution controls, and environmental policies focus on restructure economic practices towards long-term sustainability. Furthermore, price volatility[9] was pointed out as a parameter which could affect the recycling markets. Indeed, uncertainty itself is a common driver of price volatility (Stromberg, 2004).

Under the current situation, price volatility has become extremely important (see section 2) and its influence on material recycling prices is expected to decrease as soon as decisive issues will be slowed down, especially as it comes to Asian demand as well as the international crisis.

5. Conclusion

Recycling entails different environmental benefits like energy savings, natural resources conservation and reduction of waste disposal to landfill. Nevertheless, in order to determine which end of life option is also efficient, not only from an environmental point of view but also from an economic point of view, an economic analysis should be carried out.

It is necessary to assure the Ecoefficiency (environmental and economic analysis) of the recycling process selected and it is also important to consider the different market studies available, since effectiveness of recycling can be influenced by market conditions. It has been stated the influence of living in a globalized world, in which the price of recycled materials in Europe is strongly distorted by geopolitical issues in the distant Asia or in the oil producer countries. The high demand of different recovered materials by the Asiatic southwest has made prices to be increased in the last years. Furthermore, differences in prices of machinery, production and labour costs between countries compel the final price of recycled materials. Moreover, uncertainties due to the crisis have made prices to appear like a saw, and especially the effect on the great price fallings in 2008. As it comes to oil price, it influences indirectly the recycling price of plastics since it modifies the virgin material cost and, as a result, makes one of them more economically attractive. It should be

[9] Price volatility is a measure of the variation of a financial instrument (recycled materials prices in this case) over the time.

remembered that the main competitors of recycled materials are the virgin ones, and economic and environmental issues constitute the main drivers to select one of them. Thus, it is necessary to consider the recycled materials market as a whole by taking into account information from both, the regional and international perspectives. Market studies make it easy to decide whether selling a product in one or another country is better or not, and even to consider recycling as the best end of life option.

Recycling starts at the collection stage and together with transport and sorting could affect seriously the final material price. In some cases these three stages can represent the highest economic impact on recycling. A good sorting will allow increasing the recycling yield. Governments have pushed inhabitants to adopt this sub-stage at household in order to facilitate waste collection. Then, a shredding is necessary to increase the efficiency of the next stage: the recycling itself. Nevertheless, when it comes to paper and wood, shredding damages fibres and is not advised. The combination of the different steps as well as the recycling process depends greatly on the material and on its physicochemical characteristics. Specific sub-stages can be observed in the treatment of only some materials. For instance, melting is required in plastics, metal and glass recycling, whereas it is substituted by pulping in paper and board.

The economic and environmental efficiency of a recycling process is highly subjected to a series of aspects not directly related to the recycling process itself but to other aspects such as transport conditions, collection practices or the availability and the final price of the final recycled material. As a result, a non-careful study of those aspects could lead to situations where the benefits of recycling may be overshadowed by the economic and environmental impacts of the previous or posterior recycling stages.

Ecodesign comprises both improving the design for disassembling and selecting the best materials. Examples of how design for disassembling improves the recyclability have been shown and its relation with the disassembling cost in multimaterial products. It is of great interest to facilitate this point since it contributes to reduce the economic impact of this sub-stage. The selection of materials is also another way of improving the ecodesign and can constitute an advantage from the economic perspective, due to environmentally friendly waste might have a better purchasing cost over the most damaging ones. By using ecodesign, apparently trivial issues can be detected as important contributors to recycling cost. For example, it has been stated the fact that plastics usually represent less weight than glass or metals in a determined volume. But considering plastic collection in which trucks are driving and the associated costs, i.e. fuel consumption and driver salary, it is clear that the selection of materials constitutes an important point from the EOL perspective.

The combination of the environmental LCC and the LCA studies (ecoefficiency) make up a dynamic tool capable of defining the most environmentally and economically important of the recycling process. There are three types of LCCs: conventional, environmental and societal, which compose different approaches to the economic analyses and are used as tool in order to select the best alternative in EOLs and recycling processes. Since there are various types of LCCs, it should not confuse the costs and boundaries included in the economic studies. It has been stated that the environmental LCC is the approach which offers a better understanding since is similar to the LCA but in economic terms and considering both, the internal and the external costs. Environmental LCC share similar requirements with LCA like the functional unit and system boundaries. Thus, they allow obtaining a complete analysis of the recycling process from an economic and environmental point of view. However, it should be remarked that in addition to the LCA and LCC analysis, market studies are strongly recommended to assure that market conditions will not hamper or influence negatively the general efficiency and effectiveness of recycling.

The use of LCC encompasses methodological problems. On the one hand, discounting rates should be taken into account when working out an economic study on a recycling facility because of its long life span. This fact is not considered in short-life products like most of packaging. On the other hand, uncertainty due to lack of data constitutes a weak point. Such deficiency could be solved in case information is found in literature or through interviews to experts or stakeholders like in the SLCC. Nevertheless, it cannot be obtained in some cases like when deciding the inflation in the coming years.

6. References

American Foundry Society South California (AFSSOCAL) (2011). Foundry metals, 21.09.2011, Available from http://www.afssocal.org/AFS_about.html

All-recycling-facts (2009). The glass recycle process, 21.09.2011, Available from http://www.all-recycling-facts.com/glass-recycle.html

Aschilias, D. S. & Karayannidis G.P. (2004). The chemical recycling of PET in the framework of sustainable development. *Water, air and soil pollution: Focus*, Vol. 4, Numbers 4-5, pp. 385-396

Boureau of International Recycling (BIR) (2009). *Recycled materials supply 40% of the global raw material needs*, 22.09.2011, Available from http://www.bir.org/industry

Boureau of International Recycling (BIR) (2010). *World steel recycling in figures 2006 – 2010. Steel Scrap- a Raw Material for Steel Making*. Boureau of International Recycling, Ferrous Division, 15.09.2011, Available from http://www.bir.org/assets/Documents/publications/brochures/aFerrousReport Final2006-2010.pdf

Bovea M.D. & Vidal, R. (2004). Increasing product value by integrating environmental impact, costs and customer evaluation. *Resources, Conservation and Recycling*, 41, pp. 133-145

Chang, S.-Y. & Farr, E. A. (2000). Life cycle cost and multiobjective evaluation of low-level mixed waste treatment alternatives, Waste Management Conference, WM'00, 2000, Tucson, Arizona, USA, February 27-March 22

Confederation of European Paper Industries (CEPI). (2009). *Sustainability Report 2009*. CEPI, 15.09.2011, Available from http://www.cepi.org/Objects/1/Files/CEPI-Report09.pdf

Craighill, A. L. & Powell, J. C. (1995) *Life cycle assessment and economic evaluation of recycling: a case study*. CSERGE Working Paper WM 95-05. Centre for Social and Economic Research on the Global Environment University of East Anglia and University College London. ISSN 0967-8875

Dahlbo, H., Ollikainen, M., Peltola, S., Myllyma, T. & Melanen, M. (2007). Combining ecological and economic assessment of options for newspaper waste management. *Resources, Conservation and Recycling*, 51, pp. 42-63

European Plastic Recyclers (EuPR). (February 2010). *How to increase the mechanical recycling of post-consumer plastics. Strategy paper of the European Plastics Recyclers association.* European Plastic recyclers, 15.09.2011, Available from http://www.plasticsrecyclers.eu/uploads/media/eupr/HowIncreaseRecycling/1265184667EUPR_How_To_Increase_Plastics_Recycling_FINAL_low.pdf

Eurostat. (2010). *Waste statistics.* Eurostat, 15.09.2011, Available from: http://epp.eurostat.ec.europa.eu/statistics_explained/index.php/Waste_statistics

Gillespie, R. & Bennett, J. (2011). *Willingness to pay for kerbside recycling the Brisbane region.* ISSN 1835-9728. Environmental Economics Research Hub. Research Report No 97. March 2011. Crawford School of Economics and Government. The Australian National University

Glass Packaging Institute (GPI) (2010). 21.09.2011, Available from http://www.gpi.org

Greenpac, Intelligent textiles, Green Spun (2011). 21.09.2011, Avalilable from http://www.greenpac.co.uk/

Institut National de l´EnviRonnement Industriel et des riSques (INERIS) (2010). 5.1 Applied processes and techniques. 21.09.2011, Available from http://www.ineris.fr/ippc/sites/default/interactive/brefpap/bref_pap/english/bref_gb_traitement_processus.htm

Intergovernmental Panel on Climate Change (IPCC) (2006). *IPCC Guidelines for national greenhouse inventories. Volume 5: Waste. Chapter 2: Waste composition, generation and management data,* Iges, ISBN 4-88788-032-4, Japan, 21.09.2011, Available from http://www.ipcc-nggip.iges.or.jp/public/2006gl/pdf/5_Volume5/V5_2_Ch2_Waste_Data.pdf

Kim, J., Hwang, Y. & Park, K. (2009). An assessment of the recycling potential based on environmental and economic factors; case study in South Korea. *Journal of Cleaner Production*, 17, pp. 1264-1271

Kloepffer, W. (2008). Life Cycle Sustainability Assessment of Products. *International Journal of Life Cycle Assessment,* 13, (2), pp. 89-95

Letsrecycle (2011). 21.09.2011, Available from http://www.letsrecycle.com/

Lightart, T. N. & Ansems, A. M. M. (2007). *Single use cups or reusable (coffee) drinking systems: An environmental comparison.* TNO report 2006-A-R0246(E)/B, 15.09.2011, Available from http://www.prodisposables.nl/nl/file/20110104172024/3/An-environmental-comparison-of-single-use-cups-and-reusable-cups-.html

Lyrstedt, F. (2005). *Measuring eco-efficiency by a LCC/LCA ratio. An evaluation of its applicability. A case study at ABB.* Master of Science Thesis in the Master Degree Programme. International Project Management. Chalmers University of Technology. ESA Report No. 2005:11

McGregor, A. An exploration into the efficiency and effectiveness of a recycled content mandate in creating a closed-loop for plastic bottles. MSc Thesis. Cranfield University. School of Applied Sciences September 2009.

Magnaghi, G. (2009). *World recovered paper market in 2009.* Boureau of International recycling, 15.09.2011, Available from http://www.bir.org/assets/Documents/industry/MagnaghiReport2009.pdf

Maine. State Planning Office. Waste Management and Recycling Program (December 1999). Solid Waste Program Costs: A Study of Selected Maine Communities, 15.09.2011, Available from www.maine.gov/spo/recycle/docs/mswcostreport.pdf

Metafore summary report (2006). The fiber cycle technical document, 21.09.2011, Available from http://www.postcom.org/eco/sls.docs/Metafore-Paper%20Fiber%20Life%20Cycle.pdf

Nakamura, S. & Kondo, Y. (2006). A waste input-output life-cycle cost analysis of the recycling of end-of-life electrical home appliances. *Ecological Economies*, 57, pp. 494-506

Norris G.A. (2000). Integrating Economic Analysis into LCA. *Environmental Quality Management*, 10, 3, pp. 59-64

Park, P., Takara, K., Jeong, I. & Lee, K. (2006). Comparison of four methods for integrating environmental and economic aspects in the end-of-life stage of a washing machine. *Resources, Conservation and Recycling*, 48, pp. 71-85

PlasticsEurope (2010). *Plastics- The facts 2010. An analysis on European plastic production, demand and recovery for 2009*. PlasticsEurope, 15.09.2011, Available from http://www.plasticseurope.org/documents/document/20101028135906-final_plasticsthefacts_26102010_lr.pdf

Porter, R. C. (2002). The Economics of Waste, Resources for the Future, Washington, USA. ISBN 1-891853-42-2 (cloth), 1-891853-43-0 (paper)

Rebitzer G. & Hunkeler, D. (2003). et al. Life Cycle Costing in LCM: Ambitions, Opportunities, and Limitations - Discussing a Framework. *International Journal of Life Cycle Assessment*, Vol. 8, No. 5, pp. 253-256

RISI (2010). World recovered paper monitor. Analysis and forecasts of the global recovered paper markets. RISI, January, 2010

RISI (2011). Pulp and Paper International: PPI Global. Price Watch. RISI, 15.09.2011, Available from: http://www.risiinfo.com

SETAC-Europe Working Group on Life Cycle Costing (2008). *Environmental Life Cycle Costing*. SETAC, ISBN 1-880611-38-X, New York, USA

Stromberg, P. (2004). Market imperfections in recycled markets: conceptual issues and empirical study of price volatility in plastics. *Resources, Conservation and Recycling*, 41, pp. 339-364

Ungureanu, C. A., Das, S. & Jawahir, I.S. (2007). Life-cycle cost analysis: Aluminum versus steel in passenger cars. Aluminium Alloys for Transporting, Packaging, Aerospace, and Other Applications. Edited by Subodh K. Das, Weimin Yin. TMS (The Minerals, Metals & Materials Society). pp. 11-24

United States Environmental Protection Agency (EPA). (2010). *Waste – Non-Hazardous Waste – Municipal Solid Waste*, 15.09.2011, Available from: http://www.epa.gov/osw/nonhaz/municipal/

Utme, I. B. (2009) Life cycle cost (LCC) as a tool for improving sustainability in the Norwegian fishing fleet. *Journal of Cleaner Production*, 17, pp. 335-344

World Business Council for Sustainable Development (WBCSD). (October 2000). *Eco-Efficiency: creating more value with less impact*. WBSCD, 15.09.2011, Available from: http://www.wbcsd.org/plugins/DocSearch/details.asp?type=DocDet&ObjectId=Mjc5/

Wright, E., Azapagic, A., Stavens, G., Mellor, W. & Clift R. (2005). Improving recyclability by design: a case study of fibre optic cable. *Resources, Conservation and Recycling*, 44, pp. 37-50

7

Electronics Waste: Recycling of Mobile Phones

Pia Tanskanen
Nokia Corporation
Finland

1. Introduction

As the consumption of electronics products has increased the management of new type of waste, electronics waste (e-waste, WEEE), has become a global concern. Countries in the European Union are creating 17kg e-waste per capita annually and developing markets such as China and India are currently creating 1kg e-waste per capita a year. The amounts are expected to be rising in the future, and the joint disposal of e-waste together with municipal waste cannot continue (Chancerel &Rotter, 2009). E-waste contains many recyclable materials such as ferrous metals and aluminum, copper and precious metals as well as different engineering plastics. These are typically highly integrated into each other. This means that recycling of electronics products is technologically more complicated than for example glass or paper recycling. Most importantly disposal of e-waste causes loss of these valuable, non-renewable resources as electronics products contain wide range of valuable materials, many of them becoming scarce in the nature. Depletion of raw material sources together with increasing need for materials in manufacturing of new products together mean that collection and recycling of obsolete products becomes more and more important. Electronics waste recycling processes may also pose a risk to environment if electronic products are not treated in a proper manner at their end of life stage. Substances of concern may leak to the environment or cause health and safety risk at the treatment phase. Examples of improper treatment of e-waste are widely presented in the literature. The trend has been in the electronics industry to remove the potentially hazardous materials from the products so that there is smaller risk of contamination even if the improper recycling practices take place.

Product end of life process or value chain can be divided into different sub-processes that all aim at recovery of the material and energy content of obsolete products. Optimization of the whole value chain is important in order to get the best value for economy and for environment. This means that a system perspective needs to be taken into account when working to improve the parts of the recycling process as all phases have an impact on the others. For example waste collection logistics should not discredit the environmental or business benefits of recycling. To increase the e-waste recycling it must be noted that not all the recycling challenges are technical. The biggest obstacle in recycling is the lack of consumer awareness on collection and recycling possibilities, leading to low collection amounts. Without returning products for recycling the next phases, technical recycling processes, cannot take place. Leakage outside of the value chain of the products during the end of life process may lead to improper recycling practices. The cooperation and interaction with other players in the value chain becomes essential in closing the material loop. Naturally technical processes for separating and refining the recycled materials need to be in place as well as good data systems supporting the decision making processes (International Council on Mining & Metals [ICMM], 2006).

As the biggest challenge in e-waste recycling is the collection of the waste materials from the consumers, this chapter focuses on to the first step of the end of life process, on how to develop an efficient collection process to collect obsolete products from consumers. Awareness rising is currently the key to successful electronics waste management as the practices are not yet seen as everyday business. Consumer study is presented to show how people in different countries feel about recycling and case studies are presented to demonstrate Nokia's contribution in building a recycling culture. Many similarities can be seen in the results as consumers are appreciative of the information on how and where to recycle their old electronics. The success of consumer collection programs lie not only on the convenience and awareness but also on patience; systems need to be in place for years before recycling becomes a habit.

2. End of life phases

By definition recycling means processing waste (e.g. unwanted or useless materials) into new products to prevent waste of potentially useful materials, reduce the consumption of virgin raw materials, reduce energy usage, reduce air pollution (from incineration) and water pollution (from landfilling) by reducing the need for waste disposal, and lower greenhouse gas emissions as compared to virgin production. Recycling is a key component of today's waste reduction and is a component of the "Reduce, Reuse, Recycle" waste hierarchy. Recycling of products made of one material type, such as newspapers, glass or plastic bottles or metallic structures is already a common practice in many countries. More complicated products, such as cars, batteries and electronic products have entered into the recycling realm as well.

Product end of life treatment is not a single step, but rather a process consisting of process units that aiming at recovering reusable parts and recyclable materials. Every process step consists of one or more unit operations. Process units, or phases, can be classified according to their function in the recycling process (Chancerel & Rotter, 2009). More specifically end of life can be divided into three different phases which require different management methods and focus, and having different impact on recycling economics, as shown in figure 1.

The first phase is the collection and consolidation of waste, so called take back or collection in the case of the consumer recycling initiatives. This is very much a logistics challenge and requires a high awareness level of the consumers who need to return obsolete products for recycling. In the business environment the first phase can be controlled in a much more efficient way than in the post-consumer collection. The second phase is the pre-treatment phase, taken care of by recycling companies who separate the different materials in a product and then sell them further on to the third phase, recycling and recovery of materials and energy or even disposal. Every phase has a minor side flow of disposal of the fractions that cannot be further processed, such as wet cardboard packaging. Product design can make the second and third phase easier or more difficult therefore having an impact on recycling cost and efficiency. The second phase is a bit more complicated in the post-consumer waste collection compared to the business environment, as the collected material typically contain impurities and materials that are not meant to be collected. The second phase can include the usage of different techniques from manual disassembly to mechanical and chemical pre-processing. In the third phase there is not much difference where the waste is originated (Tanskanen, Takala, 2006).

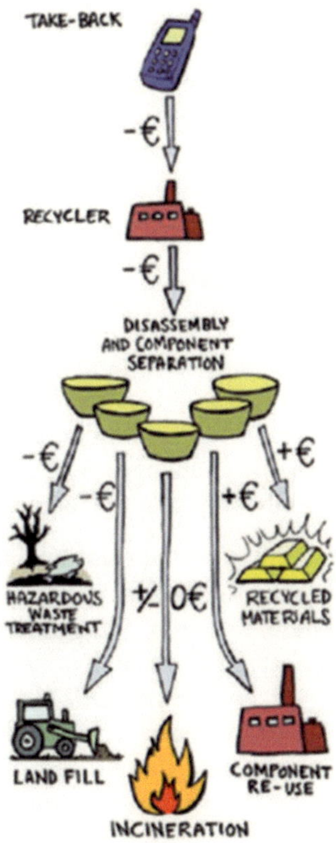

Fig. 1. Process steps of product end of life treatment with economic indicators (Tanskanen &Takala, 2006)

As the biggest challenge in e-waste recycling is the collection of the waste materials from the consumers, this chapter focuses on that phase, on how to develop an efficient take back process. Every step of the recycling process needs optimization, the challenge is not to sub-optimize the value chain but the whole system to support eco-efficient recycling process. Waste collection and management always happens locally, and case studies from different countries are presented here also to demonstrate how a global corporate strategy can be turned into local actions.

3. Challenges in e-waste recycling

Product end of life process needs to be looked as a system and the optimization should be done so that the whole value chain is taken into account rather than focusing on a part of the process. Waste management of post-consumer waste faces different kinds of challenges compared to the waste management of the office or factory waste. Waste management at the corporation's own operations is much easier to control and therefore high recovery rates for waste can be achieved. Post-consumer waste management faces different kind of challenges for post-consumer waste management are in the first phase of the end of life process, the collection.

At company premises processes can be well managed, waste amounts can be predicted and people are easier to train. For example at Nokia production sites Environmental Management Systems (EMS) and the ISO 14001 standards are used to control and manage the environmental aspects. The goal of the Nokia EMS is to improve the environmental performance, where one focus area is waste management. Most of the waste from Nokia factories is packaging waste: cardboard boxes, wood pallets and different types of plastic packaging. Typically the amount of cardboard is about 26 % and the amount of plastic 37 % of the total waste stream. E-waste amounts to only 1 % of the total waste amounts. Waste utilization rates are very high at Nokia plants: average of 92 %, so that 6 out of the 10 factories have utilization rate of over 96 %. Waste utilization rates vary because of the local infrastructure and recycling partner availability and because of the small differences in waste separation at the manufacturing sites. Most importantly all the waste that is created can be sorted and directed to dedicated waste companies for recycling. This is an activity that takes place every day. Most of the challenges that consumer e-waste recycling faces, from awareness to collection are mainly solved in the factory environment.

Consumer plays a big part in the first phase of the value chain, so the recycling behavior of the individual is crucial for the whole process. For the consumer convenience and awareness are the key points that will encourage them to start recycling. All the e-waste that can be collected needs to be directed to proper recycling facilities so that recycling happens in an eco-efficient manner to avoid contamination and ensure the efficient recovery of resources. E-waste recycling today has three main challenges that need to be improved in the way to the full recycling society. These are consumer awareness and collection, best practices in processing such as cost efficiency and value generation, and getting the material to proper recycling.

3.1 Awareness

Consumer study by Nokia shows that less than 10 % of people have recycled their old mobile phones. Most of the unused phones are still at home, making the recycling potential huge. The lack of awareness that recycling is even possible and knowledge on existing recycling programs and locations are the main obstacles for consumers. This means that the first challenge in motivating people to recycle electronics is to get them to understand that it is possible and to show how can be done.

3.1.1 Consumer behavior

There have been many studies on consumer attitudes to recycling. A schema picture is show in the figure 2 to summarize the different features that have an impact to persons recycling behavior. The first layer around the consumer describes the direct impact on the recycling decision, such as accessibility and attitude. The next layer shows the ways how to motivate the recycling behavior, like past experiences or transparency of the system, and the following layer presents the different ways to influence recycling behavior from marketing and social media activities. The outer layer shows the future values and social trends that may have an impact to consumers recycling behavior, communality and downshifting being examples of those trends. As seen from the figure, there are many factors impacting the recycling behavior, from personal attitudes and experiences to the quality of the recycling systems that are in place. Recycling is a global theme that needs local execution. Different ways to communicate, incentivize and motivate consumers are effective in different countries. To better understand what people think about phone recycling, Nokia has conducted two global consumer studies on the topic. The results for the studies are being used in planning and executing phone recycling programs.

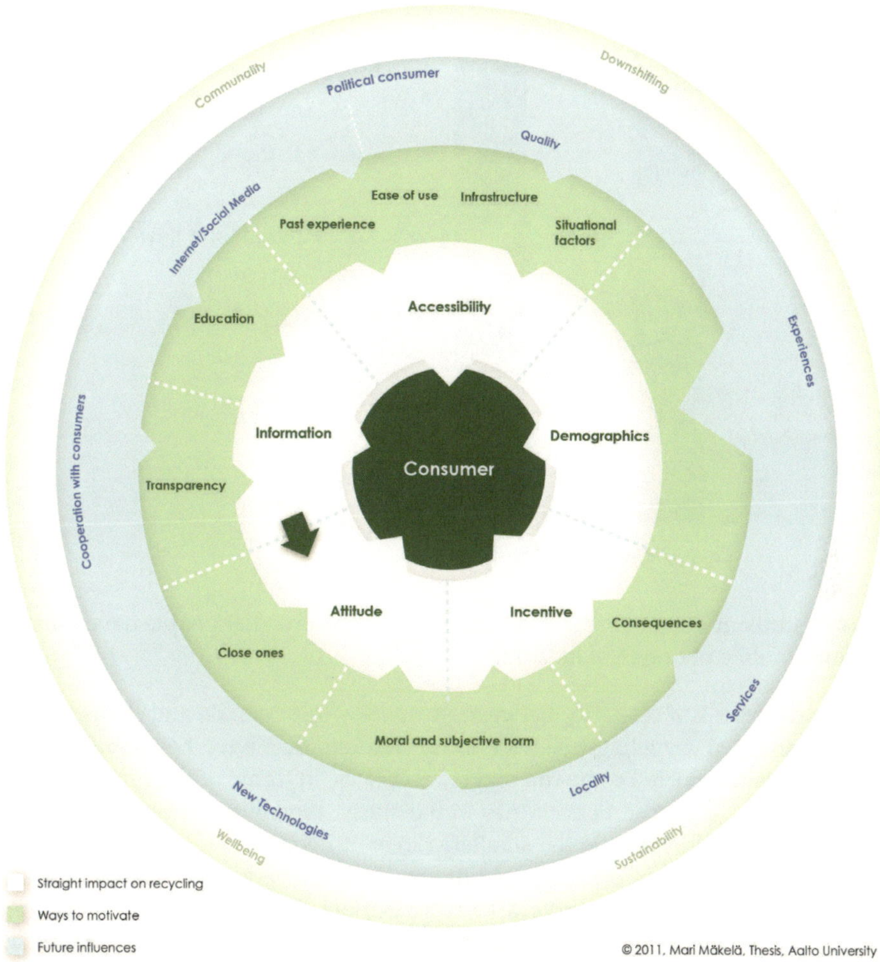

Fig. 2. Features affecting consumer recycling behavior (Mäkelä, 2011)

Based on the first consumer survey on consumer recycling behavior and attitudes at 2007 it was found out that despite the fact that households on average have each owned around five phones, very few of these have been recycled once they are no longer used. Nearly half of the consumers were unaware that it is even possible to recycle a mobile phone. Two thirds said they did not know how to recycle an unwanted device and 71% were unaware of where to do this. Only 3% said they had recycled their old phone. The survey was based on interviews with 6,500 people in 13 countries including Finland, Germany, Italy, Russia, Sweden, UK, United Arab Emirates, USA, Nigeria, India, China, Indonesia and Brazil and in the second study additionally also Argentina, Spain and Nigeria were included, but not Sweden, Brazil, Italy and Russia. Figure 3. shows what people have done with their previous phone according to the consumer study done in 2011. A majority of the old phones is kept at home or given to somebody else for further use. Fortunately the survey in 2008 showed that only 4% of the old phones had ended up into landfill, which is a concern from the environmental efficiency point of view. (Nokia, 2008, 2011).

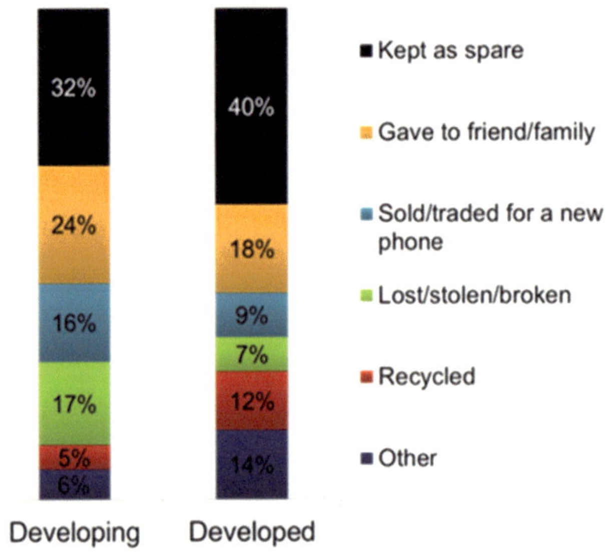

Fig. 3. Global study at developing and developed markets on what people have done to their previous mobile phone (Nokia, 2011b)

It was seen that there is a disparity between awareness of materials and items that may be recycled and reported recycling behavior in all markets that have been studied. Finland, Germany and Spain reported to be the biggest recyclers, in terms of the range of items that people usually recycle. Of the 11 countries in the study, the United Arab Emirates, Nigeria and Indonesia reported to recycle the smallest range of items. Overall, developed countries reported to be more aware of the range of different materials and items that can be recycled than developing nations, and also tend to recycle more. More barriers to recycling were reported to exist in developing countries, where awareness of both recyclable materials and items and recycling channels is lower. Developing nations have fewer recycling channels available to them, and that means that it is not as convenient to recycle as it is in many developed countries. Figure 4. shows the mean number of different items out of 11 possibility people claimed to have recycled when asked. Items that were listed were: paper/cardboard, plastic bottles, cans, glass, metals, clothes/shoes, mobile phones, batteries, televisions, refrigerators and computers. From this list the most commonly recycled materials were paper/cardboard, plastic bottles, glass and cans. More complicated products, like electronics were reported to be recycled less than other materials both in developed and developing markets.

Fig. 4. Mean number of items out of 11 that are being recycled (Nokia, 2011b)

According to the Nokia consumer survey from 2011, overall, 9% of respondents in the survey claimed to have recycled their last mobile phone, an increase of six percentage points from 2007. The developed market is at the forefront of this, although strong growth has been seen in India and China since 2007. Figure 5. shows the results for claimed mobile phone recycling in different countries.

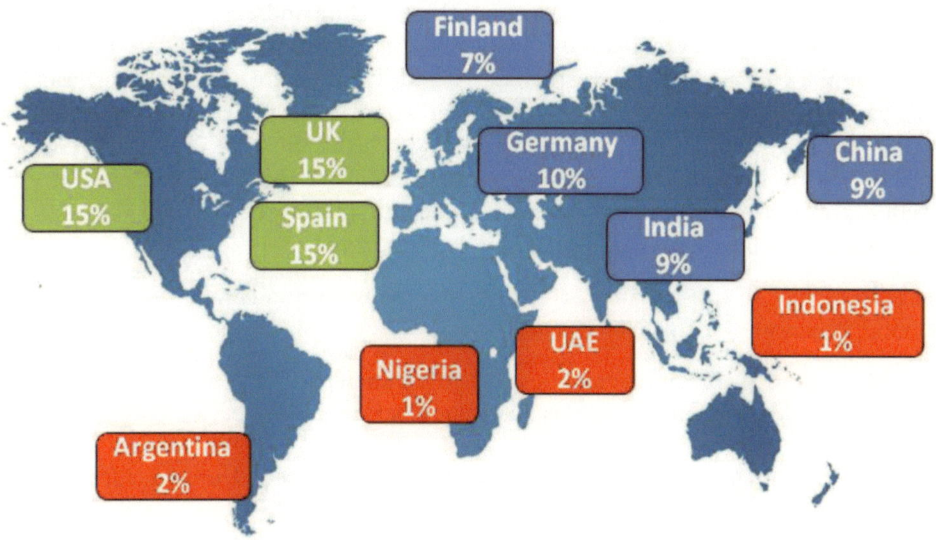

Fig. 5. Reported recycling of last mobile phone (Nokia, 2011b)

For the purpose of the study, recycling a mobile phone was defined as breaking down the phone into its composite parts to reuse in making of new products. According to the study most have an accurate understanding of what recycling a mobile phone traditionally means, although recycling can mean many other things as well, such as trading in old phones for a discount on a new phone, or giving it to someone else. This is an example of the challenges in communicating recycling messages, as the term "recycling" alone can be understood in so many different ways.

3.2 Collection

Electronics waste needs to be collected separately in order to the recycle and reuse the material content. E-waste can be collected on a voluntary basis or to fulfill legislative regulations. The motivations for collecting electronics for recycling range from economic to environmental protection and to brand enhancement motivations. The main objective for the majority of the legislations for e-waste is to prevent it to be disposed together with household waste.

Collection methods can be classified by the used models and by the persons or organizations responsible for organizing and/or financing the operations. The following models are available for collection (Chancerel, 2010, Hai-Young, 2005):

1. Drop-off program, with permanent collection centers or retailers, containers on the streets, or temporary collection events;
2. Pick-up program, where the e-waste is collected at the homes or offices
3. Distance collection, where the user sends the e-waste by post to the collector.

Collection methods each have they pro's and con's. Pick-up and mail-back programs generate higher logistics cost than drop-off programs. Collection places need space and they need to be supervised, which can be a challenge e.g. with retail and home. Event type of drop-offs can help in raising the awareness of the recycling and are simpler to operate than the permanent systems. Convenience is highest in the pick-up collection, but drop-off points are helpful in the case when the waste material is difficult to store at house because it takes lot of space or is dirty. For valuable waste materials there are typically many programs for recycling them, e.g. for charity, so the competing scheme must be even more convenient than the existing ones. A study shows the distribution of the different types of e-waste collection programs as follows: permanent collection (47%), special drop-off events (45%), and curbside collection (8%) (Hai-Youg, 2006, Jenkins, 2003).

The responsibility to finance and/or to organize the collection can be done by one of the following: (Chancerel, 2010)

1. Public authorities like municipalities and governments;
2. Private commercial organizations like manufacturers, retailers or recyclers, informal sector
3. Private non-commercial organizations like non-governmental organizations or citizen initiatives.

Generally the optimal level for collecting and recycling waste materials is difficult to determine as the recyclable material, local conditions, technical and operational issues have an impact on it. When households recycling behavior has been studied the results show that the most important factor to enhance recycling is convenience. Improvements on collection service and design are the most important factors on enhancing the recycling behavior; together with information and promotion. Other factors, such as financial incentives and socio-economical features show more mixed results. Financial incentives may play a role when non-recyclers are being activated to change their behavior, and they are the most effective when they are directed in improving the community or influencing tax rebates. (Shaw, 2008). It has been proven that the familiarity with recycling other household waste fractions, such as glass, paper, metal and plastics boosts the willingness to recycle e-waste as well. (Saphores, 2006) Improvements should focus on improving the program and information about it.

The largest environmental impact of small electronic devices recycling can easily come from the logistics related to consumer activity in recycling. This is the part where consumer takes recyclable material to a collection point, in the worst case driving to the collection location without any other reason. An example from household glass collection shows that the collection amount would triple if the collection box is placed close to the household, but at the same time environmental impact is increased because of the increased logistics. (Teerioja, 2010). These suggest that optimizing the convenience with logistics, costs and environmental benefits is important in recycling programs to keep them attractive and still eco-efficient.

3.3 Best practices in processing

Even though metal recycling as such is an ancient technology, there is room for the development. Electronics products are first of all complex, containing materials from all different material types. Materials are highly integrated into each other, they are present in low quantities are therefore not always easily separated in recycling processes. On the other hand each product group, from refrigerators to hair dryers and mobile phones are different in structure and composition. Not even all mobile phones are the same, so the variety of the product types and structures in the recycling process is huge. When a closer look is taken to the product composition, it can be seen that they vary not only on the structure, but also on recycling economics. Some of the products contain relatively large amounts of valuable materials, and recycling of these products generates value and profit. These are typically products with precious metal containing printed wiring board or large metal pars. Some products contain hazardous parts that need to be disassembled and treated in specialized waste treatment facilities, like cooling agent removal from old refrigerators. Products may also be constructed of materials with no reselling value, which means that in these latter cases recycling becomes a cost. Product design can either help or hinder the efficient pre-treatment of waste electronics products and has therefore an impact to recycling cost. There has been lot of development in removing potentially hazardous substances from the electronics products. For example materials such as lead, brominated flame retardants and PVC are no longer used in Nokia mobile phones. Typical material content of a mobile phone is presented in the figure 6.

Because electronics products are so diverse in composition and structure, it is quite challenging to develop advanced recycling technologies that would be suitable to all product types. Another important factor hindering the development is the low collection amounts of e-waste, influencing revenues of the recycling operators. These things together mean that there is not yet many facilities that could focus on specific type of e-waste recycling, but most of the processors need to take different products into their facilities. As volumes for each of the different products are low, automatization levels remain low and waste treatment remain on a relatively general level.

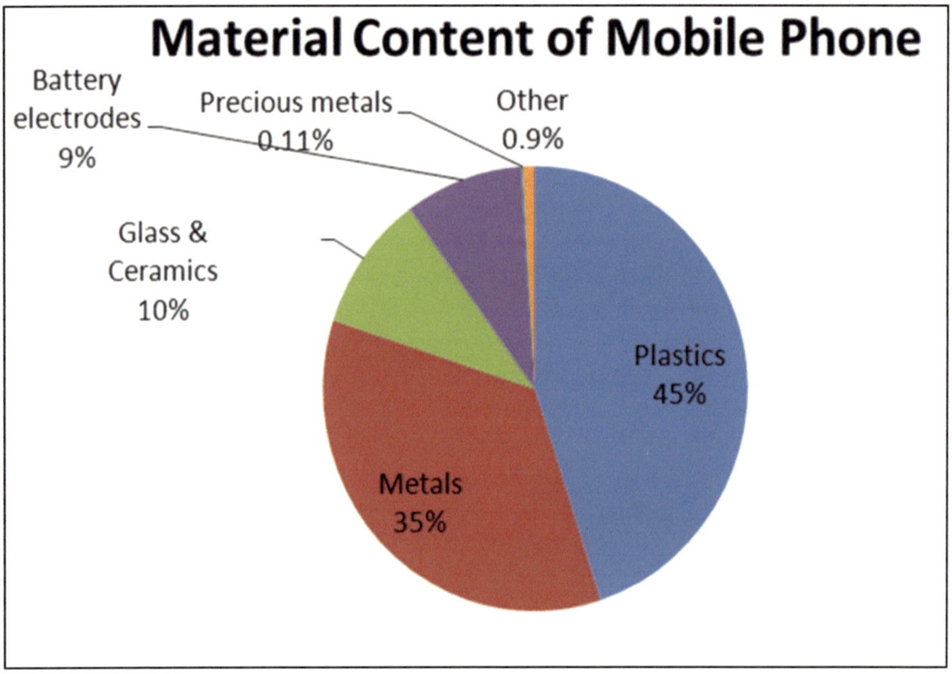

Fig. 6. Material content of a mobile phone (Nokia, 2011a)

It has been shown that products containing relatively high amounts of precious metals, such as mobile phones, separate collection and treatment, even including higher costs for logistics, will bring higher environmental and economic value. For the products containing about 250 ppm gold and 150 ppm palladium separate collection is always more eco-efficient that mixing these products with the ones containing smaller amounts of precious metals. An assumption for the calculation is that collection amounts are big enough, few tonnes rather than few hundred kilograms. (Huisman, 2003). With the use of proper recycling technologies, 100 % of the materials in mobile phone can be recycled and recovered as material or energy, and nothing is wasted. (Nokia, 2011a).

3.4 Getting material to proper recycling

One of the risks in e-waste handling is that the materials end up to be treated in improper recycling facilities, often in developing countries. These practices, like recovering copper or precious metals by uncontrolled burning and leaching processes, have been reported to not only contaminate the surrounding environment but also causing health risk to the people operating the process.

E-waste can leave the formal management system to enter the informal sector and be even exported to developing countries. This show how important it is to look at the whole recycling process and how it needs to be controlled in order to direct e-waste to proper recycling facilities. E-waste can leave the system at any point of the end of life process, creating value at that point of the process but causing harm and lost resources at some other point.

4. Mobile phone take back cases – Creating a recycling culture

The first take back pilot program for mobile phones took place in Europe in 1997. It was run in Sweden and the UK and it was executed by member companies of ECTEL (European Telecommunications and Professional Electronics Industry association) group consisting of six mobile phone manufactures (Tanskanen & Butler, 2007). Since the first program there have been many Nokia led collection initiatives globally aiming at raising the consumer awareness on mobile phone recycling. For Nokia the target for running recycling campaigns for mobile phones is to raise the consumer awareness on the recycling options. Campaigns are always supporting the existing infrastructure that is in place, so that there is continuity for recycling behavior. Nokia has for example in 2010 run co-operation programs with telecom operators and retail companies in seven countries, university events in 11 countries, direct consumer campaigns and events in 19 countries and special campaigns in Nokia repair network in eight countries. National recycling programs with Nokia's participation was done in 29 countries and four countries had a mail-back system in place in 2010. Altogether Nokia was offering permanent take back for obsolete phones in almost 100 countries in more than 5000 locations. (Nokia, 2010)

The build-up of the working recycling process for consumers can be divided into four subsequent steps. Introducing the recycling agenda and building up the infrastructure are the first steps leading to wider collection network together with partnerships. When the program has been tested and expanded through cooperation with others the next phase is the improvement and development of the program through data collection and analysis and through better understanding of the local consumer behavior. Case examples are shown here on how to raise the awareness on recycling in traditional media and in social media.

Four steps in building up a recycling program:

- Introducing recycling agenda at new markets
- Building up an infrastructure
- Expanding the program and building partnerships
- Data collection and understanding consumer recycling behavior

4.1 Introducing recycling agenda

When Nokia started to plan the first mobile phone recycling campaign in Uganda in 2010, many people were sceptical about the success of the campaign. Comments stated that "Recycling may work in Western countries, but not in Africa" "People expect money in return for their old phones, and if you do not give money you will not get any phones back." Even with the skeptical comments there was a belief that with more than 11 million mobile subscribers in Uganda, there is great potential to influence and make a difference to the environment and find a solution to increasing the problem of electronic waste.

The first recycling campaign was arranged in the Uchumi supermarket in Kampala, Uganda for two days. During those two days 459 old phones, 254 chargers and 239 batteries, in total of almost thousand items, were collected and good feedback was received: "Recycling makes a lot of sense". Consumers in Uganda see every day the problems created by the illegal dumping of electronic waste in their country, and they are ready to take action. One of the success factors of the campaign were the radio ads which Nokia ran on several radio stations to encourage people to recycle. Radio is a powerful medium in Uganda, and many people heard the message and took action. Another success factor was the fact that Nokia gave small gifts (for example, a bag made out of recycled material) to people who brought handsets back for recycling. And everyone also got the chance to enter a raffle, with the chance to win a Nokia N8 mobile phone. The short campaign in Kampala was a small, but successful step, as it is noted that creating awareness and changing people's behavior towards the environment takes time. In addition, permanent collection points have been set up at the Nokia repair centers in Kampala, with recycling bins for people to drop of their unwanted phones and accessories.

In 2010 Nokia introduced new recycling programs in Kingdom of Saudi Arabia, Lebanon, Uganda, Sri Lanka, Uruguay, Pakistan, Panama, Belarus, Bosnia, Malta and Ukraine in order to take the first steps in of creating the recycling culture. (Nokia, 2010)

4.2 Building up an infrastructure

Recycling starts with setting up a collection infrastructure. Although take back vehicles are in place for all consumers the lack of awareness of these possibilities can offer some real challenges in making the take back program successful one. Making the consumer aware of the opportunities to recycle and changing the mindset and habits of the disposal patterns in society are the true keys to any successful program. Communication together with easy take back options are seen as means to increase awareness which further leads to increased recycling of electronics waste.

Commonly used methods to collect mobile phones, batteries and accessories are special collection bins for phones, mail back envelopes and general e-waste collection locations.

A visible collection method is the take back kiosk or bin, an example of the collection bin is shown in figure 7. In this approach, challenges are many; location of the bins must be easily accessible and secure, bins need to be emptied often and finding an effective logistics solution that is economically viable is difficult in rural environments. People also tend to put all kinds of rubbish from empty cans and paper cups to apples to the recycling bins. However, bin collection can be an effective means on company premises -Nokia has implemented collection bins in all its locations globally. Bins are also effective in communicating the recycling message.

For a consumer, one of the easiest ways to recycle is to use postal services. This is especially suitable for small devices such as mobile phones. With a pre-paid postage envelope one can drop off the phone to the nearest mailbox and it will be sent to recycling. An example of a recycling envelope is shown in the figure 8. There are several ways to distribute the envelopes. Prepaid postage and addresses can be downloaded and printed from the internet, envelopes can be mailed directly to the consumers, distributed in retail shops, or can be included in the sales package of a new product. Envelopes provide an easy return process, but are costly in terms of reverse logistics and are neither efficient nor safe when larger volumes or unknown waste is in question. Results from envelope programs have shown an average of 1.2 handsets per package. This cannot be classified as low environmental impact in terms of logistics compared to programs where hundreds or even thousands of handsets can be collected in a single drop off location. However, mail back programs do offer the consumer the easiest method possible.

Fig. 7. Recycling Bins for mobile phones

Communal waste collection points can be used as take back locations for a wider spectrum of waste having special containers for mixed e-waste. In these locations there can be collection for not only different types of e-waste but also other materials such as wood, metal and glass. This is a cost efficient way of collecting waste materials, and also easy for people dropping of a different kinds of waste materials at one location.

Fig. 8. Recycling envelope for mobile phone

4.2.1 Case India

Nokia started its take-back and recycling program in India in 2008 with the key objective of raising consumer awareness on recycling of mobile devices and making it easy for consumers to recycle. Before the project could be started there was a need to build up a collection infrastructure, after that the following steps were awareness rising by using a variety of media and incentivizing consumers for recycling. The main challenge in starting the recycling program in India was a low level of environmental awareness amongst the consumers and among trade partners and retailers. This meant in practice that there was no existing recycling infrastructure (collection places and recycling companies) in place and information on e-waste recycling was also not available. In addition there was no regulatory framework or policies on e-waste management and roles of different stakeholders were not defined.

Recycling program was initiated by laying out recycling infrastructure across India. Building up an infrastructure for phone collection included the following:

- Recycling bins and information material for consumers were put in place. More than 1400 specially designed secure bins where placed at Nokia Care Centers and Branded Retail Stores.
- Reverse Logistics and IT system were developed for collection of phones from drop-off points, consolidation and shipment to recycling company. IT system was also used for capturing the number of phones and accessories dropped for recycling along with some details of the person who had dropped it.
- Local recycling companies needed to be assessed to ensure that they fulfilled the requirements from business, environmental and security point of view. Cooperation model and contract was set up with the selected company.
- Training of over 8000 employees of the Nokia Care Centers and Branded Retail Stores was done so that they could engage consumers on mobile phone recycling.
- A SMS based solution was developed to make it easier for consumers to identify the nearest location of recycling points.

After setting up the infrastructure, the recycling program was launched as a pilot in four cities - Bangalore, Delhi, Gurgaon and Ludhiana for a period of 40 days in 2009. The aim of the program was to make recycling easy and to raise consumer awareness. A key challenge during the launch was overcoming the skepticism of the trade partners and retailers. The trade partners believed that Indian consumers were not yet ready for recycling. In the second phase the program was expanded to cover an additional 28 cities. Different marketing methods were used to get more publicity. Roadshows were run at shopping malls and in the main markets. Billboards, radio and print media were also used in advertising together with Bollywood celebrities.

An incentive was added to the recycling program to get people more involved. Nokia committed to plant a tree for every phone dropped for recycling during the specified campaign period. By using plantation of trees as a declared incentive, Nokia aimed to build the right consumer attitude towards recycling wherein recycling is done for benefits to the environment and society at large rather than for personal gains.

The recycling programs have received positive feedback and results. More than 50 tons of old phones and accessories have been collected in India during the first 2.5 years of the recycling program. The collection amount has been increasing every year showing the importance of the long term planning of the recycling programs. The recycling message has reached more than 100 million people across India via various marketing media. Until the summer of 2011 Nokia has in return planted over 90.000 trees and planting continues. Collection and recycling is now put into the agenda and will continue after the active marketing campaigns.

The pilot campaign in India brought some insights on recycling behavior. An important conclusion is that people will start recycling if they are provided the right information, and it is easy for them to recycle. Cooperation in the collection efforts is also crucial as it was noticed that trade partners are important in convincing the consumers and encouraging them to recycle.

It was noticed in India that old mobile phones are used until no further value can be extracted out of them. They are then kept as back-ups, handed over to others or sold or exchanged for a discount on a new phone. Many consumers hang on to their old phones for the sake of emotional attachment.

For the majority of the consumers the concept of recycling is not clear. For most people recycling means getting some value out of something that has become useless. The findings implied that any communication around recycling needs to be very short and simple and emotional and social rather than rational. Rational messages are more likely to put consumers in a skeptical frame. Taking care of the environment should be linked to doing one's duty for society: either the future of children or healthy environment for the family. In India the results indicated that engaging women & youth should be the main focus for communication efforts. The last important learning is that the take-back initiative should make consumers understand that this initiative is a 'process', 'a way of thinking' rather than a 'scheme' to ensure the continuity of the action. (Singhal, 2010)

4.3 Expanding the program and building partnerships

After setting up a basic infrastructure for recycling and starting to raise the consumer awareness on recycling possibilities the program can be expanded in cooperation with different partners. Nokia aims to learn from each of the recycling program and develop the programs to be more attractive, convenient and efficient. Recycling programs are being done in cooperation with telecom operators, retail chains, environmental NGOs and with schools and universities. These are the partners that can help in getting the programs closer to the people so that they are convenient to use and also help to raise the awareness of recycling by education.

Nokia started a recycling program in China already at the end of the 2005. The program was launched together with the teleoperator China Mobile and with Motorola. The next year six other mobile phone manufactures joined the program. The program has been growing every year, it started with 40 big cities and 1500 recycling location and after six years it covers already 300 cities. Program has collected more than 160 tons of e-waste by the end of 2010. Different activation methods have been used over the years, from prepaid phone cards, eco-friendly shopping bags and tree planting. Experience van has also been used to spread the message to smaller cities that were not included in the program. In Latin American countries recycling programs started in 2006 in Mexico in cooperation with telecom operator Telefonica. Cooperation has expanded to other countries Peru, Brazil, Chile, Columbia, Ecuador and to Argentina (in cooperation with Claro). Latin American countries have collected 375 tons of e-waste in these programs in the first four years. These programs are good examples of how long term one must think when starting to communicate and operate recycling program for consumers and how the working concept can be duplicated from one country to another.

Nokia has also raised recycling awareness in cooperation with environmental NGO's in many countries. In Lebanon Nokia has been working in cooperation with The Association for Forests, Development and Conservation (AFDC) to conduct recycling events in universities. In United Arab Emirates there is collaboration with Emirates Environmental Group (EEG) working with schools and corporations to raise the awareness level about recycling. Posters created by a well-known film maker in UAE help to change children into "recycling heroes". Nokia India also made two educational books for children in cooperation with TERI- The Energy and Resource Institute to support awareness raising at schools. In Singapore as part of Nokia's Environmental Conservation initiative - Recycle A Phone, Adopt A Tree Program- students from Temasek Polytechnic got a change to participate in a field trip to WWF re-forestration site Rinjani Forest in Lombok Island, Indonesia that Nokia is sponsoring. Wild Ocean is a nature film that in the USA has helped Nokia to spread the sustainability message for school groups. A Wild Ocean education campaign was launched utilizing Nokia-branded educator guides and activity posters available for schools.

Cooperation with universities and researchers can also be wide ranging. In Mexico Eco Rally 2010 was a two day recycling event at the Technical University of Monterrey. During the two days not only thousand items were collected for recycling but also a workshop with students was arranged to raise environmental awareness. Nokia is also a member of United Nation University SteP (Solving the E-waste Problem) initiative. One of the SteP activities is an annual E-waste Summer School with post graduate participants from all over the world looking to find solutions on how to manage e-waste in their countries. To take a totally new perspective for recycling, Nokia together with an environmental design office S.E.O.S have organized two "Phone Liberation" workshops. These workshops have collected engineering, industrial design and economics students in Espoo Finland and in Shanghai China to brainstorm improvements for phone recycling. The second workshop was done in cooperation with Aalto University and Tongii University who hold a summer school program on public transportation. The purpose of this workshop was to innovate how recycling can be done in public transportation environment. An example of the ideas on how to communicate recycling in public media is shown in the figure 9. In the idea an old mobile phone is exchanged with the access ticket to public transport, e.g underground or train.

Fig. 9. Mobile phone used as an access ticket to public transportation

4.4 Data collection

Data collection and analysis is one of the cornerstones for measuring and improving of any activity. For recycling this is important for example for planning the logistics or collection methods, calculating the business cases for marketing campaigns and for setting targets for programs. Good and transparent data systems are also important in supporting the decision making for closing the loop in recycling society and for building partnerships. These include sharing the best practices in public forums and public reporting of performance. (ICMM, 2006). Transparent information systems are also important in preventing the illegal waste transportation from developing to developed countries leading to improper recycling practices.

4.4.1 Case Finland

In Finland, the EU WEEE directive was implemented in national legislation in 2004. As a member of producer association Elker (www.elker.fi), Nokia is taking part in maintaining take back points for the public to return old devices and the reporting of imported product and recycled waste flows to authorities. In addition to these collection points, end users may return old devices to the retailer when buying a new device. Nokia also offers a service where an old device may be returned to an authorized service point or through pre-paid mail service for proper end of life treatment. Existing channels for unused mobile phones in Finland are many, but the awareness of consumers regarding these options can be increased.

Three different recycling campaigns have been carried out in Finland in 2006, 2008 and 2009 to raise awareness of various take back options. The idea of the campaigns was to bring answers for two main obstacles in recycling consumer electronic products: awareness and incentives. On top of that the idea was to make recycling as easy and convenient as possible in order to activate typical non-recyclers. Campaigns utilized prepaid postage envelopes for collecting obsolete devices for recycling. Different distribution methods for envelopes were studied. Envelopes were distributed in the sales box for online sales, direct mailing to households, as magazine insertions, at Finnish Post offices, Nokia service points and retail and it was downloadable from Nokia internet page. Marketing efforts included campaign web pages, banners on partner web pages, posters and also newspaper ads.

Different incentives for consumers were used to get the consumers to mail back their used phones utilizing postage paid envelopes. In the two first projects Nokia donated 2 Euros for each returned Nokia phone or accessory for WWF campaigns, first to a climate program and later to protect the Baltic Sea. The third campaign promoted the use of immaterial services for mobile devices, such as games, music and navigation. When a phone was returned, the consumer received a code for free download of the selected service.

These campaigns have offered a unique opportunity to analyze consumer behavior regarding recycling The combination of advertising and incentives can be attributed to increasing consumer awareness and participation. As seen in many other recycling initiatives as well, not only private consumers were activated but also companies started asking how to return used phones, batteries and accessories for recycling.

These three campaigns together managed to collect 53.000 phones. Not all the consumers were interested in the incentives; it was interesting to see that there was a group of consumers, 16 % from total, who only wanted to recycle without any incentive at all. This supports other studies showing that financial incentives do not need to be in place for recycling activities to start- convenience is the main driver. After the campaign the mail-back system is still in use and one can download pre-paid envelope anytime at the Nokia.fi internet page.

Recycling campaigns are activating people to recycle and the large amounts of envelopes that have been distributed and downloaded are coming back to recycling. In the first campaign all the envelopes were distributed directly to people by using different channels from direct mail to magazine inserts, and 11 % of the envelopes were returned to recycling. Previous experiences showed that 2 % of the envelopes put into the sales package were returned back, so the result was really good. In the next campaigns people were directed to pick-up an envelope themselves, or even go to the internet page to seek for more information. 19 % of the people who visited the campaign internet page also downloaded the recycling envelope. From the ones who had downloaded the envelope 64 % returned it to recycling, while the envelopes that were distributed elsewhere (street, retail) 29 % were returned. These very high return rates were achieved when envelopes were given to those who showed some interest in recycling. Envelopes are being stored at home and they are being used later, when there is something to return. It has been seen that envelopes are returned even five years after the campaign has finished. When different distribution methods have been compared it can be concluded that the most efficient return rate can be achieved when people download or pick up the envelope themselves, rather that when it is provided to them.

The phone models which have been returned in recycling campaigns are rather old, typically from the 1990's. In addition to mobile phones, return envelopes have included batteries, chargers and plastic covers. This indicates consumers have been saving handsets, old parts and even plastics waiting for a proper and easy disposal option. An average of 14 % of the envelope includes more than one phone. 12 % of envelopes have not had a phone but they contained accessories and batteries and 60 % included one phone with a battery. People typically have more than one item for recycling, in 2009 envelopes contained an average 1.43 items and 1.05 phones, showing small decrease from the first campaign in 2006 with 1.52 items and 1.1 mobile phones.

4.5 Utilizing consumer insight

The main barrier for phone recycling is that people like to keep the phones as spare or back-up. Figure 10. shows the ten main reasons for not recycling a mobile phone. According to Nokia consumer survey, globally over a third of the respondents claimed to have kept their last mobile phones, see also figure 3. People in developed countries are particularly likely to keep their old mobile phone, whereas in developing nations people are more likely to seek further value from their old phone by giving it to someone else, or selling or trading it for a new phone. Following the desire to keep old mobile phones as spares, awareness is also a major barrier to recycling, with a fifth of respondents citing this as the reason for not recycling their previous phone. This is particularly relevant in the developing countries where there is the lowest awareness that mobile phones can be recycled and where they can be recycled. The study also showed that there is an opportunity to encourage more recycling through education, by building awareness that old mobile phones can be recycled, and highlighting both the environmental and personal gains associated with this. It must also be noted, that recycling amounts of e-waste are still quite low because recycling infrastructure for e-waste is not yet mature or easily available in most countries. Legislative frameworks are also important to foster this development.

Fig. 10. Reasons for not recycling last mobile phone (Nokia, 2011b)

The more opportunities people are presented with, and the more convenient it is, the more likely people are to recycle their mobile phone. Even though people feel positive towards recycling campaigns, the underlying questions "how much effort do I have to make?" and "what's in it for me?" are typically noted. More mobile phone recycling may be encouraged by seeing this as a transaction in which consumer's effort (and the value of the phone) is traded for a benefit, an incentive, that can be a soft (emotional) or hard (financial) one. This especially encourages first time recyclers into action. Through all markets the biggest interest is expressed in exchanging an old phone for discount for new handsets, and this is being followed by more altruistic initiatives such as making a donation to an environmental charity.

The results of the consumer survey showing that awareness and phone hoarding are the main barriers for not recycling has lead into the development of the Nokia recycling program "3 steps to recycle". In this digital program simple answers are given to people on how and where to recycle. These include a 3 step program with the relevant information that phone can be recycled and how the data can be saved followed by map showing the locations of the recycling points. As noted previously one of the main reasons not to recycle is an emotional attachment to an old mobile. It is quite common that people like to store the old devices as there are so many precious memories attached to them. To tackle this recycling obstacle Nokia started a recycling campaign in Twitter, called "I#recycling". The campaign utilized old iconic Nokia phones such as Cityman, Nokia 3310, Nokia 8810, Nokia 5110 and Nokia 2760 and gave them different personalities. Phones were tweeting about recycling in order to make it fun and interesting and they also gave practical hints on where and how to recycle. During the 3 weeks of the campaign time it reached 170 000 people online in 44 countries and created 2800 Facebook likes. The campaign was translated also into Chinese and run in Chinese social media channels. These digital marketing campaigns are changing recycling attitudes and help to move towards a recycling society.

5. Conclusion

Electronics waste is globally an increasing waste stream that needs to be directed to proper recycling systems in order to save precious natural resources and to avoid contamination caused by landfilling or waste incineration practices. Product end of life is a process with three main steps from consumer collection to pretreatment and material refinery. First step of the process, collection, is a prerequisite for the rest of the process. The main challenge there is to get people to return their old products for recycling when they no longer need them. Consumer attitudes and recycling behavior has been studied in many countries to improve the success of the recycling programs. The most important factors enhancing the recycling behavior are convenience and awareness on where and how to recycle. The inhibiting factors for recycling are the emotional attachment to the old phone and willingness to keep a spare product. The following steps of the recycling process, pre-treatment and refinery are the ones where possible environmental risks may take place in the cases where e-waste is not handled properly. With proper technologies 100% of the materials in a mobile phone can be recovered and nothing needs to be wasted.

First mobile phone recycling programs started in the late 90's in Europe. Today Nokia is offering recycling programs for mobile phones in almost 100 countries. At the company's own premises waste sorting can be very efficient leading to true closed loop material circles. When take back of obsolete phones from consumers is discussed, the challenge is more in the awareness raising and in building up the eco efficient collection infrastructure at different parts of the world. Financial incentives can be used in initiating the recycling behavior, but the importance of them is getting lower when recycling becomes a daily habit. The different phases for starting a recycling program have been presented from building up a recycling infrastructure to partnerships and to different marketing campaigns. One of the main learning has been the usage of right language and tonality to communicate the benefits for recycling and finding the right partners, and most importantly long-term commitment to programs. Consumers who have recycled their old mobile phone have been reported to increase from 3 % to 9 % between 2007 and 2011. There are differences between developed and developing markets in the access to recycling points and information, but the importance of the topic is becoming globally understood.

6. Acknowledgements

Big thanks to the Sustainability team members at Nokia Markets who provided information on the local recycling programs and to River Research at London for consumer insight on recycling.

7. References

Chancerel, P., Rotter, S. (2009), Recycling oriented characterization of small waste electrical and electronic equipment, *Waste Management*, volume 29, Issue 8, August 2009, pp 2336-2352

Chancerel, P. , (2010), *Substance flow analysis of the recycling of small waste electrical and electronic equipment- An assessment of the recovery of gold and palladium*, Institut für Technischen Umweltschutz, Technische Universität Berlin, ISBN 978-3-89720-555-0

Hai-Yong Kanga, Julie M. Schoenung, (2005), Electronic waste recycling: A review of U.S. infrastructure and technology options Resources, *Conservation and Recycling*, Volume 45, Issue 4, December 2005, Pages 368-400

Huisman, J. (2003), *The QWERTY/EE concept, Quantifying Recyclability and Eco-Efficiency for End-of-Life Treatment of Consumer Electronic Products*, Doctoral thesis, Delft University of Technology, Delft, the Netherlands

ICMM, (2006), *Maximizing Value- Guidance on implementing materials stewardship in the minerals and metals value chain*, ICMM International Council on Mining & Minerals

Jenkins, R.J. Martinez, S.A Palmer, K. Podolsky, M.J. (2003) ,The determinants of household recycling: a material-specific analysis of recycling program features and unit pricing, *Journal of Environmental Economics and management*, 45 (2003) 294-318

Mäkelä, M, (2011, in press), *The Complexity of Mobile phone recycling, Developing user oriented design guidelines and future concepts for mobile phone recycling*, School of Art and Design, Aalto University, Helsinki

Nokia, (2008), Global consumer survey reveals that majority of old mobile phones are lying in drawers at home and not being recycled, *Nokia Press Release*, 23.3.2010, available at: http://www.nokia.com/press/press-releases/showpressrelease?newsid=123429

Nokia, (2010), *Nokia Sustainability Report 2010*, 15.8.2011, available at: http://www.nokia.com/corporate-responsibility/reporting

Nokia, (2011a), 22.8.2011, *Materials and substances*, available at: http://www.nokia.com/environment/devices-and-services/creating-our-products/materials-and-substances

Nokia (2011b), *Consumer study 2011*, not public

Saphores, J.M, Nixon, H. Ogunseitan, O.A., Shapiro, A.A. (2006), Household willingness to Recycle electronic waste: An application to California, *Environment and Behavior*, March 2006

Shaw P.J., Maynard, S.J., (2008) The potential of financial incentives to enhance householders curbside recycling behavior, *Waste Management*, 28 (2008), pp. 1732-1741

Singhal, P., (2010), Shaping consumers' behavior for responsible recycling in India, *Proceedings of Going Green Care Innovation*, Vienna, November, 2010

Tanskanen P, Butler, E., (2007) Mobile phone take back -learnings from various initiatives, *Proceedings of IEEE Electronics and Environment*, Orlando 7-10 May

Tanskanen, P. Takala, (2006), R., A decomposition of the End of Life process, *Journal of Cleaner Production*, Vol 14, Numbers 15-16

Teerioja, N,(2009), *Kahden paperinkeräysmenetelmän ympäristövaikutusten ja kustannusten vertail*u, University of Helsinki, available at hdl.handle.net/1975/9353 [23.3.2010]

8

Design for E-Waste Recycling Deposit System and Expense Mechanism in China

Hua Zhong
School of Management and Economics, Beijing Institute of Technology
China

1. Introduction

With 50 million metric tons of e-waste disposed worldwide each year, e-waste recycling has become an increasingly important issue globally. China alone adds 2 million tons of e-waste and the number increases at a rate of more than 10% each year (Ni, 2007). American consumers replace their mobile phones every one and half years on average (Grow, 2008), producing 130 million discarded cell phones annually nationwide (Hanselman, 2007). In China, 70 million mobile phones and 10 million PCs are being discarded each year (NDRC1, 2006). Likewise, more than 50 million PCs are disposed in the U.S. each year (Hung, 2007), with this number accounting for only 10% of the disposable computers. The majorities of defunct computers are not being used and simply collect dust (Zhong, 2009).

Along with the execution of policies and regulations for e-waste recycling, the Chinese National Development and Reform Commission have launched four national pilot projects in four major cities including Hangzhou, Qingdao, Beijing and Tianjin since 2004 (NDRC2). The objectives of the pilot projects include: setting up a network for e-waste collection, supporting the development of standards and regulations for e-waste management, and developing key technologies as well as equipment for Waste Electrical and Electronic Equipment (WEEE) recycling (Yu, 2010, Streicher, 2007). Despite the advanced equipment and technologies supporting the pilot projects, the effect of these pilots was only minimal due to the high cost involved. While typical recycling companies are challenged because of the shortage of supplies, lacking efficient incentive has become a bottleneck of e-waste recycling in China.

To find the best solution to the e-waste recycling issue in China, this research carefully considers and examines recycling models such as Extended Producer Responsibility (EPR), Online Recycling System and the existing Deposit-refund System. After weighing pros and cons of each, it is concluded that we need to explore a new approach under EPR to best suit China's socio-economic situation. We then combine and tailor the concepts of the 3 above-mentioned systems based on China's situation, and proposed a customized deposit-refund system under EPR structure with the facilitation of online information technology. With the deposit-refund system under EPR, it is much easier to increase consumers' incentives to recycle and accordingly ensure the construction of a formal recycling channel.

Surrounding the proposed deposit-refund System under EPR, this research carefully analyzes the correlative monetary, information and product flow for China's e-waste special fund administration and tries to apply closed-loop supply chain optimization theory to design an expense allocation incentive mechanism that helps make all parties participate the recycling process and track the environment effect. Focusing on the issues of recycling, environmental effects and incentives, and re-evaluation of the appropriate roles of the government, this research provides a new perspective on supply chain analysis. Based on Savasakan's supply chain model, we refer the deposit parameter as a new parameter, and then examine the global supply chain from the angle of total social welfare and environment impact. Lastly, in this research, we try to define and introduce the quantitative and standardization theory, which will greatly enrich and complement the existing recycling theory system for future study.

2. Review of alternative recycling systems

2.1 EPR system

EPR (Extended Producer Responsibility) is defined as "a policy principle to promote total life-cycle environmental improvements of product systems by extending the responsibilities of the manufacturer of the product to various parts of the entire life-cycle of the product, and especially to the take-back, recycling and final disposal of the product" (Lindhqvist, 1999). EPR requests producers to take responsibility and minimize its negative environmental impacts from designing recyclable products in the first phase (eco-design) to collecting those to be disposed at the end of the products' useful lives (Zhong, 2009). Furthermore, it avoids "everyone's responsibility is no one's responsibility" because it sets forth the overall "polluter pays" principle. As an environmental legislation, EPR has provided a new means to clarify responsibility implementation through administrative, economic, and informational instruments. The concept of EPR was first formally introduced in Sweden. In 2000, the European Parliament passed a directive requiring its member countries to institute an EPR program for EOL vehicles (Forslind K.H, 2005) and an additional directive for WEEE was approved in early 2003. Other countries, such as Japan has also enacted an EPR law covering four large electrical home appliances (Spicer A, 2004). The United States established a similar system with a variation of focusing more on product responsibility, instead of product. Table 1 shows the definition and explanation of EPR under respective legislations.

The ultimate goal of controlling e-waste is minimizing the impact of end-of-lifecycle (EOL) products on the environment. Scholars have explored the practices of EPR in different regions and across different products. They reached a consensus recognizing that EPR policies can actually stimulate product innovation and environment-friendly design in reducing materials, resources and energy usage by eliminating the use of toxins, extending the useful life cycle, increasing opportunities for recovery and re-use of the product at end-of-life (Scott N., 2007,Tojo N., 2001). But all of that is under the condition of higher economic development level, the execution force of legislation and active and positive involvement from consumers. In this model, consumers' incentives to recycle disposed items to a great extent depend on their voluntary will and their sense of responsibility to comply their duties under the EPR legislation. However, in China, pure constraint from legislation alone does not form enough motivation for consumers to perform their duty in recycling process and it is practically hard to ensure the full execution of legislation.

		Europe	Japan	U.S.
legislations		WEEE Restriction of Hazardous Substance (RoHS) Package and EOL Vehicles (ELVs)	Home Appliance Recycling Law	20 states now have laws for discarded electronic products that require producers to finance or manage collection and responsible recycling
Responsibilities	Producer	Collection, recycling, disposal and charging the fee	Complete recycling rate index	In EPR system, producer, supplier and customer take on responsibility for the environmentally appropriate disposal of e-waste
	Consumer	Pay the recycling fee and insure the integrity of e-waste	Inform the retailer when home appliances retire	
	Government	Pay a part of fee of collection, disposable or make policy of fiscal subsidies of EPR organization	Enforce legislation	

Table 1. Definition and legislation of EPR in Europe, Japan and U.S.

On Feb. 25th, 2010, Chinese government promulgated "Regulation of Management of the Recycling and Disposal of Waste Electrical and Electronic Equipment" and it has taken effect as of January 1st, 2011. This scheme, quoting EPR's concept, proposes to establish special fund contributed by producer, retailer and consumer who are responsible for environment prevention and assisting e-waste recycling. However, it will still be a critical issue to enforce this finance system in real practice in the future if the problems of high cost and lack of incentives are not solved.

2.2 Online recycling system

Some scholars proposed a third-party system (Spicer A, 2004, Zhong, 2009). Under this model, Zhong introduced the concept of an online recycling system in China (Zhong, 2010). This online recycling system focuses on building an e-commerce platform to collect and analyze recycling date in real time. Similar to forward logistic shopping website, it only provides recycling related services. With this system, a consumer will only need to submit the e-waste information including the product category, brand, model, purchase date, current condition and etc. via the internet or over the phone. The system will then automatically generate a list of the items to be recycled, which will be picked up by professionals from a third-party recycler. After the transaction, third-party recycler pays the consumer via internet or in cash as compensation. The e-waste will then be transferred to an e-waste recycling center, a third firm dedicated to recovering reusable materials from EOL products and selling them in second-hand markets. Once it has been delivered to the recycling center, the e-waste will first be inspected to determine whether it should be repaired or disassembled. Repaired products can be sold in second-hand markets, while disassembled EOL products will be segregated into reusable, recyclable, and disposable materials, with each sent to its appropriate inventory. The reusable and recycling components are tested for their usability potential before they are sent to a producer or retailer, and the disposable components are sent to the disposal site. At the same time, the government and producers will provide financial support to recycling center to realize the EPR principle. This system is fast, convenient and can be easily spread once established, but it needs an original power to push this huge e-commerce recycling market and a series of policies and assessment methods such as LCA theory to support the system operation. It requires high capacity for recycling and dispatching and also needs the consistent participation from all consumers and a high reward to the participated consumers. With China's current economic level, and the fact that civic recycling consciousness hasn't been widely established throughout the whole nation, without enough recycling compensation, establishing such a big and complete system can be a long term task and needs us to seek an effective means to achieve it in gradual steps.

2.3 Deposit-refund system in the World

For most pollutants, the standard response of economists is to tax the offending activity. However, a direct tax is not easy to impose on dumping or litter, therefore, an effective alternative is to apply the deposit-refund system (DRS) (Don Fullerton, 2000). With DRS, consumers pay deposits when they purchase items and get refund when they return the items for recycle.

Take beverage container recycling system in United States as an example. Bottle bills require deposits on beverage containers, such as bottles and cans, at the time of purchase. These deposits can be totally or partially recovered by individuals who recycle these containers. When a retailer buys beverages from a distributor, a deposit is paid to the distributor for each can or bottle purchased. The consumer pays the deposit to the retailer when buying the beverage. When the consumer returns the empty beverage container to the retail store, to a redemption center, or to a reverse vending machine, the deposit is refunded. The retailer recoups the deposit from the distributor, plus an additional handling fee in most U.S. states. The handling fee, which generally ranges from one to three cents, helps cover the cost of handling the containers (Yeh, Chiou-nan, 2008).

The beverage container recycling program in the United States is an important representative of a successful deposit-refund system. The system encourages consumers to return beverage containers by providing a refund on the deposit, and is considered by many stakeholders as a successful recycling program to be expanded as a federal policy (U.S. GAO, 2006). The eleven states (California, Connecticut, Delaware, Hawaii, Iowa, Maine, Massachusetts, Michigan, New York, Oregon and Vermont) employing this system have achieved higher recycling rates and have recycled more beverage containers than the other 39 states combined (DOC, 2007). The refund value of the container (usually 5 or 10 cents) provides a monetary incentive to return the container for recycling (Ramzy, 2008).

Similarly, container deposit legislation is widespread in many countries in the world, such as Australia, Canada, Germany, Netherlands, New Zealand, Denmark, Norway, Sweden,

Finland, United Kingdom, and etc. Even in China, you can find the reverse vending machines that recycle beverage bottle or can and returns 5-10 cents as refund. However, it only provides a channel to encourage consumers' return behaviors.

Although some enterprises and organizations oppose the deposit system, government may pass container deposit legislation for following reasons:

- To encourage recycling and complement existing curbside recycling programs
- To recycle plastic PET (Polyethylene Terephthalate), aluminums that may create more economic value.
- To specifically reduce beverage container litter that pollutes the environment.
- To provide an economic incentive to control and govern environment pollution
- The system is in fact a significant source of income to some individuals and non-profit civic organizations

To summarize, Deposit-refund System can help consumers prevent from polluting the environment. It can also help fund environmental programs and cover the costs of processing returned containers with redeemed capital.

Not limited to beverage area, researchers proposed deposit application in many other fields such as e-waste, battery and auto industries and etc. In recycling area of WEEE, Ramzy tried to establish an e-market for e-waste returned deposit in the US competitive culture context (Ramzy, 2008). For China, Yu proposed a deposit system with shared responsibility to incentivize consumer participation (Yu, 2010) and He (He, 2010) developed a recycling program by constructing deposit refund scheme in battery recovery. However, neither provided elaboration on the detailed implementation of the system.

Through the above introduction, we find that EPR system emphasizes producer's responsibility and requires producer's participation. It is most feasible for developed countries with high socio-economic level and good sense of responsibility from consumers. The fact that China's economic development is still at a lower level and all consumers haven't built up a relatively good sense of responsibility dictates we cannot directly apply the EPR model without customization in China. We need to, instead, seek economic incentives and meanwhile constrain consumers' behavior in order to achieve a satisfying recycling effect.

Online Recycling System, similarly, fully operates on government and producer, but still cannot address the issue of lacking incentives from consumers. Its convenient online process may help encourage consumers to recycle, but still does not provide enough fund to boost up consumers' incentives. Besides, having all procedures online requires high cost and establishment of a series of surrounding systems to complete the process, such as assessment system and reverse logistic system and etc. Again, China's current recycling market lacks original motive to push this system to form the scale.

The existing Deposit-refund system works well in beverage products and does effectively motivate consumer to recycle disposed products. But if applied to electrical and electronic products without any other measures, it may not work as effectively as for beverage products, for e-waste products are different from beverage containers and they require more complicated process and demand more resources to recycle and reuse. Deposit alone may not be powerful enough to push the recycling process through. We still need producer's involvement to improve products' design to reduce source of pollution, extend the products' life cycle, and increase opportunities for recovery and re-use of the products.

Based on all analysis above, we propose a Deposit-refund System Under EPR, that integrates concepts from the three models and works best with China's socio-economic situation. This system takes account of economic measures and in an eased-up way delegates the recycling responsibility to consumers. It also incorporates the concept of EPR by having producers control and improves the design of the products from the beginning phase. It facilitates the process with certain online platforms and enforces the constraint of consumers' behavior with the regulations of deposit funds. With the current established "home appliance replacements" system, consumers who contact local recycling company will receive pick-up service and qualify for a direct price reduction when buying the new appliances with vouchers from the company, however, the subsidy comes from National and local finance, in this context, the deposit system under EPR is reasonably more cost effective and practically easier to implement in China.

3. E-waste deposit system under EPR

Although EPR system operates very well in western countries, one big obstacle to the adoption of EPR initiatives appears to be the establishment of the fee structure to address the costs involved in e-waste collection, disposal, and management. Take the U.S. for example, nineteen of the twenty states that have passed the e-waste legislation are adopting producer's responsibility, which requires manufactures and vendors to pay for the recycling costs and also provide free recycling to consumers. On the opposite, California switched such financial burden to the consumers (consumer responsibility), charging a $6 to $10 disposal fee on every computer and television purchased (Hanselman, 2007). An often seen dilemma is that producers are not willing to pay for recycling because such costs, very likely, were not being considered in the initial pricing model, on the other hand, consumers are reluctant to take such responsibility because in some states, there are still options to trash e-waste for free, or they may choose not to recycle at all.

Therefore it is necessary to introduce EPR implementation forms, which can be classified into three major approaches: voluntary, mandatory and economic. In terms of capital operation, in Europe, Korea and Taiwan, all funds come from the producers; while in Japan, the capital comes from the consumer. On the other side, the United States emphasizes more on economic benefits rather than legislation. Based on the situations in China, the authors believe that China should pay more attention on economic methods to design the expense mechanism with enough incentive to get all stakeholders participate in e-waste recycling.

As mentioned above, China has enforced the "Regulation of Management of the Recycling and Disposal of Waste Electrical and Electronic Equipment", the counterpart of the EU WEEE Directive in EPR system and very important for the establishment of an entire management framework for e-waste recycling in China (Liu, 2006). Following EPR principle, the regulation stipulates that e-waste should be collected by multiple channels and recycled intensively by manufacturers of WEEE. A special fund has been set up. The producer and the importer of electronic products shall perform their duty in making contributions to fund for e-waste recycling (Yu, 2010). Given this context and reference to Ramzy's model of e-market for deposit-refund (Ramzy, 2008), we propose a deposit recycling system that emphasizes on incentive collection, at the same time establishing a competitive market for reuse and recycling services, and ensuring all stakeholders to get involved in this system. This special fund should be managed by a special non-profit organization (a foundation) under the supervision of the government and the capital should come from three sources: consumer's deposit in advance, manufacture's recycling disposal fees and government's subsidy.

As illustrated in Figure 1, we proposed an e-waste deposit recycling system under EPR, called Recycling Fund Administration System. It consists of recycling fund and the fund administration information system. When consumer purchases a new product, he/she must pay a deposit to retailer, who transfers the deposit to the foundation and then receives a variable portion when consumer returns unwanted e-products to the third-party recycling company directly or via "home appliance replacements" system. The non-profit organization and the third-party recycling company are the crucial key to the proposed e-waste deposit recycling system.

Further questions we should take into account when implementing this deposit system include: 1) How to allocate the contribution to the funding? 2) How to stimulate producer to realize their duties? What measures should be taken to provide enough incentive?

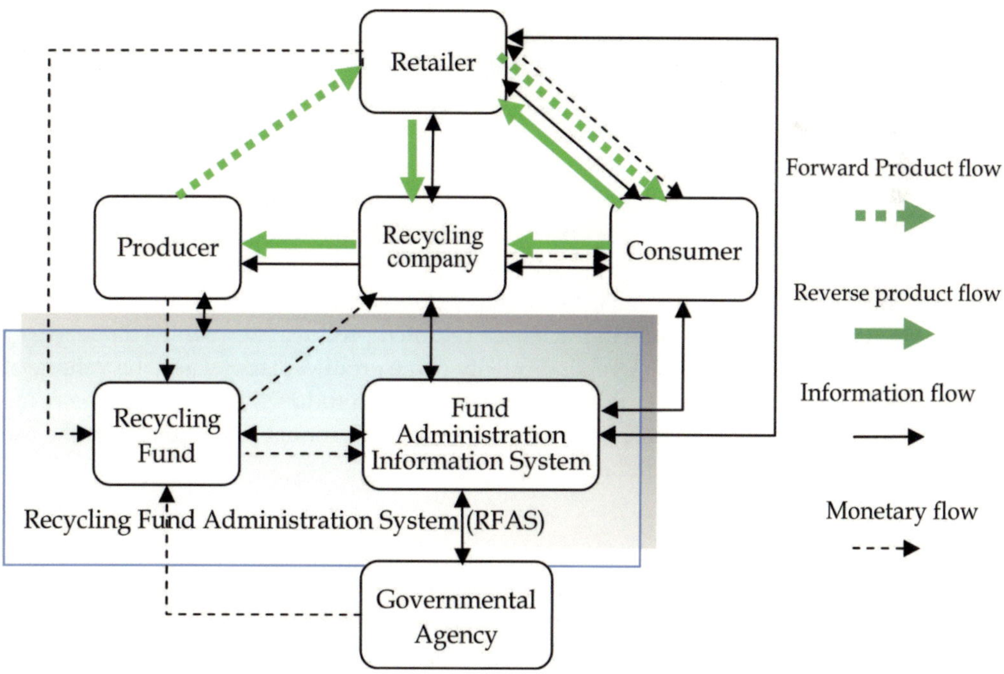

Fig. 1. E-waste deposit recycling system Under EPR

3.1 Monetary flow

To simplify the graphical presentation, we only draw the recycling related monetary flow in this framework, which starts at the point of sale with consumer's payment of the deposits. Then the deposit is transferred by retailer to the foundation, from which the third-party recycling company receives subsidies as reward. The amount varies depending on their processing capability and scalability. When consumer returns his/her unwanted products, he/she has choices to select recycling companies and receive the refund. In the meantime, to abide by the EPR principle, manufacturers and government also need to pay partial of the recycling fee to maintain the foundation's operation.

More details should be covered outside this paper by further research regarding the operations and the management of the deposit system, such as how much manufacturers should pay and how to collect and use the fund; how to determine the amount of the deposit fee; when should recycling companies get subsidy and how to control all recycling flow and manage fund.

3.2 Information flow

Fund administration throughout the deposit system should rely on advanced information technology. Based on the internet platform, the Fund Administration Information System (FAIS) ensures that all information transfers smoothly. Obviously, new electronic product sale system should connect with FAIS when the trade happens. Once consumer wants to return old products, he/she should submit information through two channels - the replace system or contacting FAIS directly. If consumer uses Replacement system of Household Electrical Appliances, he/she can input product information through replacement system which connects to FAIS. Since China has issued Implementation Measures on Replacement of Household Electrical Appliances (also called "home appliance replacements "system) and has already established the recycling database, we can utilize or modify previous information systems rather than rebuilding new ones and save money and resources. If consumer does not purchase new product or there is no replace system nearby, he/she can log into the FAIS website or make phone calls directly to select recycling company to collect their old devices. In the return process, the old electronic product information such as product make, model and year, and consumer's information such as pick-up address or contact, etc. should be inputted to FAIS. FAIS creates a unique code to track each returned product. Meanwhile, the recycling company can give feedback on the level of difficulty of the product disposal and the volume of recycling as an evaluation index of EPR performance of the producer, who also has access to this information. The FAIS's another function is to publish an annual report to ensure the use of specific funding transparent to public.

3.3 Product flow

In the close-loop supply chain, the forward product flow starts from the manufacture. To simplify the process, we do not consider the intermediate suppliers and products directly transferred to the retailer and then sold to consumer. When a consumer decides to discard the old equipment, he/she could select a disposal company by comparing different offers from the competitive recycling market. According to the signed agreement between the consumer and the disposal company, consumer can freely choose the channel to deliver unwanted product. The returning channel can function in two ways. One way is through the original replacement system channel. It integrates forward and reverse shipping system into one, delivers new product to consumer's location, collects the old returned equipment of the consumer and then finally ships the old equipment to a local collect point where recycling company can pick up later or directly send to recycling company warehouse. This is a way that can save transportation cost and increase logistics efficiency. The other way is through the FAIS channel. Consumer logs into FAIS, submits product information online such as pick-up address, prints the pre-paid shipping label of the recycling firm and drops the product to the shipping company or post office. The deliver policy will differ depending on the competitive recycling company.

4. Design of expense mechanism under e-waste recycling closed-loop supply chain with deposit system

4.1 Summary of recycling pricing coordination mechanism based on supply chain optimization

Utilizing closed-loop supply chain optimization method can help make the mechanism design efficient, establish reasonable price strategies, allocate profits and coordinate all process with all recycling stakeholders having enough participation incentives. Among the recent closed-loop supply chain researches, Savaskan's studies were widely quoted. In his theory, closed-loop supply chain was divided into four models: concentrate coordination, manufacturer collection, retailer collection and third party collection. The four models, in general, can be summarized into 2 categories: decentralized and centralized decision-making models (Savaskan, 2004). Most researches apply Stackelberg's sequential game model and compare the different decision models' efficiency of supply chain based on revenue-expense sharing contract and coordinating mechanism of all stakeholders' interests. Since collection model and supply chain's structures are distinct, different portfolio models were developed by various scholars in the past, as shown as table 2. Additionally, based on Hammod's theory, Wang (2010) comprehensively considered the behavior of the supplier, the manufacturer, the retailer, the demand market and the collector, established the variational inequality models and then developed a remanufacture closed-loop supply chain network equilibrium model. All the researches listed above concentrated on supply chain interest allocation and coordination, but government environmental decision and consumer impetus of participation recycling were seldom referred. In particular, the EPR perspective on cost incentive mechanism related studies was very much lacked. Therefore, it is necessary to extend the research from two aspects. One is to consider government policy maker and focus on environment effects to come up with optimum strategy. The other is to inspire consumer through deposit system to ensure collection effect.

4.2 Savaskan's closed-loop supply chain model by third party collecting

In the original closed-loop supply chain model, Savasakan not only considered 3rd party collecting model (in short 3P model), also compared centrally coordinated system,

<table>
<tr><th rowspan="2">Recycling model / SC structure</th><th>Centralized</th><th colspan="3">Decentralized</th></tr>
<tr><th>Integrated SC Design Models</th><th>Manufacturer collection</th><th>Retailer collection</th><th>Third party collection</th></tr>
<tr><td rowspan="3">2-tier supply chain</td><td rowspan="3">Most of reference focus on comparison between centralized and decentralized decision</td><td rowspan="2">Comparison between manufacturer participating channel, not participating channel and integrated SC (Wang,2007)</td><td>Manufacturer and retailer recycling pricing equilibrium (Ge,2008) Stackelberg and Nash equilibrium cooperation (Gu,2005,Ye,2007) Pricing of close-loop SC considering market segmentation (Wang,2009)</td><td>Retailer selling and third-party collecting (Guo,2007) Manufacturer and recycler games on random collection quantity and capacity constraints (Sun, 2008)</td></tr>
<tr><td colspan="2">Comparison between retailer and third-party collector (Wang,2010)</td></tr>
<tr><td colspan="3">Summarize and compare three recycling models (Fleischmann, 2003, Savaskan,2004, Yao, 2003)</td></tr>
<tr><td>3-tier supply chain</td><td colspan="4">Manufacturers, repair center and retailer's price incentives, Stackerlberg and Nash model comparison (Huang, 2009), Retailer and third-party joint collection, manufacturer, retailer and third-party coordinate pricing by sharing responsibility according revenue and cost share (Zhang, 2010)</td></tr>
</table>

Table 2. Summary of Recycling Pricing Coordination Mechanism Based on Supply Chain Optimization

manufacturer collecting and retailer colleting system. Among the above models, the 3P collecting model as an alternative approach where private companies assume the EOL responsibilities for products on behalf of the original equipment manufacturers and relieves both manufacturers and the general public from the responsibilities and is a promising approach in optimizing product design, specialization, immediate economic feedback and remanufacturing market development (Dan, 2008, Huang, 2006, Spicer, 2004, Xia, 2007, Zhang, 2007, Zhong, 2009). Since Savaskan's model is most widely adopted, we will focus on it in this research, as shown in Figue2.

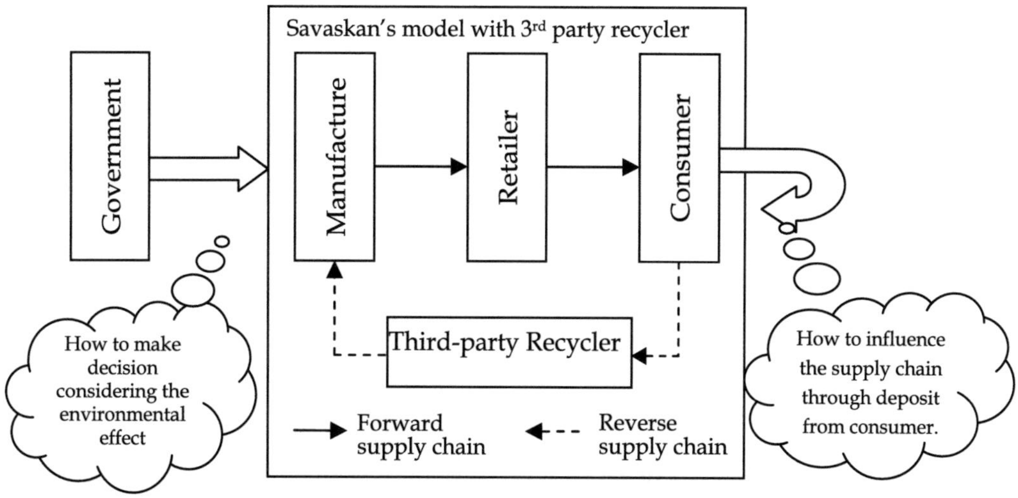

Fig. 2. Closed-loop supply chain system framework with deposit system

Below is a list of notations for Savaskan's Closed-loop Supply Chain Model.

p unit retail price
w unit wholesale price
c_m unit cost of manufacturing a new product
c_r unit cost of remanufacturing a returned product into a new one, assuming $c_m > c_r$
Δ unit cost saving from reuse $\Delta = c_m - c_r$
A fixed payment given to the consumer who returns a used product
C_L scaling parameter
b unit transfer price
$\tau(b)$ the return rate function of used products from the customers which is the expression of the transfer price
$D(p)$ demand for the new product in the market
$D(p) = \phi - \beta p$, with ϕ and β being positive and $\phi > \beta c_m$
Π_j profit function for channel member j, $j \in \{M,R,3P\}$ which denotes the manufacturer, the retailer and the third party

In this 3P model, for a given transfer price b of a used product, the third party maximizes his profits to determine the investment in used-product collection and decides the return rate τ (see Equation1).

$$\max_{\tau} \Pi_{3P} = b\tau(\phi - \beta p) - C_L\tau^2 - A\tau(\phi - \beta p) \tag{1}$$

Retailer solves for price $p(w)$ as a function of the wholesale price set by the manufacturer(see Equation 2).

$$\max_p \Pi_R = (p - w)(\phi - \beta p) \tag{2}$$

The manufacturer sets the wholesale price w of the product to Equation (3)

$$\max_w \Pi_M = (\phi - \beta p)[w - c_m + (\Delta - b)\tau] \quad (3)$$

For more detailed and complete model explanations please see Savaskan's paper (Savaskan, 2004).

4.3 The new closed-loop supply model with deposit system

Based on Savaskan's model, our new model has following changes (See Figure 2). Fistly, considering deposit impact for return rate, given d is the deposit of used product, we redefine the return rate function to $\tau(b,d)$. Secondly, we concern more about total supply chain impact when considering the government decision, which accordingly indicates an addition of government function with environment effect.

Through a survey study we can draw a reaction function of deposit change on consumer return wish and then identify the effect on collection rate $\tau(b,d)$. The deposit payment d as new parameter entering the closed-loop supply chain system model plays a key role in the new optimization supply chain model. We should firstly evaluate different deposit payment formats, for instance, a fixed fee or a proportion of the product's price, and then compare how different payment methods of the deposit can influence the rate of collection and cost of recycling. We can use statistical methods and estimate parameters to design a reasonable function to represent the model. Based on the comparison result of different combinations of different approaches and parameters, we can optimize the deposit parameters design in closed-loop supply chain system.

Then we will consider the government decision in closed-loop supply chain that indicates an additional game stage. Government's function should be clarified to satisfy the optimization of manufacture, recycler and consumer and minimize negative impacts on the human environment. For this reason, based on Savaskan's closed-loop supply chain system, we could realize global supply chain optimization by applying the Stackelberg model and maximizing the government utility function, which could be expressed by total social welfare with environment index.

The government solves the deposit d in utility function as follows (see Equation 4):

$$\max_d(\Pi_M + \Pi_R + \Pi_{3P} + \Pi_C + E(\tau)) \quad (4)$$

Π_C indicates the consumer's surplus and $E(\tau)$ is the environment index, the function of return rate.

At the same time, we should compare all stakeholders' interests and supply chain revenue under third party collection in the following situations: 1) whether the government concerns environment or not; 2) whether consumer pays deposit or not; 3) whether manufacturer pays recycling foundation or not; 4) whether manufacturer and consumer pay all together or separately. Accordingly, we can make sure that whether deposit system is practicable and know better how to share the responsibility between producer and consumer.

5. Contributions, limitations and future research

To address the serious issue of e-waste recycling in China, after comparing the different recycling systems in the world, this paper proposes to establish a deposit refund system under EPR framework suitable to China by providing a clear definition on information, product and monetary flow under the prevailing EPR policy and applying closed-loop supply chain optimization theory. Referring a deposit parameter into supply chain system and at the same time considering the government environment decision, we revise the Savaskan's supply model trying to design an expense allocation incentive mechanism under closed-loop supply chain optimization theory to have all parties participate the recycling process and track the environmental impact. In addition, it provides the quantitative and standardization theory to manage China's e-waste special fund administration. Besides the deposit system the paper also provides another glance of an alternative third-party recycling system that uses the e-commerce information platform to collect and analyze recycling data in real time, and therefore optimizes the design of the supply chain. However, we only give a rough model to solve this optimization problem since the deposit quantitative effect and environment index in government utility function are not clear. Besides, our model is limited to 3rd party collection, and we need to discuss more channels collecting to find the best solution. All these questions need to be analyzed in the future.

6. Acknowledgement

This paper was supported by the grant of humanities and social sciences under the Ministry of Education of China (10YJC630414) and New Century Excellent Talents in University (NCET-10-0048), National Science Foundation of China (No70773008，No70803004) and Fok YingDung Educational Foundation (121079). Their support is greatly appreciated. The authors would also like to acknowledge the contribution of anonymous referees whose comments would improve the paper significantly.

7. References

Dai Guilin, Zhang Yongqiang, Liu lei.(2008). Discussion to the recycle mode of waste home appliances, *Urban Problems*. pp. 78-81

Don Fullerton, Ann Wolverton. (2000). Two Generalizations of a Deposit-Refund System. *American Economic Review*, May2000, Vol. 90 Issue 2, p238, 5p

Fleischmann M ,Krikke H R ,Dekker R ,Flapper S D P.(2003). A characterization of logistics networks for product recovery, *Journal of Operations Management*. Vol.21, pp.259-279.

Forslind K.H.(2005). Implementing extended producer responsibility: the case of Sweden's car scrapping scheme, *Journal of Cleaner Production*. Vol.13 , pp.619-629.

GeJing-yan. (2008). Price decision of close-loop supply chain based on game theory, *Journal of systems engineering*. 1st ed, Vol. 23, pp.111-115.

Gu Qiaolun, Gao Tiegang, Shi Lianshuan. (2005). Price Decision Analysis for Reverse Supply Chain Based on Game Theory, *Systems Engineering-theory Practice*. 2005 March, pp.20-25.

Guo Yajun , ZHAO Li-qiang, Li Shao-jiang. (2007). Revenue and Expense Sharing Contract on the Coordination of Closed-loop Supply Chain under Stochastic Demand, *Operations and Management*. No 6, Vol.16, 2007, pp.15-20.

Grow, B., Tschang, C., Edwards, C., and Burnsed, B. (2008). Dangerous Fakes: How counterfeit, defective computer components from China are getting into U.S. warplanes and ships, *Business week* (October 02, 2008).

Lin K Catherine, Yan Linan, Davis N. Andrew. (2001). Globalization, extended producer responsibility and the problem of discarded computers in China: An exploratory proposal for environmental protection, *Georgetown International Environmental Law Review.* Spring 2001-2002,pp. 525-576.

Lindqvist Thomas. (1999). Towards an Extended Producer Responsibility——Analysis of Experiences and Proposals, *Stockholm: Ministry of the Environment and Natural Resources.* pp.29-291.

Liu XB, Tanaka M, Matsui Y.. (2006). Electrical and electronic waste management in China: progress and the barriers to overcome, *Waste Management and Research.* vol.24,pp.92–101.

Hanselman, S. E., and Pegah, M. (2007). The Wild Wild Waste: e-Waste, *Proceeding of the Association of Computing Machinery (ACM) Special Interest Group on University and College Computing Services (SIGUCCS)*, pp 157-162. Orlando, Florida, USA. October 7–10, 2007,

He Shensi. (2002). Model Design of Waste and Old Batteries Recovery System in China, *Environment prevention.* pp.40-42

Huang Ying-ying, Zhou Gen-gui, Cao Jian. (2009). Study on Price and Incentive Mechanism of Three-level Reverse Supply Chain for Electronic Products, *Industrial Engineering and Management.* No.2, Vol.14, pp.28-44

Huang Xisheng , Zhang Guopeng. (2006). Study on Extended Producer Responsibility Talking from the impetus to "Circular Economy", *Legal Forum.*Vol.21,pp.111-114.

Hung, L. Y. (2007). Go Green when Junking Gadgets, *The Straits Times (Singapore)*, April 10, 2007

National Development and Reform Commission (NDRC1). (2006). Department of Environment and Resource, Waste resource management and countermeasures in China. Dec.2006

National Development and Reform Commission (NDRC2). (2003). Notice regarding the selection of Zhejiang province and Qingdao City as pilot sites for the establishment of the national WEEE recycling and treatment system. NDRC Office: Environment and Resources.2003.

Ni Li. (2007). E-waste recycling logistics network optimization and development suggestions in China. *Market Weekly (Disquisition Edition).* pp.101-113.

Ramzy Kahhat, Junbeum Kim, Ming Xu, Braden Allenby, Eric Williams, Peng Zhang. (2008). Exploring e-waste management systems in the United States, *Resources, Conservation and Recycling.* Vol.52, pp.955–964.

Savaskan R. C, Bhattacharya S., Wassenhove Van Luk N.. (2004). Closed-loop supply chain models with product remanufacturing, *Management Sciences.* Vol.50, 2004, pp.239-252.

Scott Nicol, Shirley Thompson(2007). Policy options to reduce consumer waste to zero: comparing product stewardship and extended producer responsibility for refrigerator waste, *Waste Management Research*,25:227–233.

Spicer A J, Johnson M R. (2004). Third-party demanufacturing as a solution for extended producer responsibility, *Journal of Cleaner Production*. Vol.12, pp.37-45.

State of California Department of Conservation-Division of Recycling (DOC). (2007). Beverage Container Recycling Market Development and Expansion Grant Program, Sacramento, California(February 2007)

Streicher PM, Yang JX. (2007). WEEE recycling in China—present situation and main obstacles for improvement, In: *Proceedings of the 2007 international symposium on electronics and the environment*. pp.40–45.

Sun Hao, Da Qingli. (2008). Pricing and coordination for reverse supply chain with random collection quantity and capacity constraints, *Journal of Systems Engineering*. No.6, Vol. 23, pp. 720-726

United States Government Accountability Office (U.S. GAO). (2006). Recycling: Additional Efforts Could Increase Municipal Recycling, GAO-07-37,January 2008, Available from www.gao.gov/new.items/d0737.pdf

Wang Wenbin, Da Qingli. (2007). Distribution of profit in reverse supply chain with consumer surplus consideration, *Journal of Southeast University(Natural Science Edition)* . No4, Vol.3. pp. 726-730

Wang Wenbin, Da Qingli. (2009). Production and pricing strategy of closed-loop supply chain considering market segmentation, Control and Decision. No.5, Vol.24, pp.675-686

Wang Wenbin, Da Qingli,Hutianbing, Yang Guangfen. (2010). Remanufacturing Closed-Loop Supply Chain Network Equilibrium Model Based on Allowance and Penalty, *Operations Research and Management Science*. No1, Vol.19, pp.65-72.

Xia Yunlan, Lv Yongbo, Jia Nan, Cui Xianhua. (2007). Study on Electronic Products Reverse Logistics Mode and its Selection, *Logistics Technology*. Vol.8, pp.27-29..

Yao Weixin. (2003). Design Principle of Closed-loop Supply Chain, *Logistics Technology*. May 2003, pp. 18-20

Ye Fei, Li Yi-na. (2007). Research on Buy2back Contract Mechanism of Supply Chain Based on Stackelberg Model and Nash Negotiation Model, *Journal of Industrial Engineering/Engineering Management*, No.3, Vol.21 , pp. 39-43

Yeh, Chiou-nan; Vaughn, Percy.(2008). Consumer's Behavior under Mandatory Deposit System. *International Advances in Economic Research*, Nov2008, Vol. 14 Issue 4, p472-472, 1p; DOI: 10.1007/s11294-008-9184-5

Yu Jinglei, Williams Eric, Ju Meiting, Shao Chaofeng. (2010). Managing e-waste in China: Policies, pilot projects and alternative approaches, *Resources, Conservation and Recycling*, (Feb.2010), pp.1-9.

Zhang yulan，YAO Jian, PENG Guo-liang. (2010). Three-tier competition in the coordination of the closed-loop supply chain research, *Science-Technology and Management*，No.3, Vol. 12, pp.74-77

Zhang Heng, Da Qingli. (2007). Analysis of remanufacture logistics network model selection from strategic view, *Modern Management Science*, Vol.9, pp.28-38.

Zhong Hua, Shiller Z Shu. (2009). Recycling E-waste: A Solution Through Third Party Recycler, *Proceeding of The 9th International Conference on Electronic Commerce*. pp.965-969. Macau, Nov. 30- Dec.4, 2009

Zhong Hua, Schiller Shu, Liu Jian-chang, Wang Zhao-hua.(2010).Building a Third-party Recycling Platform for E-waste Recycling through an EPR Framework, *Proceeding of International Conference on Logistics Engineering and Management 2010*. pp.49-54. Chengdu, Oct. 8- 10, 2010

9

PET Containers in Brazil: A Logistics Model for Post-Consumer Waste Recycling

Tatiene Martins Coelho, Rosani de Castro and José Alcides Gobbo Junior
Faculdade de Engenharia de Bauru, Universidade Estadual Paulista (UNESP), Brazil

1. Introduction

Over recent decades, demographic growth, unprecedented industrialization and competitiveness for new markets have made companies face the issue of the environmental impact caused by their production processes and by the generation of post-consumption residue from their products.

In view of this, companies now have an emergent concern regarding this generation of residue, the return of which must be considered, thus providing a business opportunity through reverse logistics.

Environmental consideration is no longer regarded as a trend but a way of life for industry. For many companies over the past few decades, there has been a shift in addressing environmental issues. Actions that were taken at the operational level due to regulative demands are now being discussed at the strategic level (Nilsson, 2001). The idea of sustainability must consider the interplay of social, economic and environmental aspects with integrated and long-term perspectives.

Since the early 90s in Europe, there have been effective actions taken in order to reduce the amount of packaging, as well as its final disposal. One of the goals of the European Union was to reduce the amount of packaging waste by 70% by the year 2006. In order to achieve this goal, it is important to consider the recycling of this product since its inception (Onusseit, 2005).

Brazil is one of the fastest growing consumer markets for PET bottles, however, the country lacks an organized recycling system. This study will analyze, in an integrated manner, the best alternative to improve the recycling system by examining the PET bottle life-cycle in Brazil. Initially, Brazil's recycling of PET containers will be described. It should be noted that there are no specific laws or regulations determining responsibility for manufacturing, collection, recycling and final disposal of materials. Subsequently, an alternative system for recycling will be proposed considering social, economic, and environmental aspects with an integrated, long-term perspective.

The objective of this chapter is to outline the panorama of the post-consumption PET return system and propose a fitting application aimed at reducing the volume of residue.

2. The solid waste problematic

Countries that direct their model of development exclusively towards the economic, seek unbridled growth and an intensification of their economic activities. As a consequence, they assist in the accelerated urbanization process and the continuous transformations of nature. They also cause great disequilibrium because, if on one side they produce technological advances, on the other, they contribute towards growth in misery, environmental degradation and pollution (Cavalcanti, 1995).

According to (Tenório & Espinosa, 2004), the human being is the only one able to conduct the large scale transformation of raw materials into finished products, spewing unknown substances into the environment, which cannot be absorbed even over the long term, thus generating residue that, if not well administered, could result in pollution. From an environmental perspective, this pollution could be divided into three different classes: air pollution, water contamination and solid waste.

The Brazilian Association of Waste Treatment, Recovery and Disposal Companies reports that every year Brazil generates nearly 2,9 million tons of solid waste, 600 thousand tons, or 22%, of which receive proper treatment. Of the treated industrial rejects, 16% goes to landfills, 1% is incinerated and the remaining 5% is co-processed, that is, transformed through burning into raw materials sued for manufacturing cement (Abetre, 2008).

In face of the waste treatment, recovery and disposal context, the National Waste Policy (PNRS, 2002) has been in proceedings at the federal level since 1991. In article 12, it attributes solid waste management generated in the Federal District and cities to those respective territories.

According to (NBR 10004, 2004), solid waste may be present in solid and semi-solid states, which result in activities of industrial, household, hospital, commercial, agricultural, service and sweeping origin. This definition includes sludge from water treatment systems; sludge generated in pollution control equipment and installations, as well as certain liquids whose particularities make it unfeasible for release into the public sewage network or water bodies, or requires technical and economically unfeasible solutions for such in face of the best technology available.

In an international context, according to (Mahmood & Victor, 2001), there are many variables involved in managing solid wastes, which hamper decision making for implementing policies directed towards solid wastes.

For (Calderoni, 2003) says the definition and conceptualization of the terms "garbage", "waste" and "recycling" differ according to the situation in which they are applied. Their use in current language indeed stands out from other meanings adopted consonant to the institutional perspective or in accordance with their economic meaning; in current language, the term waste is practically a synonym of garbage. Garbage is all useless material. It designates all discarded material, put in some public location. It is the object or substance considered useless or whose existence in a given environment is considered harmful. Waste is a word often adopted to mean left over from the production, generally industrial, process.

One of the evident problems in Brazil concerns management of urban solid waste, especially when dealing with environmental impacts and the preservation of natural resources. The environmental impact caused by the unbridled production of solid waste has led the government and society to promote studies for alternatives that aim at minimizing the degradation of nature and increasing society's well-being as a whole.

According to (Scholz, 1993), final disposal and treatment of waste can be done using the following methods: landfills (disposal of household waste in the soil); energetic recycling (incineration or burning of hazardous waste, reusing and transforming the energy generated); organic recycling (composting of organic matter); industrial recycling (reuse and transformation of recyclable materials) and steam sterilization and microwave disinfection (treatment of pathogenic, septic and hospital waste).

Solid waste management activities, according to (Cunha & Caixeta Filho, 2002), can be grouped into six functional elements: generation, packaging, collection, transfer/trans-shipment stations and/or processing and/or final disposal.

According to (Machado et al., 2006), to achieve sustainable development, it is necessary to join efforts by all of society, without exclusion of any segments, characterizing the importance of the implementation study and the development of recyclable material collection cooperatives in cities as a whole, specifically at industries.

Even with the creation of measures and procedures that aim at reducing the indiscriminate use of products and actions that harm the environment, the difficulty in disposing of urban garbage is still one of the most serious environmental problems.

3. PET bottle: Recycling and reverse logistics

Recycling involves reintroducing into the system, part of the material and energy that would otherwise be waste. It is an alternative for reducing waste; however, some aspects regarding the benefits and the reuse of recyclable material have to be clarified (Grimberg & Blauth, 1998).

According to (Manzini & Vezzoli, 2005), extending the lifespan of materials means making them last longer than the products they are part of, i.e. materials can be re-processed into secondary raw materials. This is called recycling and can take place through two distinct processes:

a. **Closed loop recycling**: a production system in which the waste or byproduct of one process or product is used in making another product or derived component. In theory, this process can be used for a certain period of time without introduction of virgin materials;
b. **Open loop recycling:** A recycling system in which a product made from one type of material is recycled into a different type of product. This occurs normally in post-consumer materials. In this process, different products and manufacturers may be involved.

Recycling involves reintroducing into the system, part of the material and energy that would otherwise be waste. It is an alternative for reducing waste; however, some aspects regarding the benefits and the reuse of recyclable material have to be clarified (Grimberg & Blauth, 1998).

The main difficulty in maximizing recycling is related to the costs involved in the process, one should take into account the costs from collection, transportation and warehousing to material commercial value.

(Clinton et al., 2010) suggest that the closed-loop recycling is the best environmental option and must also take into account other factors that influence the cost of material collection and treatment. According to (Awaja & Pavel, 2005), in addition to the environmental incentive, recycling of PET as an industry is getting its driving force from the increasing value and applications of virgin and modified PET.

Various technologies have been developed for PET bottles and plastic waste. These methods include (Kawamoto, 2008):

a. material recycling in which the plastic waste is reprocessed into molded products;
b. "bottle-to-bottle" process, in which PET bottle waste is reused as raw material for new PET bottles;
c. a method in which plastic waste is used as a reducing agent in the blast furnace; and
d. a method in which plastic waste is thermally decomposed in the coke oven, producing oil and coke oven gas to be used as industrial raw materials.

Determining an overall recycling rate that takes into account social, economic and environmental aspects is a very interesting strategy for industrial organizations (McBean et al., 2005; Singh et al., 2009; Barboza et al., 2009).

Reverse logistics includes: development of products, inventories, restocking, market withdrawals and excess inventory. It also includes programs for recycling hazardous materials, and disposal of obsolete equipment and resource recovery. It can be said that recycling is the reverse channel of reverse logistics or post-consumer logistics, once it adds value after use and avoids the product life cycle to terminate at the point of the final consumer.

The main perspectives that need to be considered in relation to reverse logistics include not only networking and inventory, but also the collection of post-consumer products, its price, its use, resale and re-manufacturing through an established system. The literature on reverse logistics is, according to (Pokharel & Mutha, 2009) based on three major groups, defined by inputs, structure and processes and outputs.

Reverse logistics is the process of moving post-consumer (end-of-life industrial waste) as well as post-market (returnable due to a manufacturer defect or warranty issues) materials, from its disposal to its reinstatement in the business cycle, without causing major environmental impacts.

The reverse logistics flow must be sustainable, because this process deals with much broader issues than simple returns. The materials involved in this process can be returned to the supplier, resold, refurbished, recycled or simply discarded and replaced.

3.1 Post-consumer PET bottles in Brazil

The PET bottle is a post-consumer waste and its disposal is not regulated by any legislation in Brazil. The reintegration of these products into a new production cycle relies on the initiative of recyclable material collectors who deliver these materials to scrap dealers, who in turn refer to recycling companies.

Market trends and technological advances have spurred new applications for recycled PET. Recycling is used to produce polyester fiber for the textile industry, the manufacture of ropes and bristles of brooms and brushes, shower stalls, thermoforming and vacuum forming machinery, traffic and general signs, which adds value to business. Currently, the largest market for post-consumer PET in Brazil is the production of polyester fiber for the textile industry (Cempre, 2009).

PET is currently present in the most diverse products. Brazil has the following history of consumption/recycling over the past 10 years, as per Table 1:

Years	Recycling (Post Consumer/Índex)
1999	50,0 Ktonnes = 20,4%
2000	67,0 Ktonnes = 26,3%
2001	89,0 Ktonnes = 32,9%
2002	105,0 Ktonnes = 35,0%
2003	141,5 Ktonnes = 43,0%
2004	167,0 Ktonnes = 47,0%
2005	174,0 Ktonnes = 47,0%
2006	194,0 Ktonnes = 51,3%
2007	231,0 Ktonnes = 53,5%
2008	253,0 Ktonnes = 54,8%
2009	262,0 Ktonnes = 55,6%

Table 1. Recycling of post-consumer PET in Brazil from 1999 to 2009, Source: ABIPET (2011)

In 2009, 521,800 tons of PET resins were consumed in Brazil for the manufacture of packaging. From this, 55.6% of post-consumer packaging was effectively recycled. In order to minimize the issue of inappropriate waste disposal in the environment, recycling is encouraged through enterprise associations or individual companies who believe this to be an alternative to preserve natural resources, reduce the cost of processing raw materials and promote social inclusion (Abipet, 2011). Many Brazilian municipalities do not perform any type of collection and few have a selective waste collection system. The bottles are recovered primarily through scavengers, in addition to factories and the selective collection carried out by municipalities.

In 2006, 51.3% of PET consumed in Brazil was reintegrated into the market through recycling, and in 2009, the recycling rate was 55.6%. It is estimated that the recycling rate will continue to grow, and eventually reach that of aluminum cans recycling which is above 98% post-consumer recycling.

According to (Abipet, 2011), PET bottles in Brazil are mostly widely used as packaging material for food, which accounts for approximately 90% of the packaging industry. Most of the recycled PET in Brazil is intended for the textile industry.

It is also observed in Brazil the growth of the applications of PET in apparel, accounting for 44% of recycled PET (Abipet, 2011). The wide use of post-consumer recycled PET in tissue fibers may generate a potential environmental problem, because the material has a short life cycle and could be inappropriately disposed. The most environmentally friendly process to reintegrate this material to its production cycle is the manufacture of new bottles, even for non-food products. The intention in the use of recycled PET is presented by ABIPET (2011) who describes the intention of users to increase or maintain the consumption of resin.

Although Brazil has felt the reflexes of a global economic crisis that began in 2008, no reduction in consumption or recycling pace was observed. However, the PET recycling industry in Brazil has an idleness rate of about 30%, which makes the sector capable of quickly absorbing any volume, without any investment (Abipet, 2011).

Post-consumption recycling of PET guaranteed Brazil second place in the world for recycling the product, as in Table 2.

Countries	Post-consumer PET recycling rate (%)
Japan (2008)	77,9
Brazil (2009)	55,6
Europe (2009)	48,4
Australia (2007)	42,3
Argentina (2008)	34,0
United States (2009)	28,0
Mexico (2009)	18,1

Table 2. Post-consumer PET recycling rate around the world, Source: ABIPET - PET Recycling Census in Brazil 2009-2010, (2011)

In face of the need to reduce residue at a global level, the use of recycled PET is an alternative to increase the useful life of landfills since it is material that takes a long time to degrade in the environment (Coelho et al., 2008).

According to data from (Abipet, 2011), the destination for most recycled PET is the textile industry, as seen in Table 3.

Transformation PET	**2007**	**2008**	**2009**
Textiles	50,5%	38%	39%
Sheet and Laminates	13,1%	15%	15%
Thermoformers	12,0%	-	-
Chemical Resins	6,1%	18%	19%
Exportation	5,8%	-	-
Injection and Blow	4,1%	-	-
Tapes	3,6%	7%	6%
Engineering Plastics	3,2%	-	-
Packaging	1,7%	1%	2%
Others	-	21%	19%

Source: ABIPET – PET Recycling Census in Brazil 2007-2010, (2011).

Table 3. **Destination of PET in Brazil from 2007 to 2009**

The PET bottles in Brazil are mostly widely used as packaging material for food, which accounts for approximately 90% of the packaging industry; however, the destination for the largest volume of recycled PET in Brazil is in the textile industry (Abipet, 2011).

It is believed that the market is adapting and evolving with respect to the use of plastic waste, particularly PET.

The wide use of post-consumer recycled PET in tissue fibers may generate a potential environmental problem, because the material has a short life cycle and could be inappropriately disposed. The most environmentally friendly process to reintegrate this material to its production cycle is the manufacture of new bottles, even for non-food products. The intention in the use of recycled PET is presented by (Abipet, 2011) who describes the intention of users to increase or maintain the consumption of resin.

4. Considerations of the Brazilian PET post-consumer

It appears that effective actions are needed in Brazil to increase PET recycling; however, it is important to emphasize that collection and transportation are important steps when dealing with a process that involves costs. In the case of PET bottles, these moments are critical, since Brazil does not have a legislation requiring the manufacturer or the consumer to appropriately dispose of the post-consumer material. The key players in the process, and specifically the consumer, must be made aware of their duties and responsibilities.

The logistics flow of bottles begins in PET resin production and continues with the manufacture of pre-forms in specific companies or in beverage manufacturing companies. After filling, the product is sent to the retailer where it will be purchased by consumers who will eventually dispose of the bottle. The Brazilian reality of post-consumer PET bottles can be summarized by the individual collection performed by scavengers who survive from the economic activity provided by the trash.

In Brazil, the activity of scavengers is a reality based on the great social inequality. Struggling to survive, the scavenger becomes an important environmental agent, preserving the environment either by cleaning public places and lengthening the useful life of landfills, or by increasing the life cycle of materials (Mota, 2005).

According to data from (Cempre, 2009), in Brazil, approximately 200,000 street scavengers are responsible for collecting various materials. However, few of them live off garbage by choice, or simply to supplement the family income. Many of those living in slums and suburbs work for more than 10 hours and cover more than 12 km / day. Some work with their children and infants, tossing bags of garbage and not having regular meals (Machado, 2006). After collection, the material is sent to intermediaries, which reduces the profit of the scavengers, since it must be resold to its final destination for recycling. As a result of recycling, new products are manufactured as shown in Figure 1.

Analysis of the Brazilian system of post-consumer PET shows the need for changes in the PET bottles reintegration system. It is necessary to take into account the importance of economic, social and environmental impacts of the new proposals for post-consumer logistics in order to organize the reverse channel and increase the amount of material sent for recycling as well as reduce the rate of inappropriate disposal at landfills, dumping areas, rivers, oceans and public roads.

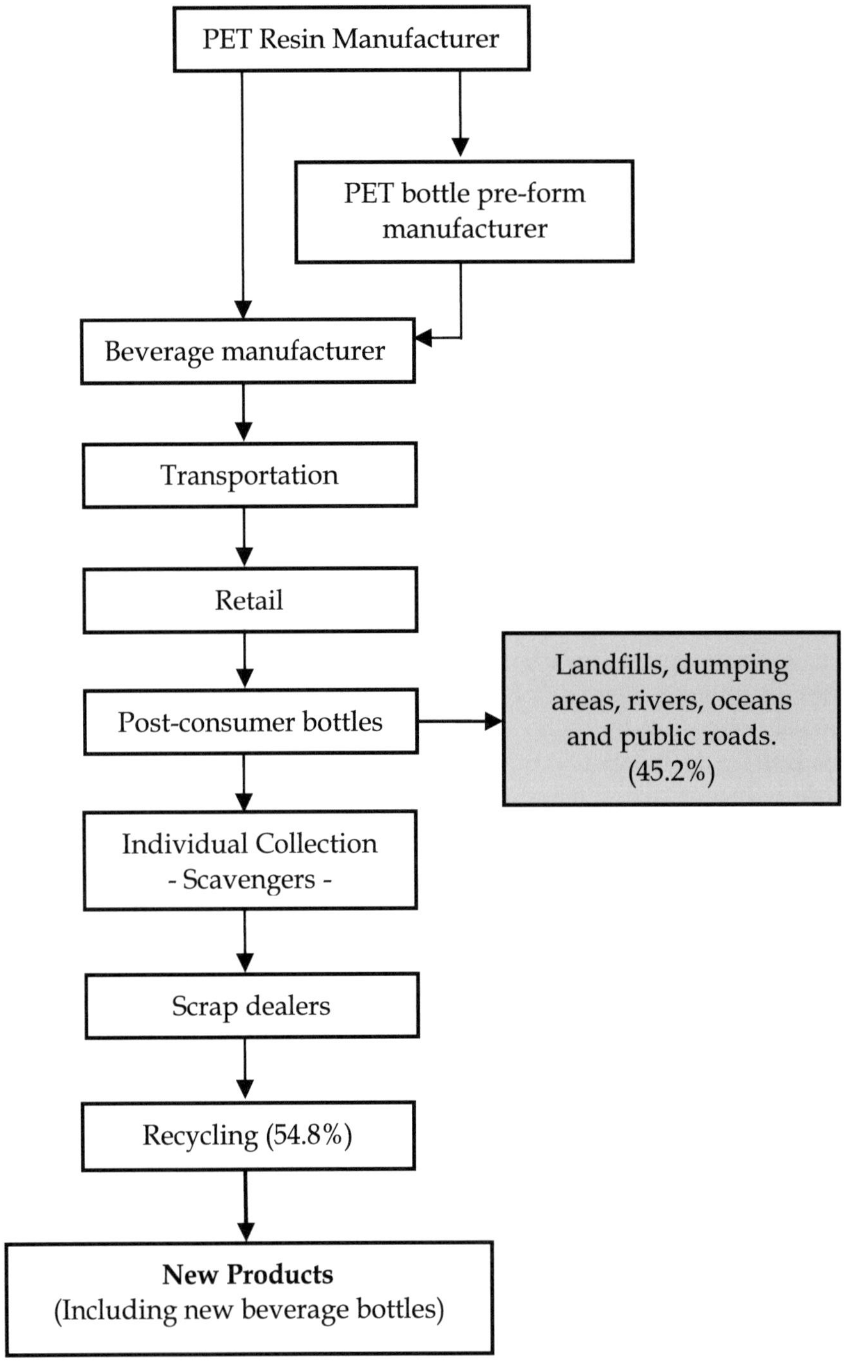

Fig. 1. Post-consumer PET Reintegration System in Brazil

Recycling is considered one of the most important tools to aid waste treatment and disposal, as well as a way of generating employment and income for a growing number of people.

4.1 Environmental aspects

According to (Boustead & Lidgren, 1984), the environmental effect of the PET bottle recycling system could be analyzed from a solid waste, energy use, pollution emissions, water pollution, hygiene, public health impacts and urban traffic congestion perspective.

The most environmentally effective strategy is one using less materials and processes with the least energy (Saphire, 1994).

According to (Boustead & Lidgren, 1984), the environmental effect of the PET bottle recycling system could be analyzed from a solid waste, energy use, pollution emissions, water pollution, hygiene, public health impacts and urban traffic congestion perspective.

The most environmentally effective strategy is one using less materials and processes with the least energy (Saphire, 1994).

In Brazil, the PET bottle is a post-consumer waste which its disposal is not regulated by any laws. The reintegration of the product to a new production cycle relies on the initiative of scavengers or recycling cooperatives who receive recyclable material through selective waste collection.

The boom of PET bottle recycling and the advances in technology, boosts new applications for recycled PET and facilitate the plastic waste removal from the landfill, reducing its impact on the environment. Only 55.6% of post-consumer PET containers were recycled effectively, a rate that can be improved if actions are effectively implemented by the government. Nowadays, recycling is accomplished through different business actions or individual companies. Recycling, on the environmental aspect, contributes to reducing soil, water and air pollution, extends landfill life, improves public sanitation, etc.

A solution to the PET waste is to maximize the reduction of the amount of such waste at the source. In Brazil, there are isolated initiatives by some manufacturing companies, but there isn't a policy established by the government to standardize environmentally friendly procedures for these companies.

4.2 Economics aspects

It is impossible to establish a sustainable product system without dealing with, not only environmental aspects, but also economic impact appropriately.

Recapturing value in recycling is the step of performing the decontamination and appropriateness of the collected material so that it can be used as raw material in manufacturing. The manufacturing industry uses the material and transforms it into another marketable product, the recycled product. The correct equation of the reverse logistics of post-consumer packaging is to make possible the recycling of PET.

The reverse logistics is the process by which recyclable material is collected, sorted and delivered to the industry. This creates an impasse, because who pays for the reverse logistics industry is neither the packaging goods companies nor the packaged goods industry, but society as a whole.

Today, society pays more for not having proper reverse logistics, different from what occurs with other countries - USA, Australia, Japan and all Europe.

PET recycling brings many economic benefits such as: reduction of the volume of garbage collected, which is transported to landfills; electrical power and oil savings, because most plastics are derived from oil; employment generation (scavengers, scrap dealers, workers, etc.); lower prices to the consumer of recycled plastic products; mechanical recycling of PET bottles requires less energy needed to produce the raw material, approximately 30% cheaper than those products manufactured with virgin raw materials; mechanical recycling of plastic beverage packaging (PET) which requires, on average, only 30% of the energy needed for raw material production.

Decreasing the amount PET waste that is sent to landfils reduces the municipality's cost of waste management. In addition, increasing its sorted collection enhances the transport from the retailer to the recycling plant. On the other hand, increasing PET waste transportation promotes increasing costs to the industry since beverage companies are responsibe for transportation of PET waste (Amano, 2004).

4.3 Social aspects

According to (Abipet, 2011), although the recycled PET packaging originates in alternative collection systems done by scavengers and their cooperatives, or by companies dedicated to this task, a large portion of the packaging ends up being sent to dumping areas. This is a common system in Brazil for the disposal of municipal solid waste. The two main reasons for this are a lack of garbage collection, mainly in poorer communities, and a habit of Brazilians for discarding trash into the rivers and streets.

According to (Abipet, 2011), although actually recycled PET packages originate in alternative collection systems, such as garbage pickers and their cooperatives or companies dedicated to that task, an important portion of the packages end up sent to garbage dumps, the most common system in Brazil for disposing of urban solid waste.

This occurs for two main reasons: first due to the lack of garbage collection, because not every Brazilian city has garbage collection for 100% of all domiciles, especially in the poorest districts and shantytowns; second, due to the Brazilian population's habit of getting rid of the package by throwing it in common garbage or, when it is not collected, in the rivers and streets.

The collection in dumping areas results in serious problems. On the one hand, it is possible to observe Brazilian adults and children in unhealthy and sub-human conditions scavenging for food, subjecting themselves to different forms of contagion and infection in an effort to find products to sell. Moreover, the product collected in landfills is heavily contaminated with grease, paints, heavy metals and dirt in general. The contaminated package requires a more expensive cleaning process, which devalues the package.

Another relevant point is that the collector receives a reduced value for the material, either because the product is contaminated, or by action of an intermediary who resells the product to the recycling industry.

In Brazil, alternative collection systems for PET packaging carried out by scavengers and cooperatives, contribute to recruiting unskilled people. By discarding the waste properly, society contributes to preserving the environment and to achieving continuous improvement in quality of life. This enables the development of standards and working conditions as well as the development of social responsibility in products and services.

System limitations focus on the consumer, who is still not fully informed about the possibility of recycling and, consequently, the economic value of the post-consumption PET bottle; with that, packaging ends up getting disposed of in common garbage. Furthermore, the lack of efficient selective collection systems impedes bottle recovery, which end up lost in landfills and garbage dumps.

5. Proposal for a Brazilian system for post-consumer PET

It appears that effective actions are needed in Brazil to increase PET recycling; however, it is important to emphasize that collection and transportation are important steps when dealing with a process that involves costs. In the case of PET bottles, these moments are critical, since Brazil does not have a legislation requiring the manufacturer or the consumer to appropriately dispose of the post-consumer material. The key players in the process, and specifically the consumer, must be made aware of their duties and responsibilities.

The logistics flow of bottles begins in PET resin production and continues with the manufacture of pre-forms in specific companies or in beverage manufacturing companies. After filling, the product is sent to the retailer where it will be purchased by consumers who will eventually dispose of the bottle. The Brazilian reality of post-consumer PET bottles can be summarized by the individual collection performed by scavengers who survive from the economic activity provided by the trash.

In Brazil, the activity of scavengers is a reality based on the great social inequality. Struggling to survive, the scavenger becomes an important environmental agent, preserving the environment either by cleaning public places and lengthening the useful life of landfills, or by increasing the life cycle of materials (Mota, 2005).

According to data from (Cempre, 2009), in Brazil, approximately 200,000 street scavengers are responsible for collecting various materials. However, few of them live off garbage by choice, or simply to supplement the family income. Many of those living in slums and suburbs work for more than 10 hours and cover more than 12 km / day. Some work with their children and infants, tossing bags of garbage and not having regular meals (Machado, 2006). After collection, the material is sent to intermediaries, which reduces the profit of the scavengers, since it must be resold to its final destination for recycling.

Analysis of the Brazilian system of post-consumer PET shows the need for changes in the PET bottles reintegration system. It is necessary to take into account the importance of economic, social and environmental impacts of the new proposals for post-consumer logistics in order to organize the reverse channel and increase the amount of material sent for recycling as well as reduce the rate of inappropriate disposal at landfills, dumping areas, rivers, oceans and public roads.

Recycling is considered one of the most important tools to aid waste treatment and disposal, as well as a way of generating employment and income for a growing number of people.

The discussion on the current reintegration system for post-consumer PET in Brazil indicates the need for a new post-consumer logistics proposal, aiming at economic, social and environmental opportunities that will provide a sustainable reverse system. Figure 2 proposes three distinct paths for post consumer PET bottles in the new reverse system.

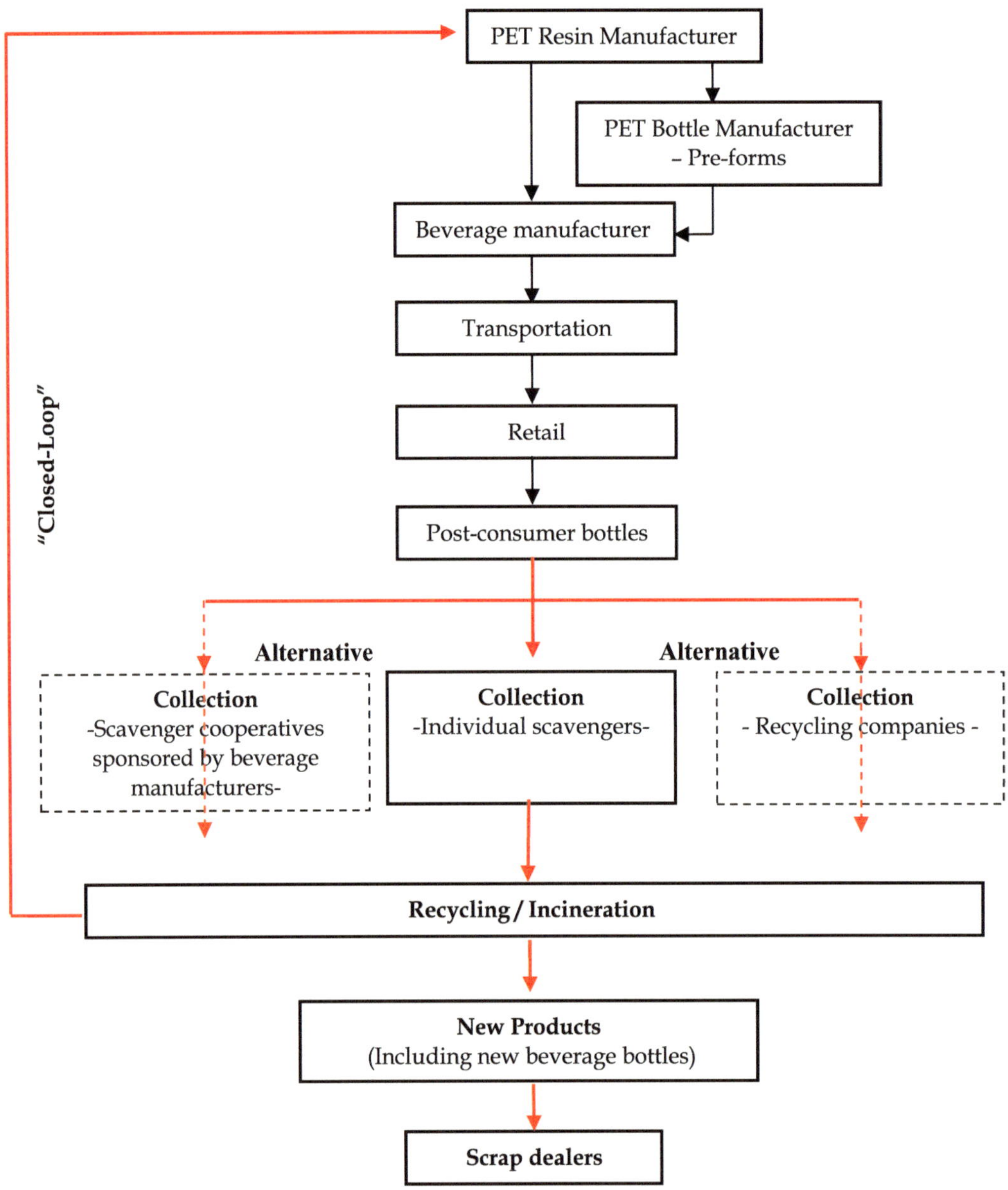

Fig. 2. Proposal for a reintegration system for post-consumer PET bottles in Brazil

The challenges of this new logistics flow for post-consumer PET bottles are concentrated in the pursuit of the following:

a. public policies concerning the responsibility for waste generated in the production process;
b. the development of strategies for the appropriate destination of PET bottles (transport mechanisms, collection, recycling and recovery of material) for the introduction of cooperatives;
c. social mainstreaming tools; and
d. the assignment of the responsibility in each cycle (production, purchase, consumption and disposal) in order to reduce the use of raw materials, waste and disposal.

All of these collection mechanisms are aimed to remove the post-consumer PET waste and send them for recycling. According to (Chilton et al., 2010), the recycling of PET results in a net reduction in the emissions of CO_2, carbon monoxide, acid gases, particulate matter, heavy metals and dioxins. This is due to the reduction in the emissions associated with manufacturing an equivalent mass of PET from virgin materials.

In the proposed logistics flow, the post-consumer bottles are obtained through: selective collection performed by scavenger cooperatives subsidized by the municipality or by the beverages manufacturers; spontaneous participation or imposed by environmental legislation; door-to-door collection (residences and points of consumption) performed by individual scavengers; and retail collection by recyclers. Collection characterization:

a. **By individual scavengers:** it is expected that 30% of the post-consumer PET bottles sent for recycling will return through these door-to-door scavengers, since they need this activity as a means of income. A successful collection relies on consumer awareness and the ability to sort out these materials.
b. **By scavenger cooperatives subsidized by the municipality or by the beverages manufacturers:** assuming that the company is responsible for the waste generated through the production process, it is possible to consider the establishment of scavenger cooperatives as a way to organize the activity and incurring gains for their own production without being exploited. This enables the development of social and health projects by achieving a better quality of life. Besides the economic and social issues related to cooperatives, the environmental aspect is also very important, because the cooperatives contribute to reducing the environmental impact of consumption. It is important that the government encourages the establishment of scavenger cooperatives, formalizing a longstanding marginalized activity, assisting with the allocation of infrastructure investments for the recovery of citizenship in the segment. In the proposed reverse system, the cooperatives are subsidized by the public and private sectors (beverage manufacturers). Investment from private companies that use PET for packaging their drinks are claimed by public policies. In this case, it is expected that 50% of the post-consumer PET bottles sent for recycling will return through these cooperatives.
c. **By recyclers:** these are companies that will benefit directly from the recycling of PET bottles. The collection points at retail would be an opportunity to expand the number of post-consumer PET bottles collected. In this case, it is expected that 20% of the post-

consumer PET bottles sent for recycling will remain the responsibility of these players. The success of the activity also relies on the consumer to play its part.

All of these collection mechanisms are aimed to remove the post-consumer PET waste and send them for recycling. According to (Chilton et al., 2010), the recycling of PET results in a net reduction in the emissions of CO_2, carbon monoxide, acid gases, particulate matter, heavy metals and dioxins. This is due to the reduction in the emissions associated with manufacturing an equivalent mass of PET from virgin materials.

Another option for the final destination of post-consumer PET bottles is incineration, because this material is capable of generating energy. According to (Chilton et al., 2010), waste PET can be recovered by collecting it with the non-recyclable household waste followed by burning it in an energy-from-waste incineration plant.

Once the PET waste is capable of generating power, the possibility exists that the collected material will be sold directly to factories to be used as a power generation source. This destination somehow does not interfere with the collection system. In addition to recylcling, it is just one more option of an appropriate destination for the material.

Jointly held shares enable the socialization of costs and contributes directly to the environment by generating other products. There is also the opportunity to transform recycled materials into new food packaging that is regulated by ANVISA.

In the proposed post-consumer logistics system, it is essential that the duties and responsibilities are divided among the various key players involved in the recycling chain. System maintenance is done through private sector investments and government intervention at the municipal, state and federal levels through public policies.

6. Conclusion

The data presented enabled the identification of the need for effective actions, aimed at increasing the amount of recycled PET. In Brazil, it is necessary to focus on post-consumer PET bottles, because the ones that are not recycled are inappropriately disposed in the environment. The need for structuring the distribution reverse channels becomes relevant, since there are limiting factors in relation to increasing the levels of PET bottle recycling.

According to (Manzini & Vezzoli, 2005), the collection and transportation stages should not be underestimated, either in terms of planning reverse logistics or in relation to environmental impact. Often, this stage affects the economy and the environmental benefits of recycling. In the case of PET bottles, the stage of recycling is fundamental, since there isn't a legislation requiring the manufacturer or the consumer to appropriately dispose of post-consumer material. The key players in the process, and specifically the consumer, must be made aware of their duties and responsibilities

The analysis of the Brazilian system of post-consumer PET bottles under environmental, economic and social aspects favored the proposal for a reintegration system for post-consumer PET with opportunities for the development of a sustainable reverse system. It also presents several challenges such as: the need for a national policy on solid waste and municipal and state actions to make the reverse logistics viable and strengthen the Brazilian recycling industry; the need to increase awareness of the key players involved in the post-consumer PET reverse chain structure; the need for consumption reduction in order to reduce the waste generated; and the need to engage industrial sectors and the government, through public policies, to support cleaner technologies along the PET bottle production chain.

Some of the difficulties related to the inclusion of the reintegration system for post consumer PET are: the costs related to the total life cycle, as the company becomes liable for its product up to the end of its lifespan; a greater integration of the production chain, assigning responsibilities and the degree of cooperation in the chain; development of larger reverse logistics management controls and information systems to facilitate the integration of reverse logistics to the normal flow of distribution.

By implementing this system, the company will be able to account for the environmental costs of their products in order to comply with environmental laws and minimize the environmental impact generated by their products. It will also allow value to be recaptured and be strategically used as a competitive edge, demonstrating the company's commitment to a sustainable development.

7. References

_____. NBR 10004. Resíduos sólidos - classificação. Rio de Janeiro, 2004.

Abipet Associação Brasileira dos Fabricantes de Embalagens PET. Reciclagem de Embalagens PET. Disponível em: http://www.abipet.com.br. Acesso: 15 ago. 2011.

Associação Brasileira de Empresas de Tratamento, Recuperação e Disposição de Resíduos Especiais - ABETRE. Disponível em: <http://www.ambientebrasil.com.br/composer.php3?base=./residuos/index.php3&conteudo=./residuos/lixo.html>. Acesso em 18 fev. 2008.

Associação Brasileira de Embalagem - ABRE. Disponível em <http://www.abre.org.br/meio_reci_brasil.php>. Acesso em 08 fev. 2010.

Amano, M. PET Bottle System in Sweden and Japan: an Integrated Analysis from a Life-Cycle Perspective. Lund, Sweden; 2004.

Awaja F., PAVEL, D.. Review Recycling of PET. European Polymer Journal 2005 41: 1453-1477

Barboza ES, Lopez DR, Amicoa SC, Ferreira CA . Determination of a recyclability index for the PET glycolysis. Resources, Conservation and Recycling, 2009 53: 122-128

Boustead, I; Lidgren, K. Problems in Packaging - the Environmental Issue. Wiley & Sons, Incorporated, John, 1984..

Brasil. Política Nacional de Resíduos Sólidos. Ministério do Meio Ambiente: Brasília, versão II, 2002. Disponível em: <www.mma.gov.br>. Acesso em: 10 jun. 2010.

Calderoni, S. Os bilhões perdidos no lixo. São Paulo: Humanitas, 2003.

Cavalcanti, C. Desenvolvimento e natureza: estudos para uma sociedade sustentável. São Paulo: Cortez, 1995.

Cempre - Compromisso Empresarial para a Reciclagem. Disponível em: <http://www.cempre.org.br>. Acesso em: 29 nov. 2009.

Chilton T, Burnley S, Suresh N. A life cycle assessment of the closed-loop recycling and thermal recovery of post-consumer PET. Journal Resources, Conservation and Recycling, 2010.

Coelho, T.M; Castro, R; Batistelle, R.A.G. Analysis of the post-consumption reverse channel of the pet bottle in brazil. POMS 20th Annual Conference. In: Anais..., Orlando, Florida U.S.A., 2009

Cunha, V., Caixeta Filho, J. V. Gerenciamento da coleta de resíduos sólidos urbanos: estruturação e aplicação de modelo não-linear de programação por metas. Gestão & Produção. São Carlos, v. 9, n. 2, ago. 2002.

Grimberg E, Blauth P. Coleta Seletiva reciclando materiais, reciclando valores. São Paulo: Polis, 1998.

Kawamoto K. Waste Recycling Technologies Required by a Sound Material-Cycle Society. Quaterly Review 2008; 27.

Machado, BA, et al. A importância social e econômica da implementação de cooperativas de materiais recicláveis. Associação Brasileira de Engenharia de Produção. XXVI ENEGEP Fortaleza, 2006. Disponível em: <http://www.abepro.org.br/biblioteca/ENEGEP2006_TR560372_8549.pdf>. Acesso em 25 maio. 2010.

Mahmood, N. Z.; Victor, D. Policy approach in life cycle of solid waste management in Malaysia. Life Cycle Management. In: 1st International Conference on Lyfe Cycle Management. Conpenhagen, Denmark, p .301-4, aug. 2001.

Mcbean EA, Del Rosso E, Rovers FA. Improvements in financing for sustainability in solid waste management. Journal Resources, Conservation and Recycling 2005 43:391-401.

Mota, AV.; Do Lixo à Cidadania. Artigo. Democracia Viva, n. 27. Instituto Brasileiro de Análises Sociais e Econômicas. Rio de Janeiro. 2005. Disponível em: < http://www.ibase.br/pubibase/media/dv27_artigo1_ibasenet.pdf>. Acesso em 25 mai. 2010.

Nilsson I. Integrating environmental management to improve strategic decision-making', Chalmers University of Technology Report, XX Environmental System Analysis 2001.

Onusseit, H. The influence of adhesives on recycling: A review. Journal Resources, Conservation and Recycling 2005 (46), 168-181.

Saphire, D. Case Reopened - Reassessing Refillable Bottles. Inform Strategies for a better environment: New York, 1994.

Singh RK, Murty HR, Gupta SK, Dikshit AK. An overviewof sustainability assessment methodologies. Ecol Indicators 2009;9:189–212.

Scholz, L.C. Coleta, tratamento e disposição final: problemas e perspectivas. In: São Paulo. Secretaria do Meio Ambiente. Coordenadoria de Educação Ambiental. Resíduos Sólidos e Meio Ambiente. São Paulo: Pini, 1993.

Tenório, J.A.S; Espinosa, D.C.R. Curso de gestão ambiental. Barueri, SP: Manole, 2004.

Section 3

Recycling Process and Optimal Production

10

Modelling of Recycling in LCA

Tom N. Ligthart and Toon (A.)M.M. Ansems
TNO, Utrecht,
Netherlands

1. Introduction

1.1 Life Cycle approach and life cycle stages

In the Life Cycle approach Life Cycle Assessment (LCA) is a systematic way to evaluate the environmental impact of systems, products, materials or activities by following a "cradle-to-grave" (a complete product chain) approach. This approach implies the identification and quantification of emissions and material and energy consumptions which affect the environment at all stages of the entire product life cycle.

The possible reasons for executing an LCA are for instance:

- To depict in as detailed as possible, the interaction of a product or activity with the environment
- To ensure a thorough insight into the interdependent nature of the environmental impacts of a human activity
- To supply decision-makers with information on the environmental impacts of activities and the possibilities for improvements

The entire product life cycle, shown systematically in Figure 1, consists of different stages:

- Extraction of raw materials
- Production of materials/semi-fabricates
- Manufacturing of product
- Service life or use phase
- End-of-Life

During all stages of the life cycle impacts exist on the surrounding environment. These impacts are caused by extracting raw materials and by emissions to air, water and soil. In LCA often several impact categories are used. The widely used Centrum voor Milieukunde Leiden/Institute of Environmental Sciences (CML) methodology (Guinee et al., 2002) for life cycle impact assessment uses the following categories:

- Abiotic depletion (ADP)
- Global warming (GWP)
- Ozone layer depletion (ODP)
- Human toxicity (HTP)
- Freshwater aquatic ecotoxicity (FAETP)

- Marine aquatic ecotoxicity (MAETP)
- Terrestrial ecotoxicity (TETP)
- Photochemical oxidation (POCP)
- Acidification (AP)
- Eutrophication (EP)
- Land competition (LC)

With the help of Life Cycle Assessment (LCA) methodologies these contributions can be quantified and aggregated as will be explained in the following section.

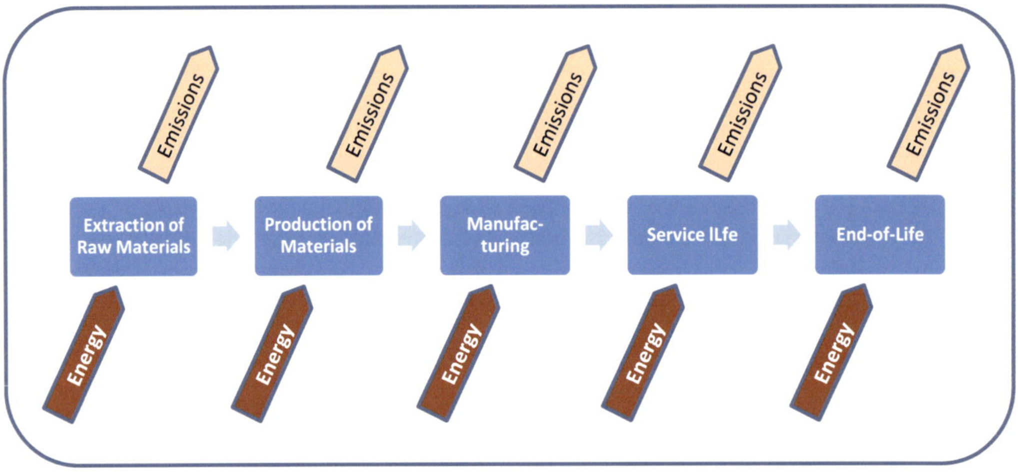

Fig. 1. The full Life Cycle of a product.

1.2 Life Cycle Assessment

The method of environmental Life Cycle Assessment (LCA) is seen as a suitable instrument for the evaluation of the environmental impacts of a product or an activity through its entire life cycle. LCA is a systematic way to evaluate the environmental impacts of product system or activities by following a "cradle-to-grave" approach. The product system consists of a set of activities (processes), all focused on the fulfilment of the required function. These activities can be executed at different places and in different time periods. Therefore it is not possible to produce results, which refer to "real" environmental effects, since this requires specific locations and specific periods.

The result of a life cycle assessment is expressed in terms of "potential" effects. These potential effects are indicators for the real effects on local, regional and global level.

The LCA methodology is structured along a framework (see Figure 2) with four main steps or phases (International Standard Organisation, 2006):

1. Goal and scope definition
2. Inventory analysis
3. Impact assessment
4. Interpretation

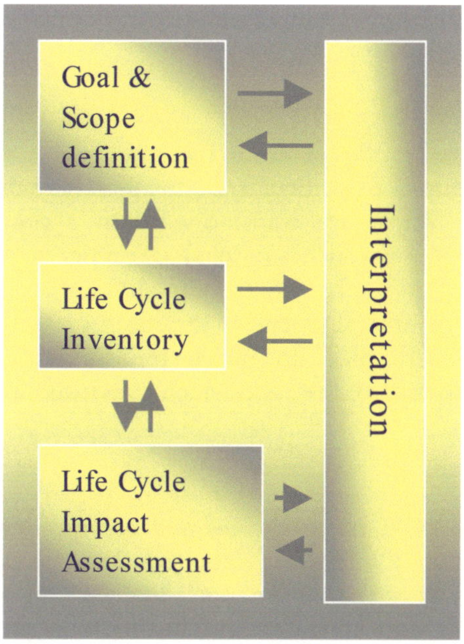

Fig. 2. The framework of LCA.

These phases are part of an iterative process; the main flow is according to the sequence mentioned before.

1. *Goal and scope definition*

This deals with the clear and unambiguous formulation of the research question and the intended application of the answer that the LCA study is supposed to provide. Important elements of the goal and scope definition are the choice of the functional unit, the selection of product (material) alternatives to be analysed, and the definition of the reference flows for each of the alternative systems.

2. *Inventory analysis*

This phase is concerned with the construction of the product systems. These systems are composed of unit processes, like industrial production, waste treatment, transport and so on.

The system boundaries and flow charts of linked unit processes are drawn for each alternative product system, and quantitative input and output data for each unit process are collected, e.g. raw materials and energy use figures, as well as emissions and waste amounts. Furthermore qualitative data for representativeness, data quality etc. are collected during this phase. For those unit processes that are multifunctional, i.e. that produce more than one product, an allocation step is made: all input and output data of the unit process are allocated to each of the products, according to chosen rule (e.g. on basis of mass ratio or economic value). A final step of the inventory analysis is the aggregation of the emissions of chemicals and the extractions of natural resources over the entire product system, in such a way that a quantitative match with the system's reference flow is achieved. The result of the inventory analysis is often a long list with disparate entries, such as carbon dioxide, nitrogen oxides, chloromethane and mercury.

3. *Impact assessment*

This phase aims to convert and aggregate the results of the inventory analysis into environmentally relevant items. In particular, we mention here the step of characterisation, in which the inventory results are transformed into a number of contributions to environmental impact categories, such as global warming, acidification, and ecotoxicity. Optionally the characterisation results may be normalised in order to relate the results to a reference value, such as the annual global or European extent of each impact. Finally, a weighting step may be performed, in which priority weights are assigned to the characterisation or normalisation results, and which may result into one final score for each alternative product system.

4. *Interpretation*

During the course of the LCA, many choices and assumptions are needed. Moreover, uncertainty may be introduced with every data item. The interpretation phase deals with the meaning and robustness of the information obtained and processed in the previous phases. The interpretation may include comparisons with previously published LCA studies on similar products, uncertainty and sensitivity analyses, data checks, external comments, and much more. It is also the place in which a final judgement and decision is outspoken.

1.3 Recycling in LCA methodology

1.3.1 Recycling as part of the life cycle

In 1.1 the different stages of the life cycle are illustrated and described. The End-of-Life stage can be filled in by following different ways and a hierarchy is followed here. The most optimal way for the End-of-Life is enlargement of the life time: Reuse. When this happens one speaks about a second life and probably the same function still exists. In this chapter we will not further discuss the effects of re-use.

The next most optimal way for the End-of-Life will be recycling with application in the same product (or group of products) without losses in quality of the observed material. It means that the physical and chemical properties of the materials or substances stay at the same quality level. In this case a closed loop recycling is considered. More on the types of recycling that are distinguished in LCA is discussed in section 1.4.

The lowest levels in the End-of-Life hierarchy are incineration, with or without energy recovery, and landfill. In this paper no attention is paid to these alternatives, however the different aspects of recycling are analysed more in depth and evaluated.

In society recycling is favourable and stimulated, because it saves primary resources and offers solutions for scarcity of raw materials. Moreover, resource efficiency mostly results in a strong reduction of the environmental impact of a product system. From that point of view recycling is stimulated by several stakeholders. In Figure 3 the specific role of recycling in the life cycle is highlighted.

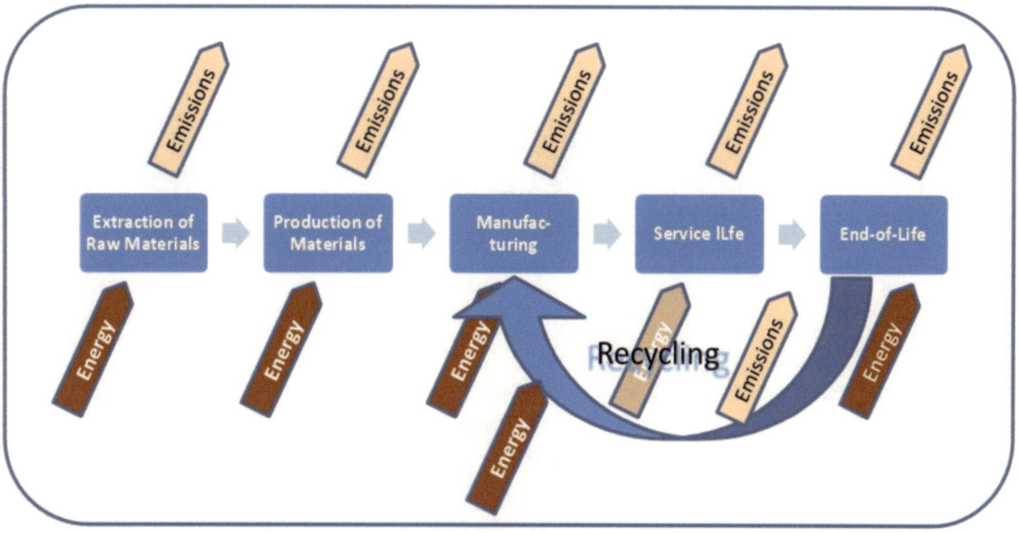

Fig. 3. The situation and place of recycling in the life cycle.

1.3.2 The difficulties of introducing recycling into LCA

Laymen to LCA often assume in a straight forward way that products are totally manufactured out of primary resources and that recycled resources, secondary resources, are applied in other products. In their view closed loop recycling does not occur. However, in practice often mixtures of primary and secondary resources are used in new products and also at the End-of-Life stage several ways of treatments exist. Combinations of closed loop recycling, open loop recycling and incineration will be applied for certain end-of-life flows.

The difficulties are now to set the right boundaries for the different flows, ending in different product systems. Which observed material flow belongs to the first observed product system and which one to the second or other observed systems? Another question is which rules or scientifically based calculation rules exist to apply broadly accepted divisions over the different product systems.

The theoretical aspects and answers of these aforementioned questions are described in the following sections.

2. Modelling of recycling in LCA

Recycling processes can be divided based on where and how the recycled material is used again:

- Closed loop recycling
- Open loop recycling
- Semi-closed loop recycling

The three schemes are shown in Figure 4 and now explained in further detail.

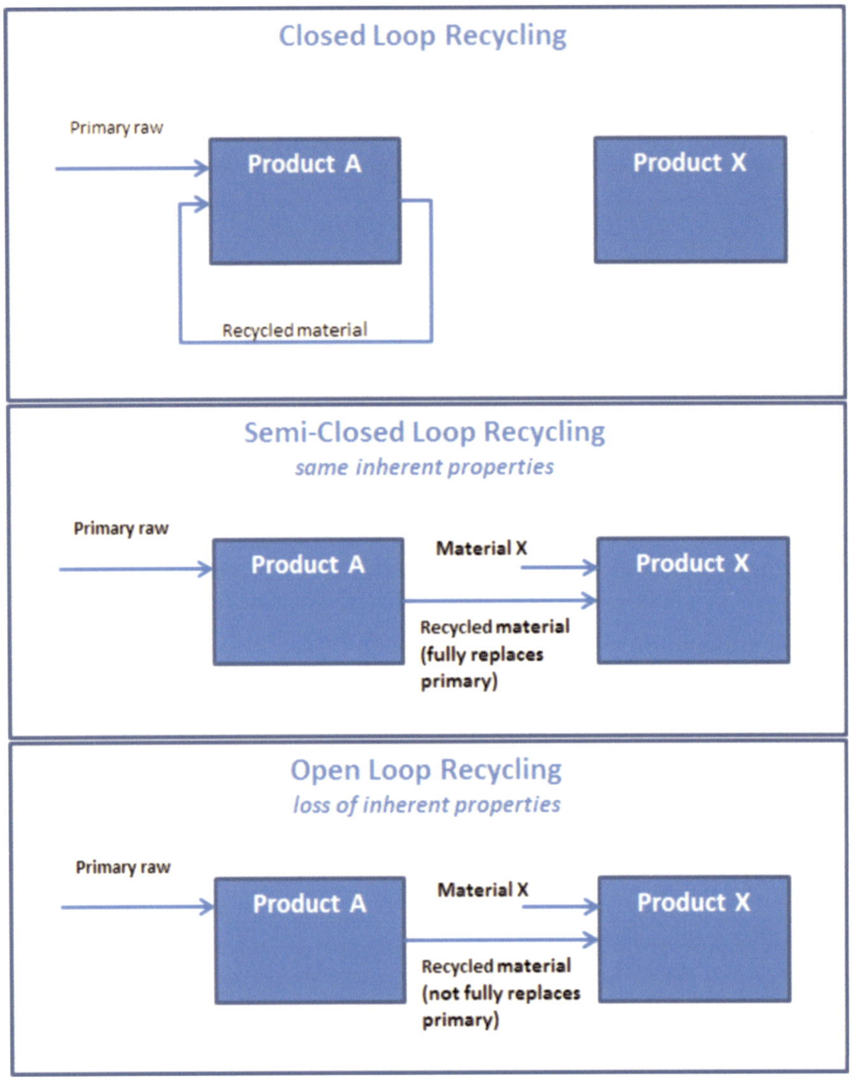

Fig. 4. The three recycling schemes.

Closed loop recycling

This occurs when the materials associated with a product are recycled and used again in the same product system. The material properties are not changed in comparison to the original primary material. Bottle-to-bottle recycling is an example of closed loop recycling.

Open loop recycling

Occurs when the recycled material goes to another product system and the inherent material properties are changed to such an extent that the recycled cannot be used in its original system. Plastic recycling is open loop recycling when the recycled plastic from i.e. packaging is recycled into other products like e.g. plastic poles. This type of recycling is also called down cycling as there is loss of quality.

Semi-closed loop recycling

When the recycled material in used in another product system, without the material's inherent properties undergoing any change. This is the case when construction steel is recycled into e.g. steel plates for ship building.

Most LCA practitioners e.g. Guinee, 2002 and the ISO 14044 standard (ISO, 2006) take semi-closed loop recycling the same as closed loop recycling (see Figure 5).

Most often a product system does not fully recycle all material that comes available after use. Even if this was the case the recycling process will normally lead to some loss of material and primary material is needed to keep the circle closed. The efficiency of the end-of-life system can be described by the following indicators (Norgate, 2004):

- Collection rate (%) $= 100 * \frac{\text{Amount of scrap collected}}{\text{Amount of scrap produced}}$
- Recovery rate (%) $= 100 * \frac{\text{Amount of scrap recovered}}{\text{Amount of scrap available}}$
- Recycling efficiency (%) $= 100 * \frac{\text{Amount of scrap reprocessed}}{\text{Amount of scrap recovered}}$
- Recycling rate (%) $= 100 * \frac{\text{Amount of scrap reprocessed}}{\text{Amount of scrap available}}$

  $= 100 * \text{Recovery rate} * \text{Recycling efficiency}$
- Recycled content (%) $= 100 * \frac{\text{Amount of scrap reprocessed}}{\text{Total Amount of material}}$

The place of these indicators in the life cycle of a system are shown in Figure 6.

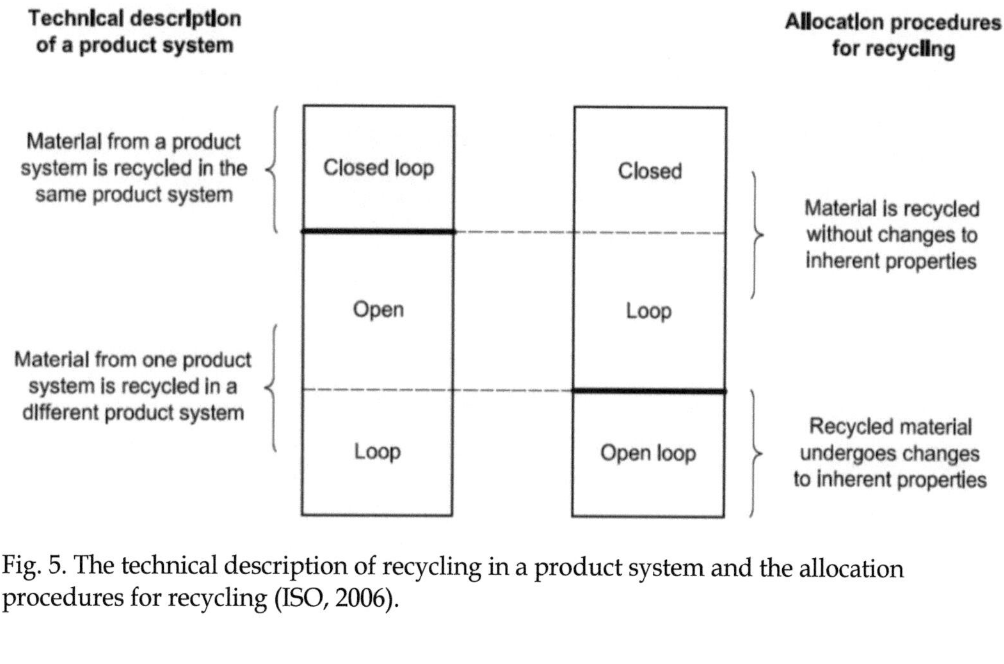

Fig. 5. The technical description of recycling in a product system and the allocation procedures for recycling (ISO, 2006).

2.1 Allocation for recycling processes

When at the end-of-life of a product system post-consumer waste occurs allocation may become necessary. This as the product system under study can provide waste materials that are used by other systems as a raw material. The question is then how to divide the impacts of e.g. recycling over the system that delivers the waste material and over the system that uses this material as input for a secondary material. This problem especially occurs in open loop recycling.

ISO 14044 (ISO, 2006) describes that the inputs and outputs shall be allocated to the different products according to clearly stated procedures that shall be documented and explained together with the allocation procedure. Whenever several alternative allocation procedures seem applicable, a sensitivity analysis shall be conducted to illustrate the consequences of the departure from the selected approach.

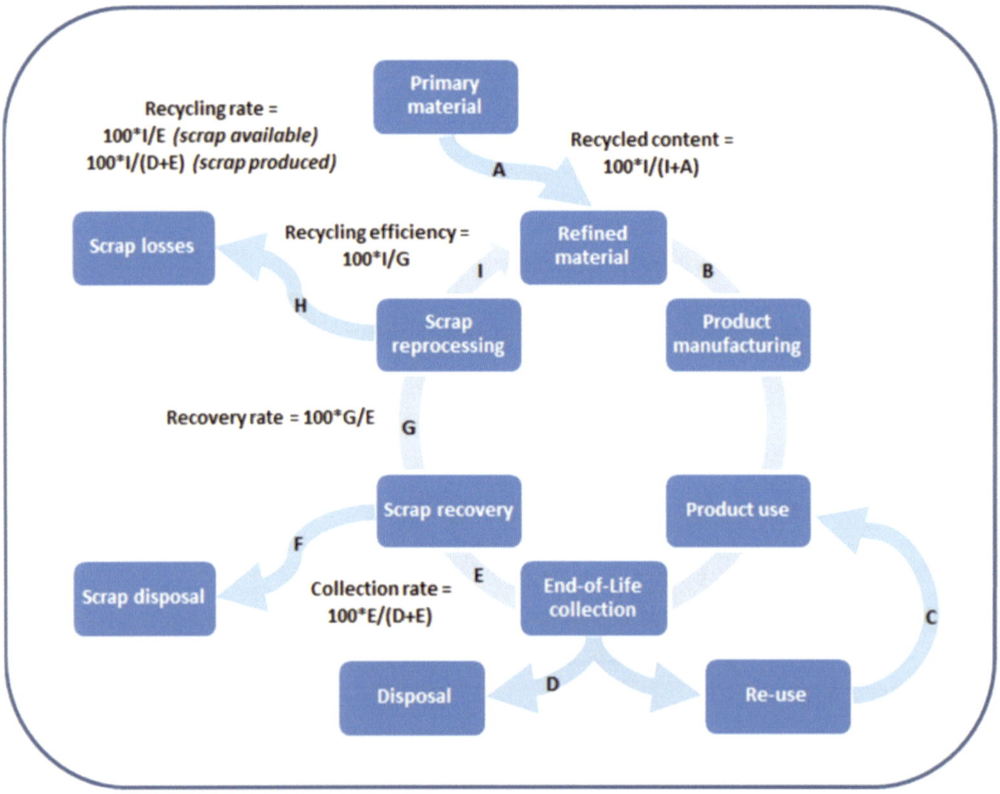

Fig. 6. Product system and recycling indicators (after Norgate, 2004).

So, what allocation procedures are applied in practice? The most common are:

1. Cut-off
2. System expansion
3. Economic allocation
4. Input oriented

5. Output oriented/Substitution
6. Value-corrected substitution
7. Multiple recycling method

These allocation procedures will now be explained and discussed.

Cut-off

One of the easiest to apply is the cut-off method (see Norgate, 2004 and Ekvall & Tillman, 1997). In this method the environmental impacts **directly** caused by the production of a product are assigned to that product, i.e. primary and secondary production are allocated to the product under study. Also an eventual waste treatment, other than recycling, is allocated to the product. No data from outside the life cycle of the investigated product are needed for this approach (see Figure 7). The extensive ecoinvent database with life cycle inventory data also uses this approach for recycling and recovery (Hischier et al., 2010).

In the cut-off approach the recycled content is of importance as this may be a way to reduce the environmental impact of the system as often the recycled material has a lesser impact than the primary material. An increase in collection rate is also beneficial as this reduces the impact of the waste disposal. The cut-off is seen to advocate the use of a high recycled content (Frischknecht, 2010).

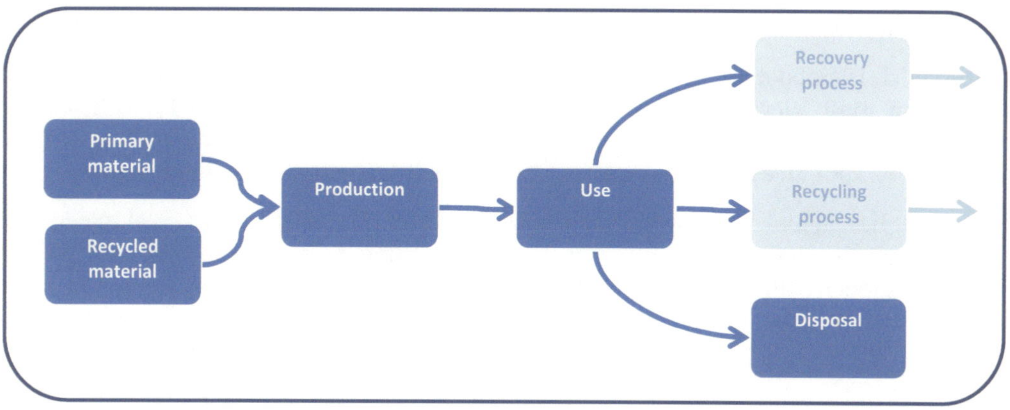

Fig. 7. The product system for the cut-off approach.

System expansion

ISO 14044 prefers to avoid allocation by expanding the system so that in fact a closed system comes into being. The aluminium industry for instance states (European Aluminium Association [EAA], 2010) that as aluminium recycling leads to no loss in inherent properties the recycling can be seen as closed loop recycling and only the replenishment of primary aluminium needed to balance the eventual losses of aluminium from the system has to be added to give a closed system (see Figure 8).

For more complex product systems application of system expansion can increase the complexity of the system by adding more functions to the system and it may even lead to new allocation issues (see e.g. Zamagni, 2008). Allocation, instead of system expansion, thus often applied in LCA studies.

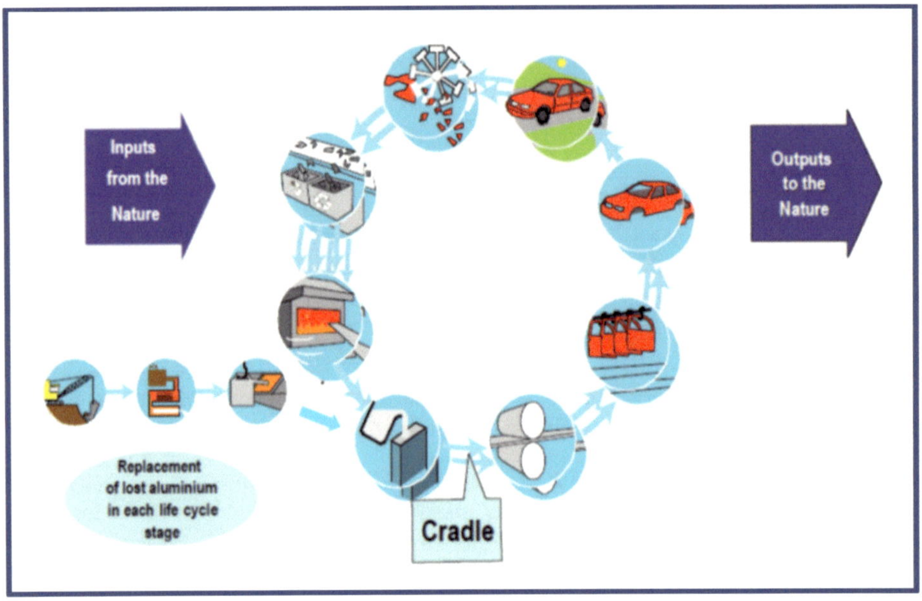

Fig. 8. Closed system of aluminium in products (EAA, 2010).

Economic allocation

In the Dutch Handbook on LCA (Guinee, 2002) economic allocation is advised as a baseline method for most allocation situations, including recycling, in a detailed LCA. Economic allocation is based on the economic value of the inputs and outputs of a recycling system. Guinee et al., 2004 give an example of the recycling of aluminium from a used engine.

Let us first have a look at an open loop recycling system. The used engine, used material in Figure 9, yields €100 for the collection and dismantling process. Another yield of this process is the scrap that is sold onto the market. The reprocessed material is used in other products than engines. The scrap yields €150, which brings the total yield of the process to €250.

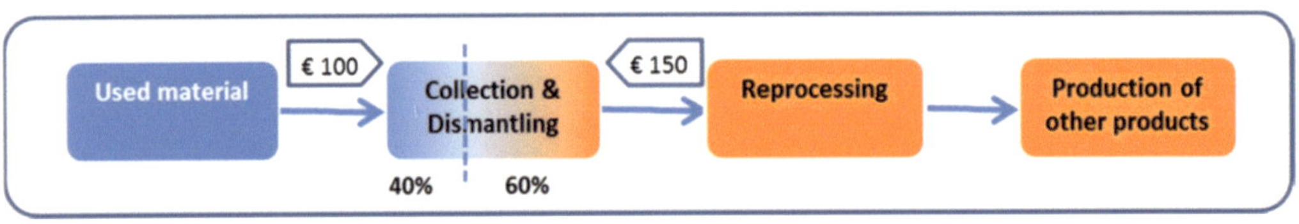

Fig. 9. Example of economic allocation of inputs and outputs of an open loop recycling system (based on Guinee et al., 2004). The 'used material' system is based on its economic value allocated 40% of the impact of the collection and dismantling.

As the used material contributes to 40% (100/250) of the collection and dismantling process, 40% of the environmental burden of the process is allocated to the system that delivers the used material. The remaining 60% is allocated to the system that subsequently uses the secondary material. This economic allocation seeks in other words where in the end-of-life the economic flows are in equilibrium or where the point of zero value lies.

In case of a (semi-)closed loop recycling scheme, shown in Figure 10, reprocessed material (partly) flows back into the original product system. Let us assume that 1 of the 5 kg material used flows back. Based on the economic allocation 40% of the collection and dismantling process is allocated to the original product system. However, also 20% (1 kg from 5 kg) of the remaining 60%, resulting in 12%, that is allocated to the reprocessed material must be added as this flows back to the original system. In total 40%+12% = 52% of the collection and dismantling process is now allocated to the original product system.

This approach of economic allocation can be seen as a modification of the cut-off method of allocation. It has in common with it that no avoided primary material production due to recycling is allocated to the product system under study.

A special situation occurs when the waste material has a positive economic value. In this case the point of zero value cannot be found. The waste material can then be seen as a co-product of the system. The relative value of the waste material can then be the basis for subtraction of part of the environmental impact from the product system. This specific case will not be discussed further in this chapter as it in practice not often occurs.

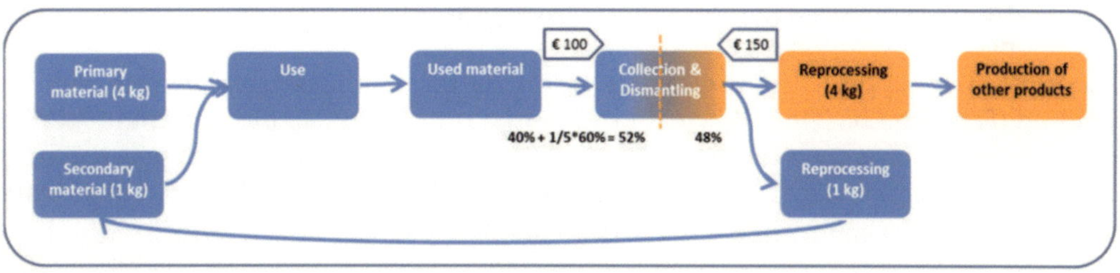

Fig. 10. Example of economic allocation of inputs and outputs of a semi-closed loop recycling system (based on Guinee et al., 2004). The used material is based on its economic value allocated 40% of the impact of the collection and dismantling.

Using economic allocation reflects that product systems are part of the economic system. As product systems are generally designed to generate economic value using economic data for allocation seems appropriate. However, the market of waste treatment may be a market distorted by regulations and fees asked for waste treatment may not always reflect real market prices. This can lead to not estimating the point of zero value fully correct.

Input oriented

The input oriented approach only takes the inputs of primary and secondary, including collection and further recycling steps, into account plus the waste disposal. The output of material going to recycling is not considered to be part of the product system. It is the same approach as the cut-off allocation (see Figure 7).

Output oriented, Substitution or Avoided burden/impact

This allocation method is based upon the substitution of primary raw material by the reprocessed (secondary) material at the end-of-life stage. It is also called the substitution method or avoided burden or avoided impact method (see Figure 11). It can especially be applied to semi-closed loop recycling in cases where the reprocessed material has the same or closely resembles the inherent properties of the primary material (see e.g. ISO, 2006, Nicholson et al., 2009). According to ISO 14044 (ISO, 2006) there is in the case of closed-loop recycling no need for allocation since reprocessed secondary material fully replaces primary material.

Effectively the product system includes or replenishes the amount of primary material that is 'leaking' from the system as it goes to the waste disposal. In this way the loop is kept closed.

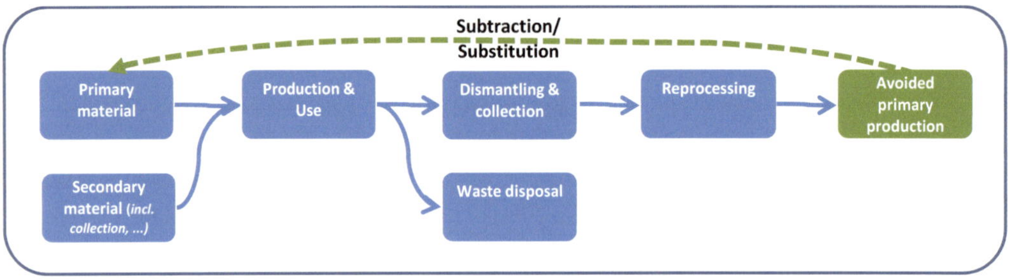

Fig. 11. The output oriented or substitution allocation approach.

The substitution approach is widely used in LCAs where recycling at the end-of-life is involved (Frischknecht, 2010). Whether in all cases it is really so that the inherent properties do not differ significantly is however the question. The substitution approach puts the focus onto the benefits of recycling. Increased recovery rates leads to more subtraction of primary material production and so lowers the environmental impacts.

Value-corrected substitution

The value-corrected substitution allocation addresses the point that the reprocessed material may not have the same inherent properties as the primary material (Guinee, 2002; EAA, 2010). This may be the case for i.e. aluminium alloys or paper. The method assumes that the proportionality of the environmental burdens caused by primary material production and the (economic) value of the recycled material (Vroonhof et al., 2002). Werner and Richter, 2000 consider the market price as an expression of the material quality within a techno-economic system. They see the relative market prices of the raw materials -not the scrap- therefore as a useful parameter to describe the "qualitative" gradation or inherent property of a material.

The method is suited for open-loop or semi-closed loop recycling. The method uses the ratio of secondary material price to the primary material price (P_{prim}/P_{sec}) as the indicator of the inherent property of the secondary material. The secondary material may substitute P_{prim}/P_{sec} of primary material.

We will clarify this method with an example (see Table 1 and Figure 12). The environmental burden of the materials is thus the sum of net difference in the value-corrected inflow and outflow of (primary) material plus the amount of material to be reprocessed in the product system. The method does include the recovery or recycling processes at the end-of-life stage of an LCA.

It must be noted that the value-corrected amounts in Table 1 do not give a closed mass balance, that is only the case for the uncorrected amounts.

Flows	Amount (kg)	Value (%/kg)	Value-correction
In: Primary	70	100	70
In: Secondary	30	90	27
In: Total	***100***	-	**97**
Out: Scrap to recovery	-80	90	-72
Out: Scrap to disposal	-20	0	0
Out: Total	***-100***	-	-72
Balance	0	-	25

Table 1. Example of value-corrected substitution.

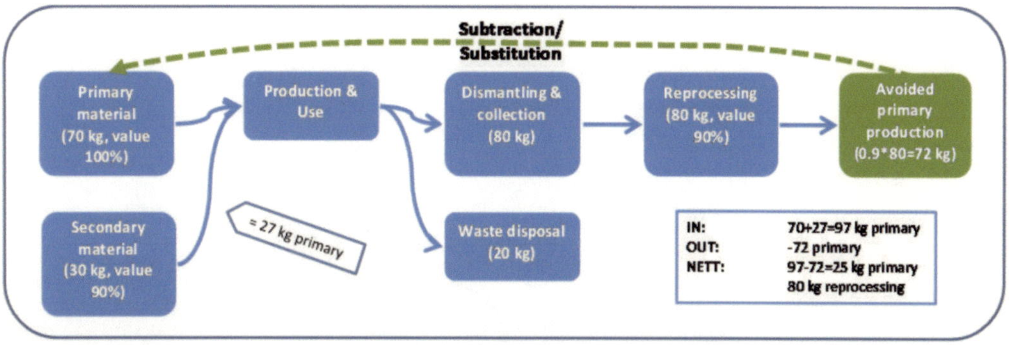

Fig. 12. Example of product system when applying the value-corrected substitution.

A precondition for this allocation procedure is a stable price relation of primary and secondary material (Werner & Richter, 2000). However, in some cases the primary to secondary material price may vary considerably during time and this will reflect in changed allocation parameters during time. The environmental impact of a system may thus vary with time while the physical characteristics of the system remain the same.

Multiple recycling method

In the steel industry scrap is used in both the primary steel route (Basic Oxygen Furnace) and in the secondary steel route (Electric Arc Furnace). Basically the primary route provides the material for the secondary route (see e.g. Thomas & Broadbent, 2007). The International Iron and Steel Institute wanted to address this interconnection for their life cycle inventory data and developed the multiple recycling method (Amato et al., 1996; Thomas & Broadbent, 2007).

The multiple recycling method uses a number of parameters (Thomas & Broadbent, 2007) to calculate the results of applying the method:

$RR = Recovery\ Rate$
$Y = Yield\ of\ recovery\ process$
$X_{pr} = Impact\ of\ primary\ production$
$X_{re} = Impact\ of\ secondary\ production$
$n = number\ of\ life\ cycles$

$Q_{ini} = Initial\ amount\ of\ primary\ product$

$Q_{cum} = Q_{ini} \times ((1 - RR^{n+1}) \div (1 - RR)) = Cumulative\ amount\ of\ product\ for\ n\ cycles$

The yield 'Y' of the recovery process is called recycling efficiency in Figure 5.

As the term $(1 - RR^{n+1})$ goes to 1 for large values of n Q_{cum} becomes $Q_{ini} \div (1 - RR)$. This means that with a Recovery Rate of e.g. 0.8 the cumulative amount of steel originating from 1 tonne primary steel is 5 tonnes. The impact of a product system can be described by $X = (X_{pri} - X_{re}) \times ((1 - RR \times Y) \div (1 - (RR \times Y)^n)) + X_{re}$

With increasing number of consecutive life cycles (see Figure 13) the environmental impact reduces as the recovered material has a lower impact than the primary material, see Figure 14. After a number of life cycles the impact reaches an almost constant value.

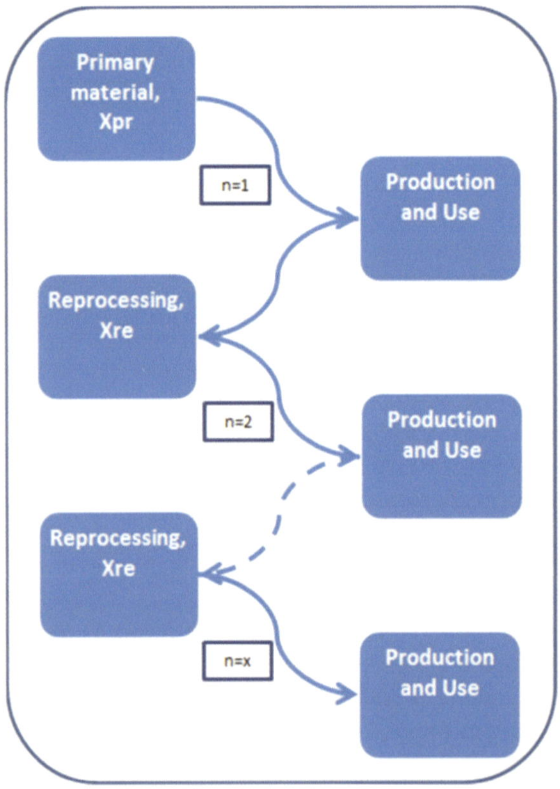

Fig. 13. Multiple recycling system with n life cycles after Thomas & Broadbent, 2007. X_{pr} Impact of primary production; X_{re} impact of recovery process.

The multiple recycling method can also be applied to other metals and materials where open loop or semi-closed loop recycling appears and the inherent properties of the material do not change or only slightly change. Recovery rates for different types of steel application can be found in the study by Five Winds International (2007).

In the next section we will discuss which methodological choices have to be made when applying these allocation methods.

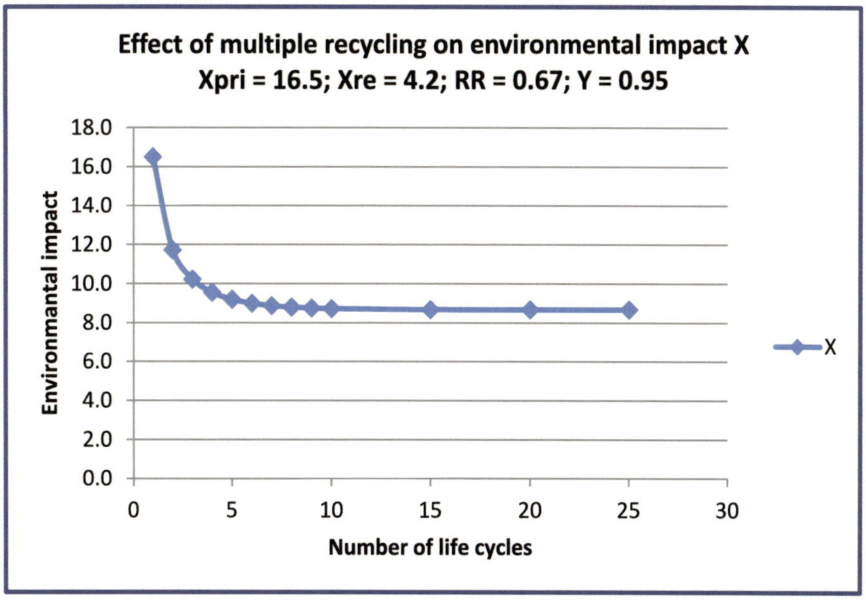

Fig. 14. The effect of multiple recycling cycles on the environmental impact assuming certain values for the parameters.

2.2 What choices are to be made?

In the case of closed loop recycling, which not often occurs in practice, methodological choices on allocation are absent. As most recycling or recovery cases are open loop or semi-closed loop these methodological choices are often to be made. Only when using the system expansion, cut-off or input oriented approaches these choices are not necessary due to the nature of these methods.

In the case of applying economic allocation the methodology is simple and the main choices occur when the revenues are not fully available or when the markets are distorted. This can be the case when fees on e.g. the recovery process is determined by regulations or by tax-like costs. Guinee et al., 2004 give strategies to overcome these issues like prices based on the actual costs made or the actual material price received by the seller.

In the output oriented or substitution method the main choice is which primary (raw) material is being avoided by the secondary material. For steel recycling the scrap substitutes pig iron. Pig iron is produced by the blast furnace and used in steelmaking in the basic oxygen furnace (see e.g. Classen et al., 2009). However, some LCA practitioners use primary steel as the avoided product although primary steel is the product of, and not a raw material for, steelmaking. This will give a higher environmental impact to be subtracted. This overestimation also happens in e.g. paper or cardboard recycling when one substitutes the primary product instead of primary pulp. Merrild et al., 2008 show the importance of making the correct choices of the recovery process for waste paper and especially the avoided type of primary pulp/paper as different choices may lead to significantly differing LCA results. In general to make the correct choice for the avoided material one has to have a close look which raw materials enter the primary production system.

Choosing the correct avoided primary material is also essential when applying the value corrected substitution method. Especially in cases of open loop recycling as then the avoided primary material may not always be clear or may in fact be more than one material. This is for instance the case with some plastic recycling schemes where the secondary material replaces concrete and/or wood. Another issue is whether one takes the price of scrap or of the secondary material itself.

The main choice to be made in the multiple recycling method is the number of times a material can be recycled. For metals an infinite number is often assumed. However, a build-up of alloying elements can reduce the number of life cycles that a material can be reprocessed (Amato et al., 1996). Applying the method to other materials like paper assuming an infinite number of loops may be incorrect and may underestimate the environmental impact of the product system.

It has been mentioned before that the cut-off approach promotes a high recycled content and the substitution approach a high recovery rate. According to Frischknecht, 2010 the cut-off approach is more appropriate when wanting to stress the preservation of natural stocks like those of metal ores. In environmental labelling the recycled content is often one of the indicators. According to Dubreuil et al., 2010 the cut-off or recycled content approach is most useful as a metric for materials that would otherwise be incinerated or landfilled as waste. The substitution approach should be more risk seeking as it depends on the actual or future recovery rate of post consumer materials.

3. Application of allocation methods in LCA

3.1 Introduction

For four of the allocation approaches that can be applied to several different type of materials the effect of applying those approaches will be shown in this section. Three different types of material (paper, metal, plastic) have been chosen. For paper the example is based on the recycling of post-consumer graphical paper. While steel has been used in the metal case. Finally, the example of post-consumer plastic recycling has been based on bottle grade PET.

3.2 Effects of applying different allocation approaches

For paper the example is based on the assumptions shown in Table 2. The results are shown for two often used single indicators; the global warming potential (GWP) and the cumulative energy demand (CED) including renewable energy sources.

For the economic allocation it has been assumed that 90% of the impact of the collection is allocated to the system under study as the value of waste paper is relatively low compared to the costs of collection. The assumption has been made that secondary pulp has a value of

90% of that of primary pulp. This value is needed for the value-corrected substitution approach.

Material and Process Flow	Amount (kg)
IN: Primary pulp	0.6
IN: Secondary pulp	0.4
Sheet production	1
OUT: Incineration	0.15
OUT: Landfill	0.05
OUT: Sorting & recycling	0.8

Table 2. Material and process flows assumed for the example for graphical paper.

Applying different allocation approaches clearly has an impact on the results of the environmental impact of the system of graphical paper (see Figure 14). Remarkable is that the highest global warming potential is found for the substitution approach which takes into account the avoided primary production. The reason for this is that the secondary pulp process is based on non-renewable energy sources while the primary pulp has a considerate input of renewable fuels of which the incineration is carbon neutral.

The two allocation approaches that include the avoided primary production (5. And 6. In Figure 15) do show a reduced cumulative energy demand. This is mainly due to the avoided production of primary pulp. The avoided primary energy production due to energy recovery from the waste send to incineration is of lesser importance in this example.

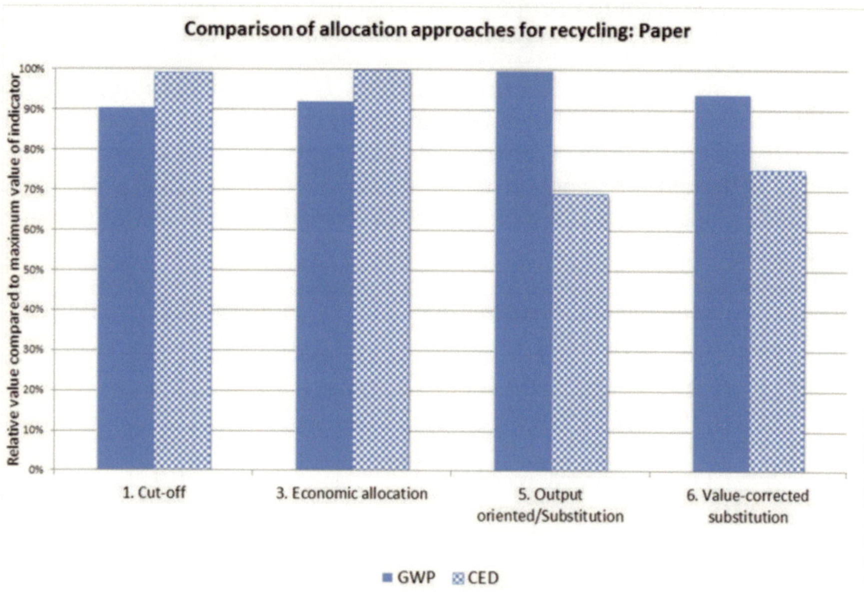

Fig. 15. Relative indicator values for both GWP and CED for the product system of graphical paper.

Specific circumstances in a product system may thus affect the outcome of a allocation approach. As seen in this example for GWP and CED the results may also differ per environmental impact. As the impact of collection is relatively small the approach of economic allocation, where a part of the collection for recycling is allocated to the system that delivers the waste, the results only very slightly differ from the basic cut-off approach.

In the case for steel the recycled content has been assumed to be 40% while 90% of the post-consumer material is going to recycling (see Table 3). As steel scrap has considerable value compared to the costs of collection only 10% of the collection impacts are allocated to the system that provides the post-consumer material. The value of secondary steel is assumed to be 90% of that of primary steel.

Material and Process Flow	Amount (kg)
IN: Primary steel	0.6
IN: Secondary steel	0.4
Sheet production	1
OUT: Landfill	0.1
OUT: Sorting & Recycling	0.9

Table 3. Material and process flows assumed for the example for steel.

Due to the limited impact of the collection of the waste the cut-off and economic allocation do not differ significantly (see Figure 16).

The allocation approaches that do include the avoided primary production show (see Figure 16) a considerable lower impact than the cut-off approach and economic allocation. Although the value corrected substitution works differently, it expresses the secondary material as a corrected amount of primary material, the results for GWP and CED do not differ greatly from those found for the 'normal' substitution approach.

The last example shown is that of PET used in bottles. The amount of secondary PET is high compared to the actual situation, but is used to enhance differences between the allocation approaches (see Table 4). For the economic allocation approach it is estimated that 15% of the impacts of collection are allocated to the system that delivers the waste. It has been assumed that the value of secondary PET is 32% of that of primary PET, this value is used by Shen et al., 2010.

Material and Process Flow	Amount (kg)
IN: Primary PET	0.6
IN: Secondary PET	0.4
Bottle production	1
OUT: MSWI	0.1
OUT: Sorting & recycling	0.9

Table 4. Material and process flows assumed for the example for PET bottles.

As seen for the other examples the results for the cut-off and economic allocation do actually not differ from each other (see Figure 17). Only in cases where the impacts of the collection itself have a considerable contribution a difference between those two approaches can be expected.

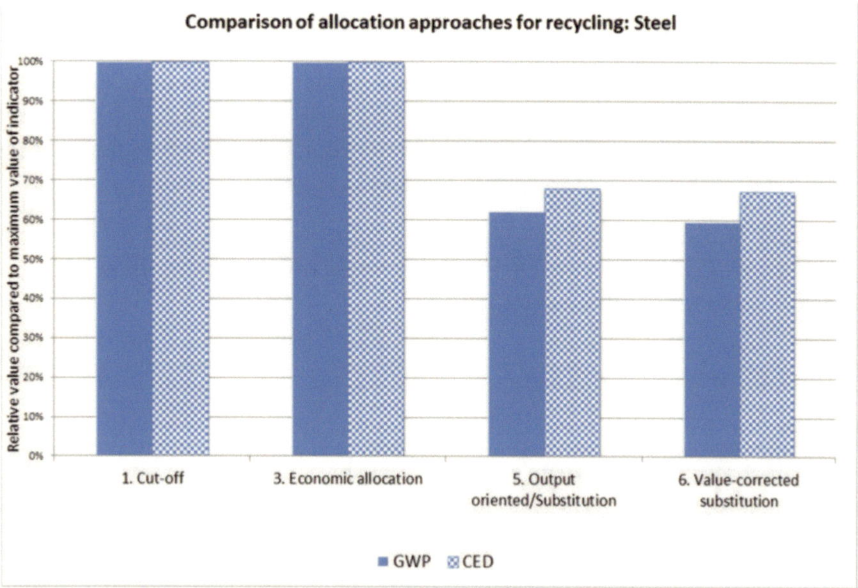

Fig. 16. Relative indicator values for both GWP and CED for the product system of steel sheet.

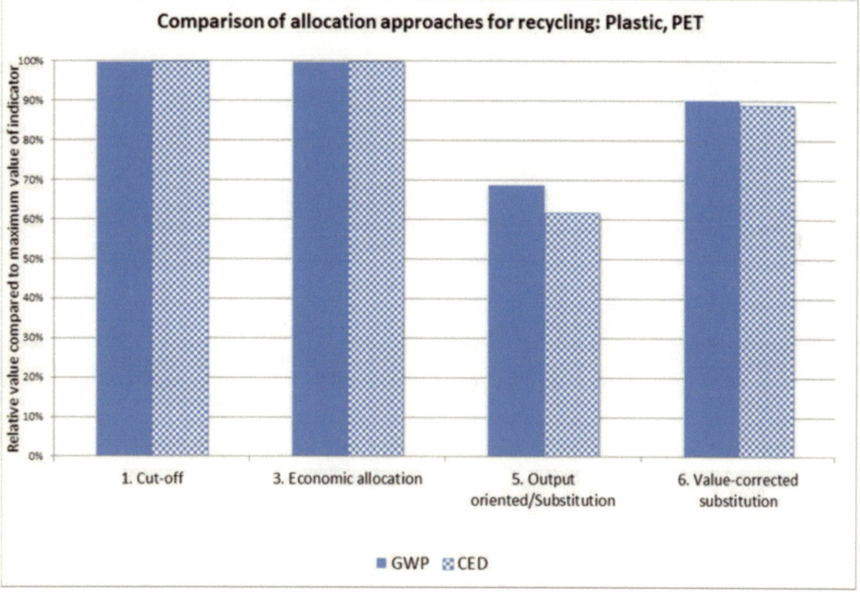

Fig. 17. Relative indicator values for both GWP and CED for the product system of PET bottles.

The two substitution approaches do indeed show a reduction of the GWP and CED results (see Figure 17). As the value of secondary PET is relatively low compared to secondary PET the value-corrected substitution approach gives a lesser amount of primary to be subtracted (32% of 0.9 kg = 0.29 kg) than the 'normal' substitution method (0.6 kg primary * recycling rate of 90% = 0.54 kg). The GWP and CED results are thus higher for the value-corrected substitution approach (see Figure 17).

The three examples show that in general those allocation approaches that do not include the avoided impacts of primary materials production show higher environmental impacts than those approaches that do include this avoided primary production. However, if the recycling process has a higher impact than the primary production the substitution approach may show higher values than the cut-off approach. This was the case for paper recycling and the GWP impact of the product system.

In case the value of the secondary material is low compared to that of the primary material, see the example of PET, the value-corrected approach will generally show higher impacts than the substitution or avoided impact approach.

4. Examples of recycling in LCA

4.1 Approach

A number of LCA studies focussing on recycling has been studied. An important source was the report issued by British Waste & Resources Action Programme (WRAP) from 2006 (WRAP, 2006) and updated in 2010 (WRAP, 2010). These reports have compared a number of LCA studies on material recycling and comparisons with other waste treatment options. The same type of materials as used in the previous section are the focus of this section.

4.2 Recycling of paper and cardboard

The recycling of paper and cardboard was in some cases environmentally preferred to incineration, but in other cases the opposite was true (WRAP, 2006, 2010). In the case of giving environmental benefits from energy recovery from waste incineration it is highly

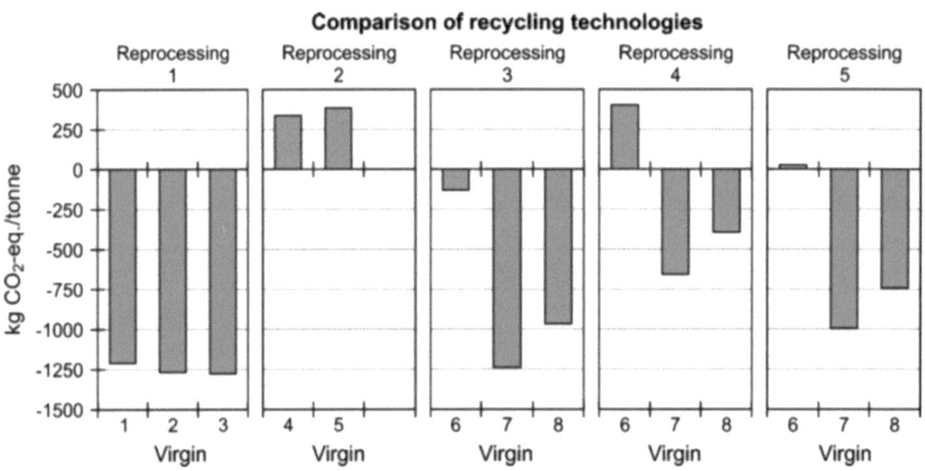

Fig. 18. Global warming potentials for recycling 1 tonne of waste paper for 14 possible recycling combinations (the top label gives a specific reprocessing technology, the bottom label gives the virgin technology that is avoided (Merrild et al., 2008). Reprocessing technologies: 1. Newspaper; 2. Fine paper; 3. Coreboard; 4. Solid cardboard; 5. Liner and fluting. Avoided technologies: 1. Newspaper; 2.,3. Printing paper; 4.,5. Fine paper; 6.-8. Cardboard.

important to choose the actual energy mix that is substituted by the energy recovered form the waste incinerator. As paper and cardboard are from renewable sources replacing energy from fossil fuels has a clear impact on the global warming impact of the system studied.

Another important point is the choice of the virgin material that is replaced by the recycled material. The recovered fibre process may deliver different kinds of products like testliner, wellenstoff and fluting. According to Arena et al., 2004 the energy requirements, which greatly influence the environmental performance, differ only slightly and the results of an LCA are thus less sensitive to the choice of the secondary material. Arena et al., 2004 compared paper and board recycling in Italy with landfill and waste incineration. They have applied system expansion in the case of packaging paper. This means that in e.g. the incineration alternative the system not only processes waste paper and board but also produces primary packaging paper. This last process takes place in Sweden the main source for imported packaging paper in Italy. In the case of Italy the energy balance of waste paper and board incineration is much better than that of recycling (Arena et al., 2004). When the avoided paper and cardboard production is allocated by substitution or value-corrected substitution the choice of the avoided product may greatly affect the LCA's results (Finnveden et al., 2000). Merild et al., 2008 conclude that the energy mix of the virgin paper production for the various paper and cardboard types is a decisive factor for the LCA's result. For the fourteen possible combinations of recycling and virgin technologies for paper and board production, which included the efficiency and quality loss of and due to the recycling processes, some combinations show a beneficial impact of recycling while others do not (see Figure 18).

Pulp is the raw material for the production of paper and cardboard. It is therefore also possible to see virgin pulp as the avoided product when using substitution allocation. It appears that most studies used in the WRAP report (WRAP, 2010) use paper as the avoided product only one study uses pulp as the avoided product (Finnveden et al., 2000). It is thus likely that the benefits of paper recycling are thus overestimated to some extent.

4.3 Plastics

The updated WRAP study (WRAP, 2010) states that mechanical recycling is the preferred waste treatment comparing it to landfill and incineration with energy recovery (see Figure 19). Important for this conclusion is the assumption that the secondary plastics replace primary plastics one to one. In case of energy recovery it is important to realize that conventional plastics stem from fossil feedstock and the incineration of them contribute to the global warming potential thus minimising the difference with the incineration of fossil fuels used in the avoided energy.

Schonfield, 2004 assumes a 1:1 substitution of primary plastic by secondary in the base case for plastic recycling. In a sensitivity analysis another scenario is tested in which only 20% primary plastic is substituted and the remaining 80% is equally divided over wood and concrete plastic poles replacing wooden and concrete poles (Schonfield, 2004). This scenario may lead to recycling from having a negative global warming impact to a positive one. For recycling of PE, PP, PET and PVC fractions a change from -620 to 439 kg CO_2-eq per tonne recycled occurred and for recycling of PE and PP fractions the change was from -464 to 415

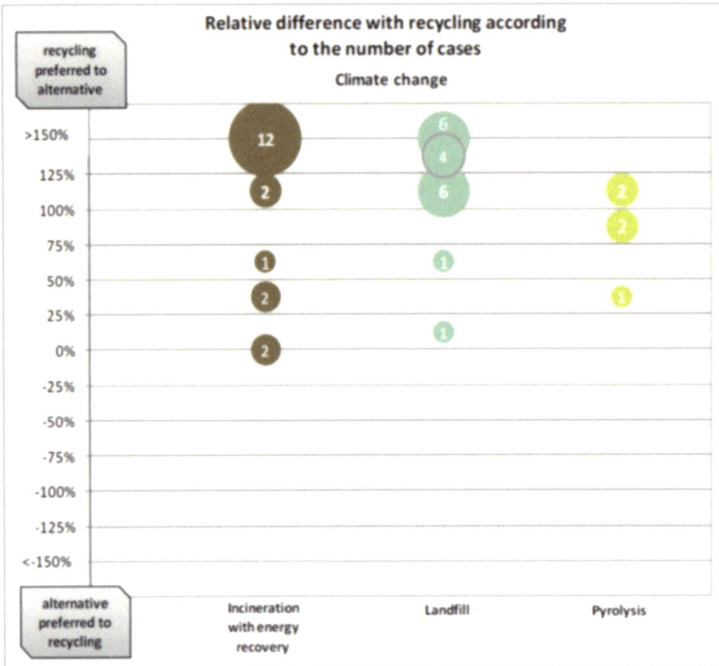

Fig. 19. Relative difference between the impacts from the different end-of-life options vs. recycling for climate change for plastics. The size of the "bubble" is proportional to the number of cases coming up with a value within the same range as another (WRAP, 2010).

kg CO_2-eq per tonne plastics recycled. The correct representation of the avoided primary products is thus crucial. Schonfield, 2004 postulates that when the substitution rate is below 70% other alternatives such as conversion to solid recovered fuel for cement kilns may be preferred.

Also other authors often assume a 1:1 substitution ratio as was found by Lazarevic et al., 2010. Only a limited number of LCA practitioners use a lower substitution ratio of e.g. 1:0.8 to address the loss in inherent properties of the plastic (Lazarevic et al., 2010). Ligthart & Ansems, 2007 used the economic value of the secondary raw material derived from waste plastic cups as a factor to correct for the quality loss using the substitution approach. This is an approach that lies between the substitution approach and the value-corrected substitution. It does not however translate the amounts of secondary raw materials entering the product system to an amount of primary raw material.

Shen et al., 2011 compared a system (reference system) in which the post-consumer waste of PET of used bottles and fibres is send to an MSWI with energy recovery with a system in which an open-loop recycling takes place. In this open loop system the secondary PET replaces part of the primary PET used for bottles and fibres. Some PET is still lost from the system and is send to an MSWI. The lost PET is replenished with virgin PET. Both systems apply system expansion to avoid allocation. The study shows that also in case when no allocation (for the recycled PET) is needed recycling reduces the environmental impact of a system (see example for NREU in Figure 20).

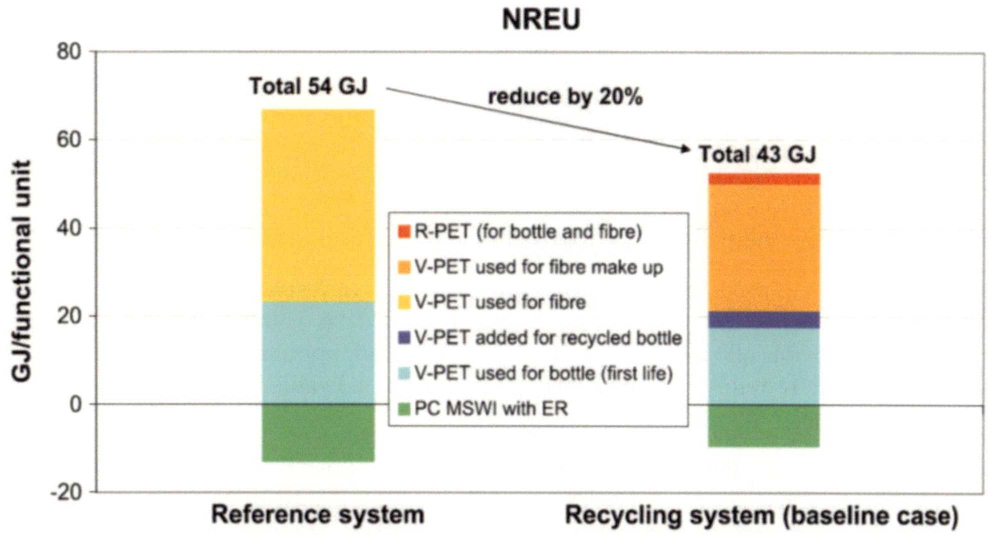

Fig. 20. Cradle-to-grave non-renewable energy use (NREU) recycling system and the reference system in which no recycling takes place (Schen et al., 2011).

Some authors like Williams et al., 2010 who studied the use of mixed plastic as a sports field drainage system are able to avoid allocation by system expansion. The system is however relatively simple as in the End-of-Life stage all waste goes to final disposal and no waste is being recycled.

4.4 Metals

The end-of-life of metals takes a special position in LCA because other than recycling there are no other end-of-life treatments, other than product re-use, that could have an environmental benefit and comparison with other beneficial treatments is not likely.

However, metal recycling may be complex as metals may be present in the form of alloys. According to Dubreuil et al., 2010 there has been until recently neither a consistent perspective nor a set of recycling indicators available for the metals industry as a whole. As the amount of available scrap would be limiting the recycled content in metal products the cut-off approach that focuses the recycled content would be less suited for modelling the end-of-life of metals (Dubreuil et al., 2010). The substitution approach would be more suited as it considers the whole life cycle of the product including its end disposition. This method is based on the premise that materials not recycled are replaced by primary materials.

Dubreuil et al., 2010 consider four metal recycling types:

1. Closed metal loop
2. Alloy loop
3. Transfer to another metal pool
4. Metallurgical re-separation

In case the inherent properties of a material are such that after recycling, the recycled material can offset the production of virgin material, a closed-loop-based calculation of environmental burden can be applied (Dubreuil et al., 2010). This approach avoids the use of allocation models. In case a metal is transferred to another metal pool, not being an alloy like brass, and the waste metal has no function it can be seen as lost from the system.

5. Conclusions

First of all it must be said that there is no best or worst allocation approach in LCA. This however under the condition that the approaches discussed here are correctly applied in LCA.

The simple cut-off approach is straightforward and so the easiest to apply. However, it does not reflect the full consequences of recycling (see e.g. Ekvall & Tillman 1997). The allocation approach of economic allocation that is based on the cut-off approach will often give results not very different from this approach. It seems that its added value is limited.

The substitution or avoided impact approach is most commonly used in LCAs in which recycling plays a role. Because it is based on the mass flow it's allocation basis is well defined. However, a change in inherent properties is not reflected in this method. A way to overcome this is to apply a correction factor for the loss in inherent properties. The main task when applying this approach is to find the proper primary material of which the production is being avoided. Heijungs & Guinee (2007) however see the specification of the avoided processes as problematic. According to them the problem lies in the fact that what is being avoided is essentially not there. Still it is likely that the substitution approach will remain the most commonly used approach for allocation of recycling in LCA.

The method of value-corrected substitution has the advantage that it is equipped to deal with changes in inherent properties of the recycled material. It works best for materials which have a well established recycling market as the market price will reflect the quality of the material. Market distortions may lead to the price of the secondary material not fully reflecting its quality leading to a distorted estimation of the avoided environmental impact.

Although allocation should be avoided according to ISO, 2006 the allocation of the environmental impacts of recycling is necessary in practice for the non-closed loop systems. The system expansion needed to avoid allocation would include many other, not always well-known, product systems using the recycled material. Because well applied allocation uses clear rules and system boundaries it ensures coherency between different LCAs.

As the choice of the allocation approach for the End-of-Life may influence the ranking of alternatives (Nicholson et al., 2009) LCA studies should, in these cases were such an influence may be expected, assess the effects of using an other allocation approach in a sensitivity analysis.

6. References

Amato, A.; Brimacombe, L. & Howard, N. (1996). Development of quantitative methodology for assessing embodied energy of recyclable and reusable materials/products *Ironmaking and Steelmaking*, 23 (3), pp. 235-241.

Arena, U.; Mastellone, M.L.; Perugini, F. & Clift, R. (2004). Environmental assessment of paper waste management options by means of LCA methodology. *Industrial and Engineering Chemistry Research*, 43 (18), pp. 5702-5714

Classen M.; Althaus H.-J.; Blaser S.; Doka G.; Jungbluth N. & Tuchschmid M. (2009). *Life Cycle Inventories of Metals*. Final report ecoinvent data v2.1

Dubreuil, A.; Young, S.B.; Atherton, J. & Gloria, T.P. (2010). Metals recycling maps and allocation procedures in life cycle assessment *International Journal of Life Cycle Assessment*, 15 (6), pp. 621-634.

EAA, (2007). Aluminium Recycling in LCA; http://www.eaa.net/upl/4/default/doc/Alu_recycling_LCA.pdf

Ekvall, T. & Tillman, A.-M. (1997). Open-loop recycling: Criteria for allocation procedures; *International Journal of Life Cycle Assessment* 2(3) pp. 155-162

Finnveden, G.; Johansson, J.; Lind, P. & Moberg, Å. (2000). *Life cycle assessment of energy from solid waste*. Forskningsgruppen för mijöstrategiska studier, Stockholm, Sweden.

Five Winds International, (2007). *Recycling Allocation: Summary of Best Practices for Use in Life Cycle Assessment*; http://www.zinc.org/resource_documents/recycling_lca.pdf

Frischknecht, R. (2010). LCI modelling approaches applied on recycling of materials in view of environmental sustainability, risk perception and eco-efficiency *International Journal of Life Cycle Assessment*, 15 (7), pp. 666-671.

Guinée J.B. (Ed.); Gorrée M.; Heijungs R.; Huppes G.; Kleijn R.; Wegener Sleeswijk A.; Udo de Haes H.A.; de Bruijn J.A.; van Duin R & Huijbregts M.A.J. (2002). *Handbook on Life Cycle Assessment: Operational Guide to the ISO Standards*. Kluwer Academic Publishers. Dordrecht http://cml.leiden.edu/research/industrialecology/researchprojects/finished/new-dutch-lca-guide.html

Guinée, J. B.; Heijungs, R. & Huppes, G. (2004). Economic Allocation: Examples and Derived Decision Tree. *International Journal of Life Cycle Assessment* 9 (1) pp. 23 - 33 http://www.p2pays.org/ref/41/40525.pdf

Heijungs R. & Guinee J.B. (2007). Allocation and "what-if" scenarios in life cycle assessment of waste management systems. *Waste Management*;27(8) pp. 997-1005. http://www.sciencedirect.com/science/article/B6VFR-4NDDSY4-3/2/1287345b1ec18bffb2b2e65077374ec0

Hischier R.; Weidema B.; Althaus H.-J.; Bauer C.; Doka G.; Dones R.; Frischknecht R.; Hellweg S.; Humbert S.; Jungbluth N.; Köllner T.; Loerincik Y.; Margni M. & Nemecek T. (2010). *Implementation of Life Cycle Impact Assessment Methods*. Final report ecoinvent v2.2 No. 3. Swiss Centre for Life Cycle Inventories, Dübendorf, CH.

ISO, (2006). *ISO 14044:2006 Environmental management -- Life cycle assessment -- Requirements and guidelines*.

Lazarevic D.; Aoustin, E; Buclet, N. & Brandt, N. (2010). Plastic waste management in the context of a European recycling society: Comparing results and uncertainties in a life cycle perspective. *Resources, Conservation and Recycling*, Volume 55, Issue 2, December 2010, pp. 246-259

Ligthart, T.N. & Ansems, A.M.M. (2007). *Single use Cups or Reusable (coffee) Drinking. Systems: An Environmental Comparison*. Date. October 2007. TNO report 2006-A-R0246(E)/B

Merrild, H.; Damgaard, A. & Christensen, T.H. (2008). Life cycle assessment of waste paper management: The importance of technology data and system boundaries in assessing recycling and incineration. *Resources, Conservation and Recycling*, 52 (12), pp. 1391-1398.

Nicholson, A.L.; Olivetti, E.A.; Gregory, J.R.; Field, F.R. & Kirchain, R.E. (2009). End-of-life LCA allocation methods: Open loop recycling impacts on robustness of material selection decisions. *Sustainable Systems and Technology, 2009. ISSST '09. IEEE International Symposium on*, pp.1-6, 18-20 May 2009

Norgate T.E. (2004). *Metal Recycling: An Assessment Using Life Cycle Energy Consumption as a Sustainability Indicator*, December 2004 CSIRO Minerals Report DMR-2616

Schonfield, P. (2008). *LCA of Management Options for Mixed Waste Plastics*. WRAP report June 2008

Shen, L.; Worrell, E. & Patel, M.K. (2010). Open-loop recycling: A LCA case study of PET bottle-to-fibre recycling *Resources, Conservation and Recycling*, 55 (1) pp. 34-52

Shen, L.; Nieuwlaar, E.; Worrell, E. & Patel, M. (2011). Life cycle energy and GHG emissions of PET recycling: change-oriented effects. *International Journal of Life Cycle Assessment* 16(6) pp. 522-536

Thomas, J.-S. & Broadbent C. (2007). *Method to take into account material recycling and end of life in LCA*. LCA Forum Lausanne, 22 November 2007. http://www.lcaforum.ch/Portals/0/DF_Archive/DF33/Thomas%20-%20ArcelorMittal%20-%20LCA%20DF33.pdf

Vroonhof, J.; Schwencke, A.; Croezen, H. & Potjer, B. (2002). *Legislation using LCA concerning Aluminium* Delft, CE, November, 2002

Werner, F. & Richter, K., (2000). Economic Allocation in LCA: A Case Study About Aluminium Window Frames. *International Journal of Life Cycle Assessment* 5(2) pp. 79-83. 2000. http://dx.doi.org/10.1007/BF02979727

Williams, T. G.J.L.; Heidrich, O. & Sallis, P.J. (2010). A case study of the open-loop recycling of mixed plastic waste for use in a sports-field drainage system. *Resources, Conservation and Recycling* 55 pp. 118–128

WRAP (2006) *Environmental benefits of recycling*. An international review of life cycle comparisons for key materials in the UK recycling sector. May 2006.

WRAP (2010) *Environmental benefits of recycling* – 2010 update. March 2010

Zamagni A. (2008) *Critical review of the current research needs and limitations related to ISO-LCA practice*. Deliverable D7 of work package 5 of the CALCAS project. 2008 http://www.leidenuniv.nl/cml/ssp/publications/calcas_report_d7.pdf

11

Optimal Production Decision in the Closed-Loop Supply Chain Considering Risk-Management and Incentives for Recycling

Takashi Hasuike
Graduate School of Information Science and Technology, Osaka University
Japan

1. Introduction

Production companies in the real world face many decision making situations, such as logistics, scheduling, and data mining in the supply chain. The optimal production decision is also major and important when deciding the optimal rate of product allocation, taking into consideration customers' demands, resource costs, and many conditions surrounding the real market. Many researchers have considered optimal production decision models with the total cost-minimization or the total profit-miaximization arising from the production processes of firms, and recently many mathematical models have been proposed (for example Letmathe & Balakrishnan, 2005; Li & Tirupati, 1997; Morgan & Daniels, 2001; Mula et al., 2006a).

In production processes, there are many uncertain factors in terms of total volume and customer's demands, such as the occurrence probability of machine breakdown and human error based on historical data, and the ambiguity that derives from the quality of information received and decision makers' intuition. Some recent articles have elaborated on studies of production planning problems including such ambiguous situations (for instance, Mula et al., 2006b, 2007; Vasant, 2000). We (Hasuike & Ishii, 2009a, 2009b) considered optimal production decision models, under both randomness for the return of each product and fuzziness for coefficients of constraints in production processes.

Particularly, in a forward supply chain, the customer is typically the end of the process. However, a closed loop supply chain includes the returns processes and the manufacturer has the intent of capturing additional value and further integrating all supply chain activities. Therefore, closed-loop supply chains include traditional forward supply-chain activities and the additional activities of the reverse supply chain such as product acquisition and reverse logistics. Then, in terms of reverse logistics and closed loop supply chain management, it is important to improve the sustainability of production companies' business considering the reintegration of their returned products into their own production network. The increase of the reintegration rate can be achieved at multiple levels also called recovery paths: reuse, repair, remanufacturing, recycling, etc. (in detail, Lebreton, 2007). Therefore, an incentive, which is the leverage for reintegrating valuable products, is a key factor in the closed roop supply chain management. If the buyback incentive is set higher than the market price, the production company will pay more than it. If the incentive is too low, customers will prefer to discard old products or resell on the secondary market rather than to return to the production company. Thus, it is most important how high incentives the production company provides to customers.

On the other hand, Corporate Social Responsibility (CSR) is currently the most important measure to sustain continuous developments of companies by performing environment-friendliness and suitable social activity, and it is also essential for avoiding the latent risk. Most recently, the CSR is not only a prominent research theme but also it can also be found in corporate missions and value statements (Svendsen et al., 2001). Companies increasingly realize that their actions in purchasing and supply chain management strongly affect their reputation and long-term success (Castka & Balzatova, 2008). Recently, many famous companies have faced damaging media reports, external pressure from activists, and internal pressure from investors. Consequently, companies start expanding their responsibility for their products, and managing the CSR of their partner within the supply chain (Bloemhof-Ruwaard et al., 1995; Emmelhainz & Adams, 1999; Kolk & Tudder, 2002).

CSR has been considered as a theme of many researchers since Carroll's study (1979). Wartick and Coghran (1985) traced the evolution of the corporate social performance model by focusing on three concepts of CSR: economic responsibility, public responsibility, and social responsiveness. They examined the management of social issues as a dimension of corporate social performance and concluded that the corporate social performance model is valuable for business and society.

CSR issues surrounding supply chains have only recently come to the fore, in the context of conceptual and survey studies (for instance, Carter & Jennings, 2002, 2004). Murphy and Poist (2002) stated that although supply chain practitioners have been slow to adopt CSR considerations, social responsibility concepts in the supply chain are increasing in importance. Carter and Jennings (2002, 2004) empirically established primary supply chain CSR categories of environment, diversity, human rights, philanthropy, and safety. Some researchers have examined individual elements of CSR in the supply chain. In response to growing CSR concerns, researchers have begun to deal with environmental risks, labor practices, procurement, and affirmative action purchasing. Moreover, some researchers have consider various types of practical models with CSR such as forest industry (Panwer et al., 2010), hospitality management (Kang et al., 2010), fashion industry (Perry & Tower, 2009), etc..

Nevertheless, decision support models to integrate CSR into supply chain management are surely needed. Within recent business and social environments, trade-offs between various objectives while providing resources to CSR activities are becoming increasingly complex. Most recently, Cruz and Wakolbinger (2008) and Cruz (2009) have considered various impacts and effects of CSR in supply chain management, and proposed the multi-criteria decision making approach. However, this research field of supply chain management with CSR is now developing, and there are few studies, particularly under uncertainty.

In this paper, we develop a multi-criteria model in the supply chain framework that captures the economic and CSR activities of manufacturing, retailer, and demand market under uncertainty. Particularly, we consider the evaluation of environmental conservation activities and resource recycling as CSR activities, and propose a new optimal production policy as well as the maximization of total profit and effects of CSR activities considering the minimization of latent risk using the advanced risk measure Conditional Value at Risk (CVaR) proposed by Rockafellar and Uryasev (2002).

CSR are generally presented as linguistic and ambiguous information including several types of subjectivities. Therefore, it is hard to consider the value of CSR as a fixed value. In this paper, in terms of linguistic property and subjectivity, we formulate effect of CSR activities as fuzzy numbers, and so our proposed model is formulated as an uncertainty programming problem. By performing deterministic equivalent transformations and considering the application in practice, the analytical and effective solution algorithm is developed.

This paper is organized as follows. In Section 2, we introduce and formulate the proposed optimal production model considering the maximization of total profit, recycling rate, and cost for CSR. In Section 3, we construct the analytical solution algorithm using the equivalent transformations. Finally, in Section 4, we conclude this paper and discuss future research problems.

2. Mathematical formulation and notation of parameters

In this section, we introduce the notation used in our proposed models integrating optimal production decision and CSR activities considering the closed-loop supply chain. Furthermore, as a risk measure, we introduce Conditional Value-at-Risk proposed by Rockafellar and Uryasev (2002). CVaR is known as a useful risk measure which is coherent, consistent with the second (or higher) order stochastic dominance, and the consistency with the stochastic dominance implies that minimizing the CVaR never conflicts with maximizing the expectation of any risk-averse utility function.

Through the whole paper, to simplify, we consider the following cases:

1. All products are produced in one manufacture and these products are delivered to p retailers.
2. We assume that the leadtime of production is larger than that of consumption, and so the production company must predict the demand in the next period and the amount of production items previously.
3. The shortage of each product at the retailers is forbidden as much as possible.

2.1 Notation of parameters

Sets

$i \in \{1,2,...,p\}$: index of retailers

$j \in \{1,2,...,n\}$: index of products

$t \in \{1,2,...,T\}$: index of production periods

$k \in \{1,2,...,m\}$: index of resource constraints

Demand and cost parameters

$d_{ij}^{(t)}$: demand volume for item *j* at destination *i* in period *t*

$r_j^{(t)}$: profit for item *j* in period *t*

$c_j^{(t)}$: production cost for item *j* in period *t*

$a_{kj}^{(t)}$: necessary resource volume for item *j* at *k*th resource constraint in period *t*

h_j : inventory holding cost per unit of item *j* at the manufacture, which is constant value to each period *t*

$u_{ij}^{(t)}$: transportation cost for item *j* from the manufacture to destination *i* in period *t*

v_j : recycling cost for item *j*, which is constant value to each period *t*

$R_C(t)$: expenses for CSR activities in period *t*

$b_k^{(t)}$: maximum volume of *k*th resource in period *t*

W : maximum total level of inventory

C : maximum value of total fund cost.

$g_j\left(p_j^{(t)}\right)$: recycling volume of item *j* in period *t* characterized by a continuous function on promotion cost $p_j^{(t)}$ for recycling

Decision variables

$x_j^{(t)}$: production volume of item j at the manufacture in period t

$y_{ij}^{(t)}$: transportation volume for item j from the manufacture to destination i in period t

$p_j^{(t)}$: promotion cost for recycling item j in period t

$w_j^{(t)}$: inventory level at end of period t for item j at the manufacture.

With respect to the relation between demand $d_{ij}^{(t)}$ and expenses of CSR activities $C_{CSR}^{(t)}$, consumers are generally well-affected in the case that production companies actively perform CSR activities. Then, their purchasing powers are certainly increasing but ambiguity due to including consumer's subjectivity. Therefore, we assume that this relation as the following linear relation:

$$d_{ij}^{(t)} = D_{ij}^{(t)} + \tilde{\alpha} C_{CSR}^{(t)} \tag{1}$$

where $D_{ij}^{(t)}$ is the random demands independent of CSR activities and $\tilde{\alpha}$ is assumed to be a fuzzy number characterized by the L-shape membership function $\mu_{\tilde{\alpha}} = (\bar{\alpha}, \delta)_L$ where $\bar{\alpha}$ is the center value and δ is the spread. Therefore, demand $d_{ij}^{(t)}$ is presented as a hybrid variable with both randomness and fuzziness.

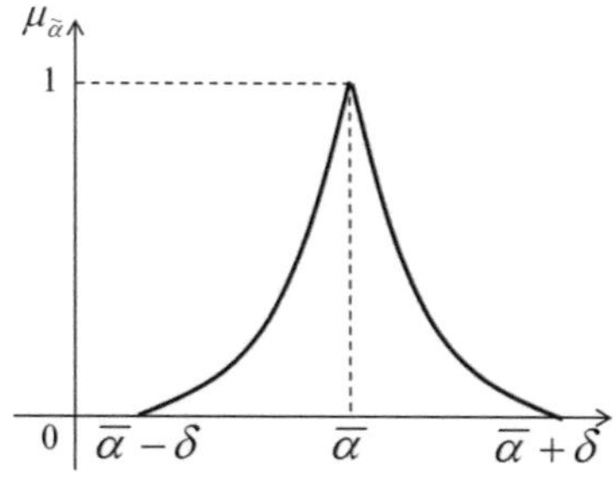

Fig. 1. L-fuzzy number $\tilde{\alpha}$

2.2 Identification of cost function

2.2.1 Identification of cost function

The objects of our proposed model are to maximize the total cost and recycling rate derived from customers' demands, the recycling volume, and the expenses of CSR activities. The total profit function is associated with total profit, transportation costs, production costs, and holding costs. Each profit or cost function is formulated as follows in the multi-item integrated model using the above-listed parameters:

a. Profit: $\sum_{t=1}^{T}\sum_{i=1}^{p}\sum_{j=1}^{n} r_j^{(t)} d_{ij}^{(t)}$

b. Production cost: $\sum_{t=1}^{T}\sum_{j=1}^{n} c_j^{(t)} x_j^{(t)}$

c. Transportation cost: $\sum_{t=1}^{T}\sum_{i=1}^{p}\sum_{j=1}^{n} u_{ij}^{(t)} y_{ij}^{(t)}$

d. Holding cost: $\sum_{t=1}^{T}\sum_{j=1}^{n} h_j w_j^{(t)}$

e. Recycling cost: $\sum_{t=1}^{T}\sum_{j=1}^{n} v_j g_j^{(t)}\left(p_j^{(t)}\right)$

f. Expenses of CSR activities: $\sum_{t=1}^{T} C_{CSR}^{(t)}$

Therefore, the total profit function $f(\boldsymbol{x},\boldsymbol{y},\boldsymbol{w},\boldsymbol{p},\boldsymbol{D})$ is obtained as the following form:

$$f(\boldsymbol{x},\boldsymbol{y},\boldsymbol{w},\boldsymbol{p},\boldsymbol{D})=\sum_{t=1}^{T}\sum_{i=1}^{p}\sum_{j=1}^{n} r_j^{(t)} d_{ij}^{(t)}-\sum_{t=1}^{T}\sum_{j=1}^{n} c_j^{(t)} x_j^{(t)}-\sum_{t=1}^{T}\sum_{i=1}^{p}\sum_{j=1}^{n} u_{ij}^{(t)} y_{ij}^{(t)}$$
$$-\sum_{t=1}^{T}\sum_{j=1}^{n} h_j w_j^{(t)}-\sum_{t=1}^{T}\sum_{j=1}^{n} v_j g_j^{(t)}\left(p_j^{(t)}\right)-\sum_{t=1}^{T}\sum_{j=1}^{n} p_j^{(t)}-\sum_{t=1}^{T} C_{CSR}^{(t)}$$

Furthermore, each constraint with respect to inventory, shortage, recycling volume, and general production process is defined as follows:

i. Inventory constraint:

$$x_j^{(t)} + w_j^{(t-1)} - \sum_{i=1}^{p}\left(y_{ij}^{(t)} - d_{ij}^{(t)}\right) = w_j^{(t)},\ \sum_{j=1}^{n} w_j^{(t)} \le W,\ \left(j = 1,2,\ldots,n,\ t = 1,2,\ldots,T\right)$$

ii. Shortage constraint:

$$y_{ij}^{(t)} \ge d_{ij}^{(t)},\ \left(i = 1,2,\ldots,p,\ j = 1,2,\ldots,n,\ t = 1,2,\ldots,T\right)$$

iii. Recycling volume constraint:

$$g_j^{(t)}\left(p_j^{(t)}\right) \le \sum_{l=1}^{t}\sum_{i=1}^{p} d_{ij}^{(l)} - \sum_{l=1}^{t-1} g_j^{(l)}\left(p_j^{(l)}\right),\ \left(j = 1,2,\ldots,n,\ t = 1,2,\ldots,T\right)$$

iv. General production process constraint:

$$\sum_{j=1}^{n} a_{ij}^{(t)} x_j^{(t)} \le b_k^{(t)} + \sum_{j=1}^{n} a_{ij}^{(t)} g_j^{(t-1)}\left(p_j^{(t-1)}\right),\ \left(k = 1,2,\ldots,m,\ t = 1,2,\ldots,T\right)$$

v. Limited total fund cost constraint:

$$\sum_{t=1}^{T}\sum_{i=1}^{p}\sum_{j=1}^{n} r_j^{(t)} d_{ij}^{(t)} - f\left(\boldsymbol{x},\boldsymbol{y},\boldsymbol{w},\boldsymbol{p},\boldsymbol{D}\right) \le C$$

Consequently, the proposed integrated production decision and CSR model considering the closed-loop formulated as the following problem:

$$\begin{aligned}
&\text{Maximize } f\left(\boldsymbol{x},\boldsymbol{y},\boldsymbol{w},\boldsymbol{p},\boldsymbol{D}\right)\\
&\text{Maximize } R\left(\boldsymbol{p},\boldsymbol{D}\right) = \frac{\sum_{t=1}^{T}\sum_{j=1}^{n} g_j^{(t)}\left(p_j^{(t)}\right)}{\sum_{t=1}^{T}\sum_{j=1}^{n} d_j^{(t)}}\\
&\text{subject to } y_{ij}^{(t)} \ge d_{ij}^{(t)},\ \left(i = 1,2,\ldots,p,\ j = 1,2,\ldots,n,\ t = 1,2,\ldots,T\right)\\
&\qquad x_j^{(t)} + w_j^{(t-1)} - \sum_{i=1}^{p}\left(y_{ij}^{(t)} - d_{ij}^{(t)}\right) = w_j^{(t)},\ \sum_{j=1}^{n} w_j^{(t)} \le W,\ \left(j = 1,2,\ldots,n,\ t = 1,2,\ldots,T\right)\\
&\qquad \sum_{j=1}^{n} a_{ij}^{(t)} x_j^{(t)} \le b_k^{(t)} + \sum_{j=1}^{n} a_{ij}^{(t)} g_j^{(t-1)}\left(p_j^{(t-1)}\right),\ \left(k = 1,2,\ldots,m,\ t = 1,2,\ldots,T\right)\\
&\qquad g_j^{(t)}\left(p_j^{(t)}\right) \le \sum_{l=1}^{t}\sum_{i=1}^{p} d_{ij}^{(l)} - \sum_{l=1}^{t-1} g_j^{(l)}\left(p_j^{(l)}\right),\ \left(j = 1,2,\ldots,n,\ t = 1,2,\ldots,T\right)\\
&\qquad \sum_{t=1}^{T}\sum_{i=1}^{p}\sum_{j=1}^{n} r_j^{(t)} d_{ij}^{(t)} - f\left(\boldsymbol{x},\boldsymbol{y},\boldsymbol{w},\boldsymbol{p},\boldsymbol{D}\right) \le C\\
&\qquad x_j^{(t)} \ge 0,\ y_{ij}^{(t)} \ge 0,\ w_j^{(t)} \ge 0,\ p_j^{(t)} \ge 0,\ \left(i = 1,2,\ldots,p,\ j = 1,2,\ldots,n,\ t = 1,2,\ldots,T\right)
\end{aligned} \tag{2}$$

where we assume that initial inventories $w_j^{(0)}$ and recycling volume $g_j^{(0)}$ are given as constant values. Then, we set the following feasible solutions set as Φ.

$$
\boldsymbol{x},\boldsymbol{y},\boldsymbol{w},\boldsymbol{p}\in\Phi=\left\{\boldsymbol{x},\boldsymbol{y},\boldsymbol{w},\boldsymbol{p}\left|\begin{array}{l}
y_{ij}^{(t)}\ge d_{ij}^{(t)},\ (i=1,2,\ldots,p,\ j=1,2,\ldots,n,\ t=1,2,\ldots,T)\\
x_j^{(t)}+w_j^{(t-1)}-\sum_{i=1}^{p}\left(y_{ij}^{(t)}-d_{ij}^{(t)}\right)=w_j^{(t)},\ \sum_{j=1}^{n}w_j^{(t)}\le W,\ (j=1,2,\ldots,n,\ t=1,2,\ldots,T)\\
\sum_{j=1}^{n}a_{ij}^{(t)}x_j^{(t)}\le b_k^{(t)}+\sum_{j=1}^{n}a_{ij}^{(t)}g_j^{(t-1)}\left(p_j^{(t-1)}\right),\ (k=1,2,\ldots,m,\ t=1,2,\ldots,T)\\
g_j^{(t)}\left(p_j^{(t)}\right)\le\sum_{l=1}^{t}\sum_{i=1}^{p}d_{ij}^{(l)}-\sum_{l=1}^{t-1}g_j^{(l)}\left(p_j^{(l)}\right),\ (j=1,2,\ldots,n,\ t=1,2,\ldots,T)\\
\sum_{t=1}^{T}\sum_{i=1}^{p}\sum_{j=1}^{n}r_j^{(t)}d_{ij}^{(t)}-f(\boldsymbol{x},\boldsymbol{y},\boldsymbol{w},\boldsymbol{p},\boldsymbol{D})\le C\\
x_j^{(t)}\ge 0,\ y_{ij}^{(t)}\ge 0,\ w_j^{(t)}\ge 0,\ p_j^{(t)}\ge 0,\ (i=1,2,\ldots,p,\ j=1,2,\ldots,n,\ t=1,2,\ldots,T)\\
d_{ij}^{(t)}=D_{ij}^{(t)}+\tilde{\alpha}C_{CSR}^{(t)}
\end{array}\right.\right\} \tag{3}
$$

Substituting demand relation (1) and performing the equivalent transformation, the feasible solutions set Φ is equivalently transformed into the following set:

$$
\boldsymbol{x},\boldsymbol{y},\boldsymbol{w},\boldsymbol{p}\in\Phi=\left\{\boldsymbol{x},\boldsymbol{y},\boldsymbol{w},\boldsymbol{p}\left|\begin{array}{l}
y_{ij}^{(t)}\ge D_{ij}^{(t)}+\tilde{\alpha}C_{CSR}^{(t)},\ (i=1,2,\ldots,p,\ j=1,2,\ldots,n,\ t=1,2,\ldots,T)\\
\sum_{j=1}^{n}a_{ij}^{(t)}x_j^{(t)}\le b_k^{(t)}+\sum_{j=1}^{n}a_{ij}^{(t)}g_j^{(t-1)}\left(p_j^{(t-1)}\right),\ (k=1,2,\ldots,m,\ t=1,2,\ldots,T)\\
\sum_{l=1}^{t}g_j^{(l)}\left(p_j^{(l)}\right)\le\sum_{l=1}^{t}\left(\sum_{i=1}^{p}y_{ij}^{(l)}-x_j^{(l)}+w_j^{(l)}-w_j^{(l-1)}\right),\ (j=1,2,\ldots,n,\ t=1,2,\ldots,T)\\
\sum_{j=1}^{n}w_j^{(t)}\le W,\ (j=1,2,\ldots,n,\ t=1,2,\ldots,T)\\
\sum_{t=1}^{T}\left(\sum_{j=1}^{n}\left(c_j^{(t)}x_j^{(t)}+\sum_{i=1}^{p}u_{ij}^{(t)}y_{ij}^{(t)}+h_jw_j^{(t)}+v_jg_j^{(t)}\left(p_j^{(t)}\right)+p_j^{(t)}\right)+C_{CSR}^{(t)}\right)\le C,\\
x_j^{(t)}\ge 0,\ y_{ij}^{(t)}\ge 0,\ w_j^{(t)}\ge 0,\ p_j^{(t)}\ge 0,\ (i=1,2,\ldots,p,\ j=1,2,\ldots,n,\ t=1,2,\ldots,T)
\end{array}\right.\right\} \tag{4}
$$

Then, the main problem (2) is also equivalently transformed into the following problem:

$$
\begin{aligned}
&\text{Minimize } \sum_{t=1}^{T}\left(\sum_{j=1}^{n}\left(c_j^{(t)}x_j^{(t)}+\sum_{i=1}^{p}u_{ij}^{(t)}y_{ij}^{(t)}+h_jw_j^{(t)}+v_jg_j^{(t)}\left(p_j^{(t)}\right)-p_j^{(t)}\right)\right.\\
&\qquad\qquad\left.+\left(1-\tilde{\alpha}\sum_{i=1}^{p}\sum_{j=1}^{n}r_j^{(t)}\right)C_{CSR}^{(t)}-\sum_{i=1}^{p}\sum_{j=1}^{n}r_j^{(t)}D_{ij}^{(t)}\right)\\
&\text{Maximize } \frac{\sum_{t=1}^{T}\sum_{j=1}^{n}g_j^{(t)}\left(p_j^{(t)}\right)}{\sum_{t=1}^{T}\sum_{j=1}^{n}D_{ij}^{(t)}+\tilde{\alpha}C_{CSR}^{(t)}}\\
&\text{subject to } \boldsymbol{x},\boldsymbol{y},\boldsymbol{w},\boldsymbol{p}\in\Phi
\end{aligned} \tag{5}
$$

This problem is equivalent to a multi-criteria linear programming problem if each demand $d_{ij}^{(t)}$ is fixed or the expected value $E\left(d_{ij}^{(t)}\right)$ of random variable. However, in reality, the wild swing of demands associated with drastic changes of social and economic conditions often happens. In this case, it is obviously impossible that each demand is generally not fixed, and it is rough to degenerate the random distribution into the expected value by neglecting important factors considering fluctuation ranges such as the variance. Therefore, we must consider more advanced risk management approaches to avoid futile supply chain disruptions, particularly, focusing on the downside risk to decrease uncertainty up to the high-cost as much as possible.

3. Analytical and efficient solution algorithm based on CVaR

First, we introduce the standard CVaR in the stochastic programming problem. Standard CVaR for randomness was proposed by Rockafellar and Uryasev (2002) as follows:

$$CVaR_{\beta}(\chi)=E\left[L(\chi,\xi)\middle|L(\chi,\xi)\geq VaR_{\beta}(\chi)\right]=\frac{1}{1-\beta}\int_{L(\chi,y)\geq VaR_{\beta}(y)}L(\chi,y)f(y)dy$$
$$\left(VaR_{\beta}(y)=\inf\left\{r\middle|\Pr\left\{L(\chi,\xi)\leq r\right\}\geq\beta\right\}\right) \tag{6}$$

where $L(\chi,\xi)$ is the loss function with fixed value χ and random variable ξ, and β is the confidence level. Then, $f(y)$ is the density function of random variable ξ. $VaR_{\beta}(\chi)$ is Value at Risk (VaR) and the standard risk measure in economic and financial fields and used in many practical risk management. Furthermore, we introduce the following function:

$$F_{\beta}(\chi,u)=r+\frac{1}{1-\beta}E\left[\left[L(\chi,\xi)-r\right]^{+}\right] \tag{7}$$

Rockafellar and Uryasev (2002) proved that the CVaR (10) can be minimized by minimizing the auxiliary function (7). Since the loss function $L(\chi,\xi)$ is convex due to the linearity for fixed demand ξ, we can obtain the optimal order quantity of our proposed model by solving the following CVaR minimization problem:

$$\begin{aligned}&\underset{\chi,r}{\text{minimize}}\ r+\frac{1}{1-\beta}E\left[\left[L(\chi,\xi)-r\right]^{+}\right]\\ \Leftrightarrow\ &\underset{\chi,r}{\text{minimize}}\ r+\frac{1}{1-\beta}\sum_{s=1}^{S}p_{s}\left[L(\chi,\xi_{s})-r\right]^{+}\end{aligned} \tag{8}$$

where probability ξ is assumed to be a finite discrete random variable like assumptions in this paper, i.e.,

$$\Pr\left\{\xi=\xi_{s}\right\}=p_{s},\ \sum_{s=1}^{S}p_{s}=1$$
$$\xi_{s}=\left\{\xi_{sij}^{(t)},\ (i=1,\ldots,p,\ j=1,\ldots,n,\ t=1,\ldots,T)\right\},\ (s=1,\ldots,S)$$

we can perform the transformation from second problem to third problem in (6). Furthermore, introducing parameters $\omega_{k},\ (k=1,2,\ldots,K)$, this optimization problem is equivalently transformed into the following problem:

$$
\begin{aligned}
&\underset{\chi,r}{\text{minimize}} \;\; r+\frac{1}{1-\beta}\sum_{s=1}^{S} p_s\omega_s \\
&\text{subject to } \;\omega_s \geq L(\chi,\xi_s)-r,\; \omega_s \geq 0,\; (s=1,2,\ldots,S)
\end{aligned} \tag{9}
$$

Using CVaR model (9) for randomness and setting loss functions for both objective functions as $L_1(\chi,\xi)=-f(\boldsymbol{x},\boldsymbol{y},\boldsymbol{w},\boldsymbol{p},\boldsymbol{D})$ and $L_2(\chi,\xi)=-R(\boldsymbol{p},\boldsymbol{D})$, we reformulate our proposed closed-loop supply chain model under hybrid CVaR as follows:

$$
\begin{aligned}
&\underset{\boldsymbol{x},\boldsymbol{y},\boldsymbol{w},\boldsymbol{p},\gamma_1}{\text{minimize}} \;\; \gamma_1+\frac{1}{1-\beta_1}\sum_{s=1}^{S} p_s\left[-f(\boldsymbol{x},\boldsymbol{y},\boldsymbol{w},\boldsymbol{p},\boldsymbol{D}_s)-\gamma_1\right]^+ \\
&\underset{\boldsymbol{p},\gamma_2}{\text{minimize}} \;\; \gamma_2+\frac{1}{1-\beta_2}\sum_{s=1}^{S} p_s\left[-R(\boldsymbol{p},\boldsymbol{D}_s)-\gamma_2\right]^+ \\
&\text{subject to } \boldsymbol{x},\boldsymbol{y},\boldsymbol{w},\boldsymbol{p}\in\Phi_d
\end{aligned} \tag{10}
$$

where Φ_d is re-defined as the following feasible solutions set using the demand scenario of deterministic values $\boldsymbol{D}_s=\left\{D_{sij}^{(t)},(i=1,\ldots,p,j=1,\ldots,n,t=1,\ldots,T)\right\},(s=1,\ldots,S)$:

$$
\boldsymbol{x},\boldsymbol{y},\boldsymbol{w},\boldsymbol{p}\in\Phi_d=\left\{\boldsymbol{x},\boldsymbol{y},\boldsymbol{w},\boldsymbol{p}\;\middle|\;
\begin{array}{l}
y_{ij}^{(t)} \geq D_{sij}^{(t)}+\tilde{\alpha}C_{CSR}^{(t)},\;(i=1,2,\ldots,p,\; j=1,2,\ldots,n,\; t=1,2,\ldots,T, s=1,2,\ldots,S) \\
\sum_{j=1}^{n} a_{ij}^{(t)}x_j^{(t)} \leq b_k^{(t)}+\sum_{j=1}^{n} a_{ij}^{(t)}g_j^{(t-1)}\left(p_j^{(t-1)}\right),\;(k=1,2,\ldots,m,\; t=1,2,\ldots,T) \\
\sum_{l=1}^{t} g_j^{(l)}\left(p_j^{(l)}\right) \leq \sum_{l=1}^{t}\left(\sum_{i=1}^{p} y_{ij}^{(l)}-x_j^{(l)}+w_j^{(l)}-w_j^{(l-1)}\right),\;(j=1,2,\ldots,n,\; t=1,2,\ldots,T) \\
\sum_{j=1}^{n} w_j^{(t)} \leq W,\;(j=1,2,\ldots,n,\; t=1,2,\ldots,T) \\
\sum_{t=1}^{T}\left(\sum_{j=1}^{n}\left(c_j^{(t)}x_j^{(t)}+\sum_{i=1}^{p} u_{ij}^{(t)}y_{ij}^{(t)}+h_jw_j^{(t)}+v_jg_j^{(t)}\left(p_j^{(t)}\right)+p_j^{(t)}\right)+C_{CSR}^{(t)}\right) \leq C, \\
x_j^{(t)} \geq 0,\; y_{ij}^{(t)} \geq 0,\; w_j^{(t)} \geq 0,\; p_j^{(t)} \geq 0,\;\; (i=1,2,\ldots,p,\; j=1,2,\ldots,n,\; t=1,2,\ldots,T)
\end{array}
\right\} \tag{11}
$$

Furthermore, introducing parameters η_s,ξ_s, problem (10) is equivalently transformed into the following problem:

$$
\begin{aligned}
&\underset{\boldsymbol{x},\boldsymbol{y},\boldsymbol{w},\boldsymbol{p},\gamma_1}{\text{minimize}} \;\; \gamma_1+\frac{1}{1-\beta_1}\sum_{s=1}^{S} p_s\eta_s \\
&\underset{\boldsymbol{p},\gamma_2}{\text{minimize}} \;\; \gamma_2+\frac{1}{1-\beta_2}\sum_{s=1}^{S} p_s\xi_s \\
&\text{subject to } \;\eta_s \geq \sum_{t=1}^{T}\left(\sum_{j=1}^{n}\left(c_j^{(t)}x_j^{(t)}+\sum_{i=1}^{p} u_{ij}^{(t)}y_{ij}^{(t)}+h_jw_j^{(t)}+v_jg_j^{(t)}\left(p_j^{(t)}\right)-p_j^{(t)}\right)+\left(1-\tilde{\alpha}\sum_{i=1}^{p}\sum_{j=1}^{n} r_j^{(t)}\right)C_{CSR}^{(t)}-\sum_{i=1}^{p}\sum_{j=1}^{n} r_j^{(t)}D_{ij}^{(t)}\right)-\gamma_1, \\
&\eta_s \geq 0, \\
&\xi_s \geq -\frac{\sum_{t=1}^{T}\sum_{j=1}^{n} g_j^{(t)}\left(p_j^{(t)}\right)}{\sum_{t=1}^{T}\sum_{j=1}^{n} D_{sij}^{(t)}+\tilde{\alpha}C_{CSR}^{(t)}}-\gamma_2,\; \xi_s \geq 0, \\
&\boldsymbol{x},\boldsymbol{y},\boldsymbol{w},\boldsymbol{p}\in\Phi_d
\end{aligned} \tag{12}
$$

This problem includes only linear constraints, but it is hard to find an unique optimal decision for problem (12) analytically due to fuzzy number $\tilde{\alpha}$ and multi-criteria. Therefore, in order to solve this problem analytically, we introduce fuzzy goals as the fuzzy programming approach.

Let $Z_{CVaR}^1 = \left(\gamma_1 + \frac{1}{1-\beta_1} \sum_{s=1}^{S} p_s \eta_s \right)$, $Z_{CVaR}^2 = \left(\gamma_2 + \frac{1}{1-\beta_2} \sum_{s=1}^{S} p_s \xi_s \right)$. We consider the necessity measure for fuzzy constraints including $\tilde{\alpha}$, and the following linear fuzzy goals $\mu_{G_1}(\omega)$ and $\mu_{G_2}(\omega)$ for multi-criteria Z_{CVaR}^1 and Z_{CVaR}^2, respectivly:

$$\mu_{G_i}(\omega) = \frac{Z_{CVaR}^{i,\max} - \omega}{Z_{CVaR}^{i,\max} - Z_{CVaR}^{i,\min}},\ i = 1,2 \tag{13}$$

where all max and min values are constant determined by the decision maker. Using these fuzzy goals and necessity measure $\inf \max\left\{1 - \mu_{\tilde{\alpha}}(\omega), \mu_{G_i}(\omega)\right\}$, problem (12) is transformed into the following model maximizing satisfaction level h:

$$\begin{aligned}
&\text{maximize } h \\
&\text{subject to } \gamma_1 + \frac{1}{1-\beta_1} \sum_{s=1}^{S} p_s \eta_s \le (1-h) Z_{CVaR}^{1,\max} + h Z_{CVaR}^{1,\min}, \\
&\gamma_2 + \frac{1}{1-\beta_2} \sum_{s=1}^{S} p_s \xi_s \le (1-h) Z_{CVaR}^{2,\max} + h Z_{CVaR}^{2,\min}, \\
&\eta_s \ge \sum_{t=1}^{T} \left(\sum_{j=1}^{n} \left(c_j^{(t)} x_j^{(t)} + \sum_{i=1}^{p} u_{ij}^{(t)} y_{ij}^{(t)} + h_j w_j^{(t)} + v_j g_j^{(t)} \left(p_j^{(t)} \right) - p_j^{(t)} \right) \right. \\
&\left. + \left(1 - \left(\bar{\alpha} - L^*(1-h) \right) \sum_{i=1}^{p} \sum_{j=1}^{n} r_j^{(t)} \right) C_{CSR}^{(t)} - \sum_{i=1}^{p} \sum_{j=1}^{n} r_j^{(t)} D_{ij}^{(t)} \right) - \gamma_1, \\
&\eta_s \ge 0, \\
&\xi_s \ge - \frac{\sum_{t=1}^{T} \sum_{j=1}^{n} g_j^{(t)} \left(p_j^{(t)} \right)}{\sum_{t=1}^{T} \sum_{j=1}^{n} D_{sij}^{(t)} + \left(\left(\bar{\alpha} - L^*(1-h) \right) \right) C_{CSR}^{(t)}} - \gamma_2,\ \xi_s \ge 0, \\
&\boldsymbol{x}, \boldsymbol{y}, \boldsymbol{w}, \boldsymbol{p} \in \Phi_d
\end{aligned} \tag{14}$$

where $L^*(h)$ is the inverse function of menbership function $L(\omega)$. This problem is equivalent to a standard linear programming problem if parameter h is fixed. Therefore, it is easy to solve this problem with fixed value $\bar{h}$ analytically and efficiently using linear programming approaches such as the Simplex method and Interior point method. Furthermore, using the bisection algorithm on h, we obtain the strict optimal production volume of the main problem (10). Consequently, we have proposed the versatile model for the closed-loop supply chain management considering the CSR activities, and developed the analytical solution algorithm based on the standard linear programming approaches.

4. Numerical example

In order to compare our proposed closed-loop supply chain model considering CSR with the previous standard models, we introduce the following simple numerical example. We assume that two periods (t=2), three products (n=3), and three retailers (p=3). To simplify, these products consists of one resource (k=1). Tables 1 and 2 show data of parameters with respect to returns, production cost, expenses for CSR, and demands in all retailers in each period which are assumed to be uniform distributions. Transportation, holding, and recycling costs are assumed to be same in two periods. Furthermore, we assume that $\tilde{\alpha}$ is a triangle fuzzy number characterized by $\mu_{\tilde{\alpha}} = (0.05, 0.02)_L$.

	Product A	Product B	Product C
$r_j^{(1)}$	20	40	20
$c_j^{(1)}$	3	10	5
$D_{1j}^{(1)}$	[25,35]	[50,60]	[37,43]
$D_{2j}^{(1)}$	[24,26]	[54,56]	[39,41]
$D_{3j}^{(1)}$	[32,38]	[40,60]	[39,41]
$r_j^{(2)}$	40	20	20
$c_j^{(2)}$	10	3	5
$D_{1j}^{(2)}$	[49,51]	[25,35]	[35,45]
$D_{2j}^{(2)}$	[50,70]	[19,21]	[35,45]
$D_{3j}^{(2)}$	[49,51]	[19,21]	[30,50]

Table 1. Data of parameters in each period

	Product A	Product B	Product C
u_j	0.5	1	0.5
h_j	4	1	3
v_j	10	8	8
a_j	1.5	2	4

Table 2. Data of parameters for common costs

We set b=2000, W=50, C=10000, and the following fuzzy goals:

$$\mu_{G_1}(\omega)=\frac{12000-\omega}{12000-10000},\mu_{G_2}(\omega)=\frac{-0.5-\omega}{-0.5-(-0.7)}$$

Then, we assume function of recycling volume is a linear function, i.e., $g_j^{(t)}\left(p_j^{(t)}\right)=2p_j^{(t)}$, and solve the proposed model (14) and the standard model without CSR object, and obtain the following optimal solution.

	Product A		Product B		Product C	
Period	1	2	1	2	1	2
With CSR	149	124.5	176	79.5	125	142.5
Not CSR	149	122	176	77	125	140

Table 3. Optimal solution of production volume

From this result, we find that the whole production volume tends to be larger than the standard model due to consideration of CSR. Furthermore, volumes of all products in period 2 of proposed model are much larger than standard model. Results of CSR activities in our proposed model $C_{CSR}^{(1)}$ and $C_{CSR}^{(2)}$ are 0 and 16.4, respectively. This means that the CSR activity for promotion and recycling is mainly performed in the second period.

Furthermore, we consider some cases of $g_j^{(t)}\left(p_j^{(t)}\right)$; linear function $g_j^{(t)}\left(p_j^{(t)}\right)=2p_j^{(t)}$, convex functions $g_j^{(t)}\left(p_j^{(t)}\right)=\left(p_j^{(t)}\right)^2, \exp\left(p_j^{(t)}\right)$, and concave function $g_j^{(t)}\left(p_j^{(t)}\right)=\sqrt{p_j^{(t)}}$, respectively. Then, we solve the proposed model (14) considering CSR for each function, and obtain the following optimal production volume and the optimal profit as Table 4.

$g_j^{(t)}\left(p_j^{(t)}\right)$	Product A		Product B		Product C		Profit
	1	2	1	2	1	2	
$2p_j^{(t)}$	149	124.5	176	79.5	125	142.5	11606
$\left(p_j^{(t)}\right)^2$	149.4	122.1	176.4	77.1	125.4	140.1	12496
$\exp\left(p_j^{(t)}\right)$	149.1	122.3	176.1	77.3	125.1	140.3	12493
$\sqrt{p_j^{(t)}}$	154.5	129.8	181.5	84.8	130.5	147.8	9106

Table 4. Optimal solution of production volume to each function $g_j^{(t)}\left(p_j^{(t)}\right)$

From Table 4., in the case of concave function, whole production volume tends to be larger than the linear and convex-based proposed models. On the contrary, the total profit in the case of concave function is much smaller than linear and convex functions. This means that the cost derived from the CSR activity is not used as promotion activity to increase the customers' demands, but used as the recycling activity to collect old-products.

5. Conclusion

In this paper, we have considered the optimal production decision model with CSR and recycling of old-products in the closed-loop supply chain. We have integrated the evaluation of environmental conservation activities and resource recycling as CSR activities into the closed-loop supply chain management, and proposed the new optimal production policy based on CSR. In mathematical programming, the proposed model including CSR activities have been formulated as stochastic and fuzzy programming problem due to linguistic information and subjectivities as well as random demands, and we have transformed the initial proposed model into the deterministic equivalent problem using CVaR as the advanced risk measure and fuzzy goals for the multi-criteria model. Consequently, we have developed the analytical and effective solution algorithm based on linear programming. Furthermore, in order to represent useful features and advantages of the proposed model, we have provided a numerical example, and obtained that the whole production volume for the CSR-based model was larger than the standard model and that the cost derived from the CSR activity is used as promotion activity or the recycling activity according to the continuous function for the recycling volume.

As the future works, we will apply this proposed model to practical systems of actual production companies, and evaluate the performance or advantage of the proposed model compared with the previous standard and useful models using real numerical and linguistic data.

6. References

Bloemhof-Ruwaard, J.M., Beek, P., Hordijk, L., VanWassenhove, L.N., "Interactions between operational research and environmental management", *European Journal of Operational Research*, 85, pp. 229-243, 1995

Carroll, A.B., "A three-dimensional conceptual model of corporate social performance", *Academy of Management Review*, 4(4), pp. 497-505, 1979

Carter, C.R., Jennings, M.M., "Social responsibility and supply chain relationships", *Transportation Research Part E*, 38E(1), pp. 37-52, 2002

Carter, C.R., Jennings, M.M., "The role of purchasing in corporate social responsibility: a structural equation analysis", *Journal of Business Logistics*, 25(1), pp. 145-186, 2004

Castka, P., Balzarova, M., "ISO 26000 and supply chains - on the diffusion of the social responsibility standard", *International Journal of Production Economics*, 111, pp. 274-286, 2008

Cruz, J.M., "The impact of corporate social responsibility in supply chain management: Multicriteria decision-making approach", *Decision Support Systems*, 48, pp. 224-236, 2009

Cruz, J.M., Wakolbinger, T., "Multiperiod effects of corporate social responsibility on supply chain networks, transaction costs, emissions, and risk", *International Journal of Production Economics*, 116, pp.61-74, 2008

Emmelhainz, M., Adams, R., "The apparel industry response to "sweatshop" concerns: A review and analysis of codes of conduct", *The Journal of Supply Chain Management*, 35(3), pp. 51-57, 1999

Hasuike, T., Ishii, H., "On flexible product-mix decision problems under randomness and fuzziness", *Omega*, 37(4), pp. 770-787, 2009a

Hasuike, T., Ishii, H., "Product mix problems considering several probabilistic conditions and flexibility of constraints", *Computers & Industrial Engineering*, 56(3), pp. 918-936, 2009b

Kang, K.H., Lee, S., Huh, C., "Impacts of positive and negative corporate social responsibility activities on company performance in the hospitality industry", *International Journal of Hospitality Management*, 29, pp.72-82, 2010

Kolk, A., Tudder, R.V., "The effectiviness of self-regulation: Corporate codes of conduct and child labour", *European Management Journal*, 20(3), pp. 260-271, 2002

Letmathe, P., Balakrishnan, N., "Environmental considerations on the optimal product mix", *European Journal of Operational Research*,167, pp. 398-412, 2005

Lebreton, B., *Strategic Closed-Loop Supply Chain Management*, Lecture Notes in Economics and Mathematical Systems 586, Springer, 2007.

Li, S., Tirupati, D.. "Impact of product mix flexibility and allocation policies on technology", *Computers & Operations Research*, 24(7), pp. 611-627, 1997

Morgan, L.O., Daniels, R.L., "Integrating product mix and technology adoption decisions: a portfolio approach for evaluating advanced technologies in the automobile industry", *Journal of Operations Management*, 19, pp. 219-238, 2001

Mula, J., Poler, R., Garcia-Sabater, J.P., Lario, F.C., "Models for production planning under uncertainty: a review", *International Journal of Production Economics*, 103, pp. 271-285, 2006a

Mula, J., Poler, R., Garcia-Sabater, J.P., "MRP with flexible constraints: a fuzzy mathematical programming approach", *Fuzzy Sets and Systems*, 157, pp. 74-97, 2006b

Mula, J., Poler, R., Garcia-Sabater, J.P., "Material requirement planning with fuzzy constraints and fuzzy coefficients", *Fuzzy Sets and Systems*, 158, pp. 783-793, 2007

Murphy, P.R., Poist, R.F., "Socially responsible logistics: an exploratory study", *Transportation Journal*, 41(4), pp. 2335, 2002

Panwer, R., Han, X., Hansen, E., "A demographic examination of societal views regarding corporate social responsibility in the US forest products industry", *Forest Policy and Economics*, 12, pp. 121-128, 2010

Perry, P., Towers, N. "Determining the antecedents for a strategy of corporate social responsibility by small- and medium-sized enterprises in the UK fashion apparel industry", *Journal of Retailing and Consumer Services*, 16, pp. 377-385, 2009

Rockafellar, T.R., Uryasev, S., "Conditional value-at-risk for general loss distributions", *Journal of Banking and Finance*, 26, pp. 1443-1471, 2002

Svendsen, A., Boutilier, R.G., Abbott, R.M., Wheeler, D., "Measuring the business value of stakeholder relationships (part one)". *CAmagazine* August, pp. 29-63, 2001

Vasant, P., "Fuzzy decision making of profit function in production planning using S-curve membership function", *Computers & Industrial Engineering*, 51, pp. 715-725, 2000

Wartick, S.L., Coghran, P.L., "The evolution of the corporate social performance model", *Academy of Management Review*, 10(4), pp. 758-769, 1985

12

Research on Multi-Step Active Disassembly Method of Products Based on ADSM

Zhifeng Liu, Xinyu Li, Huanbo Cheng and Yifei Zhan
Hefei University of Technology
China

1. Introduction

This chapter discusses the principles of multi-step active disassembly, proposes the method of products multi-step active disassembly and divides the step of product parts according to the step division principle of multi-step active disassembly products. In addition, this chapter also proposes step division process of multi-step active disassembly products and determines parts in each step according to the process. Materials which have the same trigger medium and different trigger strength are used as active disassembly material, ensuring that trigger strength (such as temperature, magnetic field strength, etc) of active disassembly device which is in different disassembly step forms gradient. Trigger strength of active disassembly device increases along with the disassembly step from low to high. The joints which are in the same disassembly step use the active disassembly devices which have the same trigger strength. In different disassembly steps, install active disassembly parts according to the gradient. Lastly, disassemble the products by sending it to the different work areas.

2. Active disassembly

As the concept of circular economy and sustainable development is gradually extended, the requirement of recycling of the life cycle terminal electronic products after they are discarded becomes increasingly stringent. At present, electronic products are disassembled by means of manual disassembly or mechanical crushing. But the efficient of manual disassembly is very low and mechanical crushing will damage reusable parts or components. To solve these problems, the concept of active disassembly is proposed. The product is disassembled when it is heated to a specific temperature by using active disassembly device instead of traditional connecting device.

2.1 The concept of active disassembly

Active Disassembly uses active disassembly device instead of traditional device like the snap -fit, rivet and thread connection. Active disassembly device deforms when it is in certain external trigger conditions. At last, the parts of products are disassembled. This technique mainly uses the properties of shape memory materials. It is also known as active disassembly using smart materials (Chiodo et al., 1998, 1999, 2002; Suga & Hosada, 2000).

2.2 The characteristics of active disassembly device

ADSM (Active Disassembly using Smart Materials) uses the principle of SMA (Shape Memory Alloy) or SMP (Shape Memory Polymer) that they can automatically return to its original shape under certain circumstances to make active disassembly devices using SMA or SMP, such as smart driver and active separation snap-fit. Active disassembly device is placed to the products during the stage of the design and assembly of the products. These products are placed on the trigger conditions of active disassembly device, such as temperature, when they are recycled. The products will be disassembled by these active disassembly devices. So ADSM makes disassembly and recycling of products very convenient.

The deformation force of SMA is very large when SMA return to its original shape. So SMA can be used to make active disassembly device which provides driving force, such as coil spring, splint pin, ribbon, cotter pin, rivet, tube and so on. The deformation force of SMP is very low, but its deformation is 400%, so SMP can be used to make active disassembly device, such as screw, snap-fit, rivet, washer, zipper snap-fit and so on. Table 1 shows several typical active disassembly devices (Chiodo et al).

Active disassembly device	Before trigger	After trigger
SMA coil spring		
SMA rivet		
SMA ribbon		
SMA tube	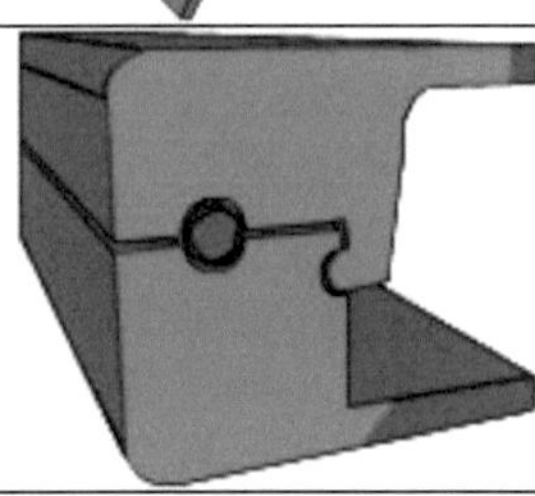	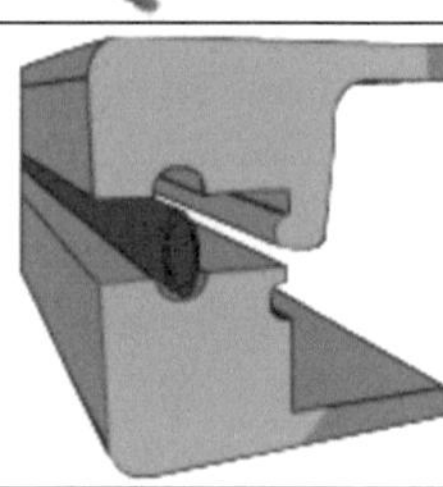

SMA splint pin	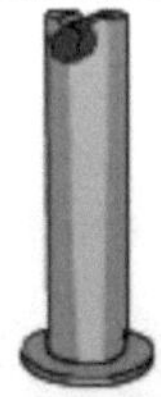	
SMP screw	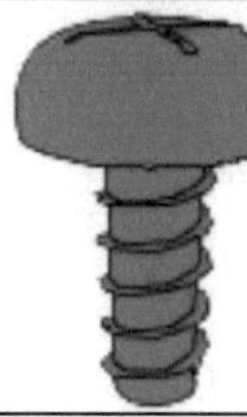	
SMP cotter pin		
SMP Snap Fastener 1		
SMP Snap Fastener 2	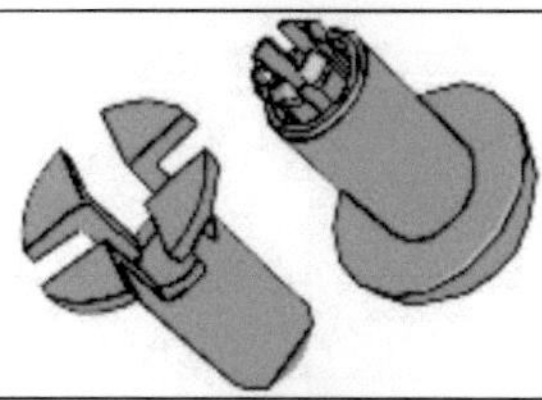	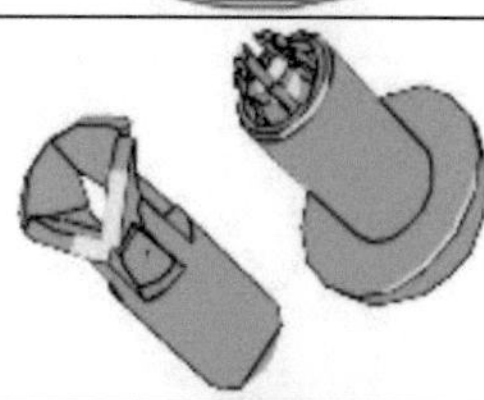
SMP Snap Fastener3	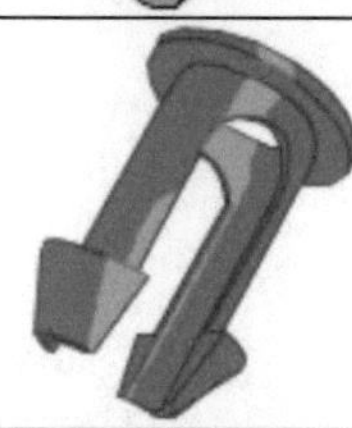	
SMP clamping washer		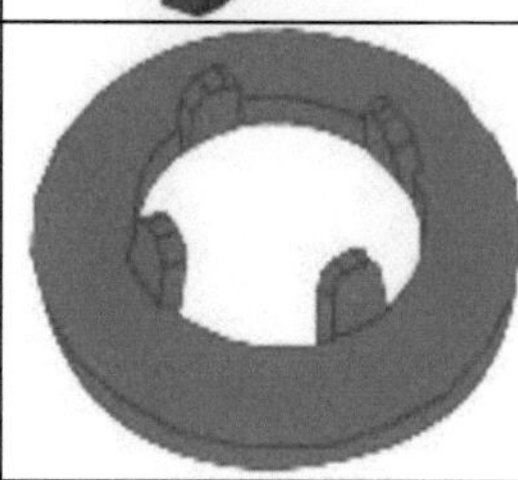

Table 1. Typical active disassembly devices (Chiodo et al,)

The most common two broad categories of active disassembly devices in Table 1 are snap-fit and screw. Screw type is generally made of shape memory polymer. The threads disappear or soften after screw is triggered by heat. So SMP screw loses its connecting function. The advantages of SMP screw are that the structure is simple and the cost is low. SMP screw is usually used in places where connection strength are not high. The connection strength of snap-fit is very high. But its disadvantages are complicated structure and high cost. SMP snap-fit is usually used in places where connection strength is high or where special requirements are needed appropriate active disassembly device will be chosen according to the needs of products device when designing products.

When a product is designed using active disassembly device, the initial device should be designed by traditional method at first. Then the material will be chosen according to initial device and the environmental conditions and the active disassembly device will be designed. After that the active disassembly device will be arranged in the suitable place to achieve the active disassembly. At last the device and façade should be optimized.

2.3 Trigger methods of typical active disassembly device

Trigger ways of active disassembly device as shown in Table 2 (Willems, 2005).

Trigger ways	Figure	Work instructions	Requirements	principle
Centrifugal force	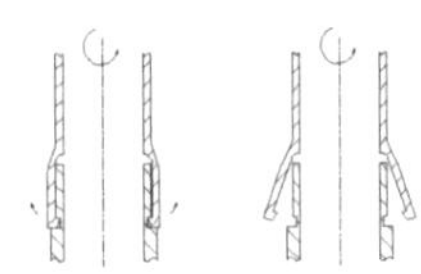	Use centrifugal force to make snap-fit release	Axisymmetric device	special device
Vibration	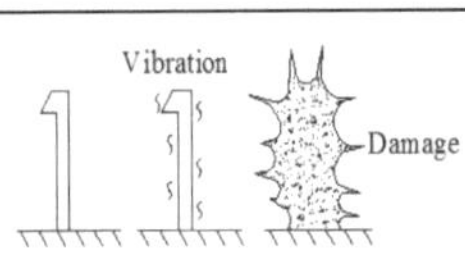	Fasteners collapse in a particular vibration frequency	Device with default/ modify characteristic frequency	Special device or material
Barometric pressure	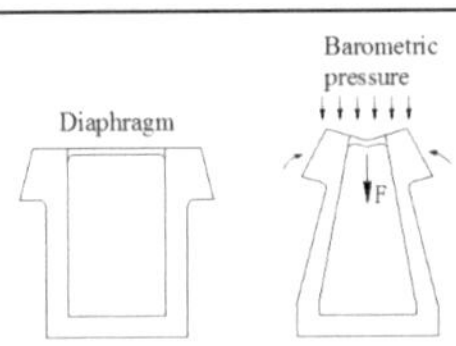	Increase air pressure make fasteners deform so that connection failures	Closed cavities filled with air	Special device
Current	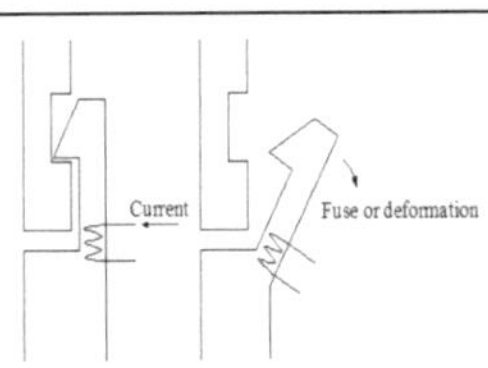	The internal resistance wire in connector melt the plastic around when it is energized and make connec- -tion failure	Bury the resistance wire during the manufacture of connectors	Special material

Dissolve	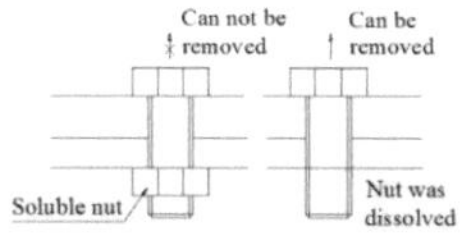	Fasteners which are made of soluble materials are dissolved away during disassembly	Suitable soluble materials	Special material
Heating	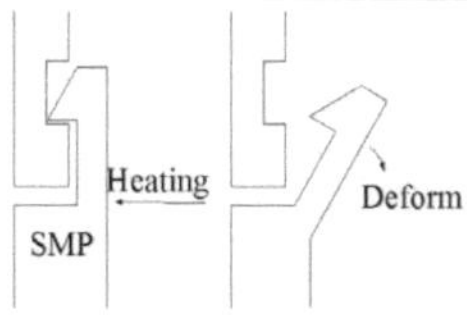	Deformation at elevated temperatures of shape memory materials / composites of two materials (such as bimetal)	Shape memory material or two bistable composite material	Special material
Magnetic field	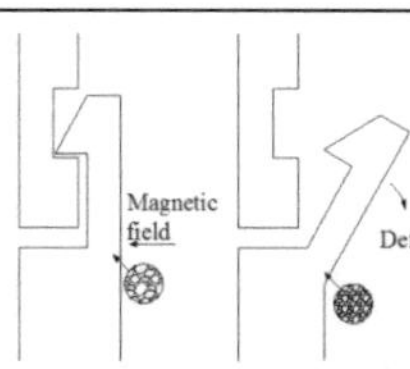	Magnetoresis- -tive materials deform under the magnetic field trigger	Magnetoresistive particular material with Special shape	Special material

Table 2. Trigger ways of active disassembly devices (Willems, 2005).

The general trigger way of active disassembly device is heating.There are five main heating trigger methods to stimulate active disassembly device: air convection heating such as air bath, immersion heating such as water bath heating, microwave heating, far infrared heating and induction heating. Five methods have advantages and disadvantages separately, as shown in Table 3 (Li, 2008).

Aimed at the active disassembly device, 5 heating ways are normal used. Water bath or air bath heats the whole product, which may damage the components and parts that have bad heat resistance, besides heat is dispersed as a result of costing more energy, and it takes a long time for heat to transmit from the surface to the inside. So the trigger time is uncontrollable and the disassembly efficiency reduces. To avoid the problems mentioned above, the active disassembly device triggered by electric heating has been studied deeply. This active disassembly device can be classified into 3 steps as follows:

The active disassembly drive parts made from SMA can be electrified directly to generate applicable force after deformation by electric heating to achieve the separation between components and parts. The trigger time can be controlled by the electric power of active disassembly drive parts (Z. Liu et al., 2011).

The snap-fit made from SMP can be electrified through the electric heater band glued at root of SMP snap-fit or heating wire in the root. When electric heater band or heating wire is electrified, the electric heat generates the enough deformation to make the connection lost its effect. The trigger time of SMP snap-fit can be controlled by the electric power of electric heater band or heating wire (Z. Liu et al., 2011).

Heating method	Advantage	Disadvantage	Scope
Air convection heating	Clean and pollution-free, simple equipment	Heating efficiency is low, the internal heating effect of the product is bad	Smaller products
Immersion heating	High heating efficiency, simple equipment	Erosion or pollution of components	Suitable for parts of products no longer reuse
Microwave heating	High heating efficiency, no affecting for the volume	Unable to inspire the SMP, may damage or even burnt parts	Suitable for SMA active dis-assembly device
Far infrared heating	Low energy consumption, heating uniformity	The internal heating effect of the product is bad	Suitable for smaller products
Induction heating	High heating efficiency, uniform heating	Energy consumption, unable to inspire the SMP, electronic devices cannot be reused	Suitable for electronics products no longer reuse

Table 3. Advantages and disadvantages of 5 heating methods and their applications (Li, 2008).

The snap-fit made from the same metrical above can be electrified through the electric heater band glued at root of it. When electrified, the root of the SMP snap-fit will be fused by electric heating to achieve the separation between snap-fit and slot, so that the connection of components and parts will lose effectiveness to achieve the active disassembly. The fused time can be controlled by the electric power of heating wire (Z. Liu et al., 2011).

2.4 Application object of ADSM method

ADSM method is primarily applied to the following types of electrical and electronic products:

1. The products which need non-destructive disassembly, because components can be reused or pollute the environment after they are damaged, such as PC, LCD display, LCD TV and so on.
2. The products which are small and whose manual disassembly are extremely inconvenient, such as remotes, mobile phone, MP3, MP4 and so on.
3. The products which have complicated device or many components cause that the efficiency of manual disassembly is very low, such as precision instruments.
4. The products which have large output, serious shortage of recycling capacity and whose disassembly efficiency need to increase significantly.

The products which are not suitable for ADSM method are as follows.

1. The products which only recycle materials instead of recycling parts and recycle through overall fragmentation instead of disassembly, such as refrigerator.
2. The products which are mainly composed of metallic materials and have requirements of high structural strength, such as Air conditioner outdoor unit.

For example, based on the design theory and method of ADSM, some remote controls are redesigned to put the active disassembly device in. The active disassembly experiments of redesigned remotes show that the cost of life circle reduces while the disassembly property is obviously enhanced. The disassembly efficiency is 300 times enhanced with no pollution and damage for the environment. So the active disassembly devices are suitable for large-scale industrial production (G. Liu et al., 2008), as shown in the Fig.1.

Fig. 1. The active disassembly efficiency of the shells of remote controls (G. Liu et al., 2008).

For medium-sized electronic products such as monitors, television sets, the results are unsatisfactory. Because these products are larger and the time for heat transmitting to the internal active disassembly device of product is longer. As a result, heating time has to be extended. It is difficult to achieve complete dismantling through one trigger. Efficiency of disassembly is very low and components with poor heat-resistant are damaged. Because medium-sized electronic products have a lot of components, all completely disassembled parts are mixed together and create difficulties for subsequent sorting. So it increases the workload of sorting parts. In a word, it is important to research on the method of products multi-step active disassembly deeply and achieve step active disassembly of products components.

3. The multi-step active disassembly of products

The multi-step active disassembly methods mainly aim at the condition that the distance between the internal active disassembly devices and the shells of products is so far that it is difficult to be disassembled by single-step active disassembly. On the other hand, the parts from single-step active disassembly are mixed up. That goes against the follow-up material recycling, so disassembly efficiency is low. By using the multi-step active disassembly, the products can be divided into multi-step for disassembling. In each step, one or several kinds of components will be disassembled from the products, which can save time of disassembly and be benefit to the follow-up material sorting, and it reduces the cost of products disassembly. Therefore, it can realize the automation of product recovery and assembly line work.

3.1 The conception of the multi-step active disassembly products

According to the properties of the shape memory materials, when different kinds of material are in the effect of certain trigger medium(such as infrared ray heating, microwave heating, water bath, etc), the trigger strength they need is not the same. Thus products can be designed in different kinds of trigger medium or increasing trigger strength in proper order in the same trigger medium. When these recycled products are sent to different disassembling areas, one or several kinds of components are disassembled in each step, until the products are completely disassembled, in order to achieve the purpose of grading disassembly. This method is called multi-step active disassembly (Zhao et al., 2011a)

When it is disconnected between the internal active disassembly devices and the housings of products, multi-step active disassembly can save the disassembling time and improve the disassembling efficiency observably. For some products that have complex device or large volume, the trigger speed of the internal active disassembly devices is low, which leads to the decrease of the efficiency of entire disassembly. In this case, different kinds of active disassembly devices with different trigger strength can be used in the stage of products design to achieve the active disassembly of components and parts.

3.2 The principle and method to achieve multi-step active disassembly products

According to the principle of multi-step active disassembly, there are two kinds of methods or their combination to achieve multi-step active disassembly (Zhao et al., 2011a).

1. Different trigger media: In different disassembling steps, active disassembly devices belong to the connection location use different trigger media. While in the same disassembling step, active disassembly devices belong to the connection location use the same trigger medium. The active disassembly device in different disassembling steps is triggered by different excitation ways, and multi-step active disassembly can be achieved.

For example, the method of air heating can be used to trigger SMP snap-fit to achieve the separation of the components and parts. And the field effect can be used to trigger SMA driver parts to achieve the separation. In this way the purpose of multi-step active disassembly can be achieved.

2. Same trigger medium, but the trigger strength increases in gradient way: Using shape memory materials with same trigger medium but different trigger strength as the active

disassembly devices in the design phase of products, it makes trigger strength(such as temperature, magnetic field strength) of active disassembly devices in different disassembling steps form a gradient. According to the disassembling step from low to high principle, the trigger strength of active disassembly device appears a gradient rise. In the same disassembling step, the trigger strength of active disassembly device is the same. Install the active disassembly devices in different disassembling step according to the gradient, and then send the products to each step of working area to complete the disassembly.

For example, by physical cross-linking irradiation, PVC can obtain property of shape memory, when the irradiation dose is 100kGy, the trigger temperature is 65°C, but the irradiation dose of the shape memory alloy is 75°C. Therefore, we can disassemble adapting piece made of shape memory polymer materials in 65 temperature field, and disassemble adapting piece made of shape memory alloy in 75 temperature field, in order to achieve the purpose of multi-step active disassembly.

According to the property of PVC, if the irradiation dose is 100kGy, the trigger temperature of PVC is 65°C, and if the irradiation dose is 4kGy, the trigger temperature of it is 80°C. So in the design phase of products, active disassembly devices with different trigger temperature can be used in different disassembling step of the products, in order to achieve the purpose of multi-step active disassembly.

Using temperature as the trigger medium of the active disassembly devices of products, connection location in products is divided into different disassembling steps according to the position and the heat transfer performance of materials around the products. The inner or worse heat transfer performance of materials around, the higher of the disassembling step. Active disassembly devices located in the same disassembling step has the same trigger temperature. The trigger temperature of the active disassembly device appears gradient rise according to the disassembling steps from low to high. The temperature difference between the trigger temperature of adjacent disassembling step shouldn't be less than 5°C.

Check the strength and the deformability of the shape memory materials in the choice of materials, and then choose the trigger temperature. Generally speaking, the trigger temperature is higher than the temperature the product is normally used, but it shouldn't be too high to damage the parts which don't have high temperature resistant during disassembly. In a general way, the SMP materials whose trigger temperature is between 60°C and 150°C are selected. The SMP materials' trigger temperature by irradiation modification is related to their molecular crosslinking degree, the molecular crosslinking degree increases with the radiation dose, until reach a steady value.

The shape memory performance of the SMP mainly depends on its material type and the molecular crosslinking degree. The shape memory performance of the commonly used polymer material PVC which goes through irradiation modification is the best; the mechanical strength and corrosion resistance ability of PVC are also well, it is appropriate for being used as material of active disassembly devices. For the same polymer material, the bigger of the radiation dose, in other words, the greater of the molecular crosslinking degree, the lower of the trigger temperature. In the same disassembling step, the connect parts use active disassembly devices with the same trigger temperature. On the contrary, in different disassembly steps, the connect parts use active disassembly devices with different trigger temperature, which can achieve the purpose of multi-step active disassembly.

Another way to trigger the active disassembly device by temperature is through the heating wire method (Z.Liu et al, 2011), by using the controllability of the trigger time, the multi-step active disassembly products can be achieved. Considering the disassembly sequence of active disassembly parts in each step in the design phase of products, electric heating elements with different power are chosen according to the trigger time, then the fully automatic disassembly of products can be achieved. Electric heating elements with different power are arranged in connection of active disassembly devices, and the time that the active disassembly devices are triggered is controlled by adjusting the power of the electric heating elements. In the same disassembling step, the power of electric heating elements which is needed in active disassembly device that belongs to the connect parts is the same, on the contrary, in different steps, it forms a certain power gradient, which can achieve the purpose of multi-step active disassembly by using the difference of the power between each trigger level.

4. The design of multi-step active disassembly products

The key of multi-step active disassembly design is to classify the products into several disassembly steps, and to set the trigger intension of active disassembly device in connections in every step. The trigger intension of active disassembly device in different disassembly steps is different, while that in the same disassembly step is the same. The components and parts in every disassembly step should be set by the gradient, and disassembled in the work area of every step.

4.1 The principle of division of multi-step active disassembly products

The principle of division of multi-step active disassembly products is shown in the Fig.2.

The purpose of multi-step active disassembly is to save time. It is conducive to recycle the materials if the components and parts with the same or compatible materials are disassembled in the same step. When classify the steps of multi-step active disassembly products, the below principles should be followed (Z. Liu et al., 2010; Zhao et al., 2011a).

1. The principle of disassembly

There are some differences between disassembling properties of multi-step active disassembly and the traditional disassembling properties. The later only consider the restraint relationship among all the parts, as shown in Fig.3. Part 1 is chosen as the basic part (the part will be disassembled last), and the disassembly property of every part from direction (+X, +Y, +Z, -X, -Y, -Z) can be determined by the method of interference-freedom matrix (Wang et al., 2005a, 2005b, 2006).

$$DAW(2)=(F_{1,2,1}, F_{2,2,1}, F_{3,2,1}, F_{4,2,1}, F_{5,2,1}, F_{6,2,1}) \wedge(F_{1,2,3}, F_{2,2,3}, F_{3,2,3}, F_{4,2,3}, F_{5,2,3}, F_{6,2,3})$$

$$\wedge(F_{1,2,4}, F_{2,2,4}, F_{3,2,4}, F_{4,2,4}, F_{5,2,4}, F_{6,2,4})$$

$$=(1,1,1,1,1,0) \wedge(0,1,1,1,1,0) \wedge(0,1,1,1,1,1)= (0,1,1,1,1,0) \quad (1)$$

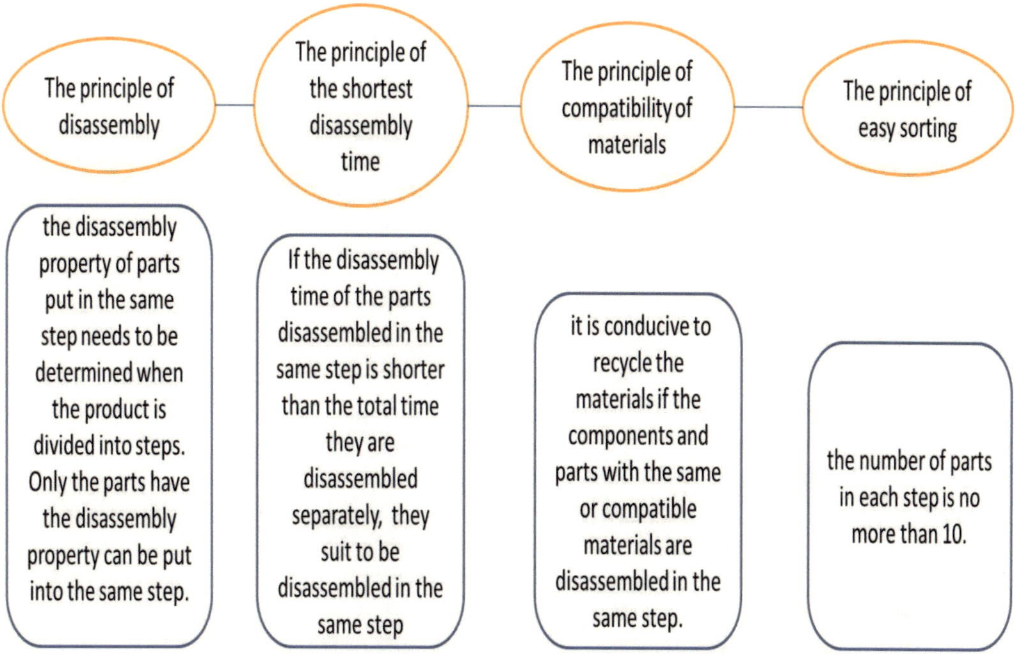

Fig. 2. The principle of division of multi-step active disassembly products

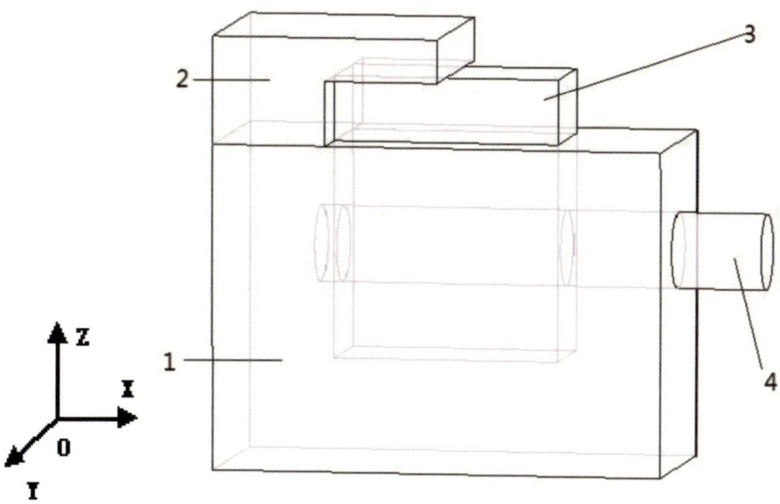

1. the foundation 2. top cover 3. the carriage 4. pin

Fig. 3. The model and its frame diagram (Zhao et al, 2011a)

$$DAW(2)=(F_{1,3,1}, F_{2,3,1}, F_{3,3,1}, F_{4,3,1}, F_{5,3,1}, F_{6,3,1}) \wedge (F_{1,3,2}, F_{2,3,2}, F_{3,3,2}, F_{4,3,2}, F_{5,3,2}, F_{6,3,2}) \wedge (F_{1,3,4}, F_{2,3,4}, F_{3,3,4}, F_{4,3,4}, F_{5,3,4}, F_{6,3,4})$$

$$=(0,0,1,0,0,0) \wedge (1,1,0,0,1,1) \wedge (0,0,0,1,0,0)= (0,0,0,0,0,0) \quad (2)$$

$$DAW(3)=(F_{1,4,1}, F_{2,4,1}, F_{3,4,1}, F_{4,4,1}, F_{5,4,1}, F_{6,4,1}) \wedge(F_{1,4,2}, F_{2,4,2}, F_{3,4,2}, F_{4,4,2}, F_{5,4,2}, F_{6,4,2})$$
$$\wedge(F_{1,4,3}, F_{2,4,3}, F_{3,4,3}, F_{4,4,3}, F_{5,4,3}, F_{6,4,3})$$
$$=(1,0,0,0,0,0) \wedge(1,1,0,1,1,1) \wedge(1,0,0,0,0,0)= (1,0,0,0,0,0) \quad (3)$$

Where, DAW denotes the disassembly property. For example, DAW(1) denotes the disassembly property of part 1.

By the calculation, we can see that the part 2 can be disassembled from the direction +Y, +Z, -X, -Y. The part 3 cannot be disassembled. Part 4 can be disassembled from the direction +X. That is, the part 2 and part 4 have the disassembly property, while the part 3 hasn't. So the part 3 can only be disassembled after disassembling part 2 and part 4 by the traditional disassembly method. However, whether the part can be disassembled or not needs to consider other parts probably in the same step when multi-step active disassembly divides steps. For instance, when the disassembly property of the part 3 is checked, the parts probably disassembled with it in the same step will be considered first. When the part 3 is arranged in independent disassembly step, by the interference-freedom matrix, the part 3 is calculated to have not the disassembly property because the parts 2 and 4 have interference for it. If the part 2, 3 and 4 is disassembled in the same disassembly step, the interference that the part 2 and 4 is to the part 3 should be wiped off when the disassembly property of the part 3 is calculated with the method of the interference-freedom matrix. Thus we can calculate that the part 3 has the disassembly property by the interference-freedom matrix. So the parts 2, 3 and 4 can be disassembled in the same step. In the same way, if the parts 2 and 3 are disassembled in the same step, the interference the part 2 is to the part 3 should be wiped off when the disassembly property of the part 3 is calculated with the method of the interference-freedom matrix. By the interference-freedom matrix, we can calculate that the part 3 has not the disassembly property because of the interference the part 4 to the part 3. So, the part 2 and 3 cannot be disassembled in the same step.

In summary, the disassembly property of parts put in the same disassembly step needs to be determined when the product is divided into disassembly steps. Only the parts which have the disassembly property can be put into the same disassembly step.

2. The principle of the shortest disassembly time

In order to reduce disassembly cost and raise disassembly efficiency, the disassembly time is expected to be as short as possible. Therefore, when classify the steps of product, ensure the disassembly time of the parts as they are disassembled in the same disassembly step is shorter than the time needed when they are disassembled separately. When two parts are disassembled in the same step, it will cost extra time for heat to pass to the other parts after one of them reaches the trigger temperature. Thus make the disassembly time of the parts disassembled in the same step longer than the total time they are disassembled separately. For example, when battery and battery compartment cover of cellphone are heated to disassemble in the same step, active disassembly of the battery compartment cover reaches the trigger temperature first, and then it will cost a period of time to pass heat from battery compartment cover to the battery. Thus make the disassembly time of the parts disassembled in the same step longer than the total time they are disassembled separately. So, the number of work area and the cost can be reduced for making the number of parts disassembled in the same step as more as possible. If the disassembly time(t_1) of the parts disassembled in the same step is shorter than the total time(t_2) they are disassembled separately, they suit to be disassembled in the same step. For example, when a product will be multi-step active disassembled by the way of air heating, the total time they are disassembled separately is t_1=25s, while the disassembly time of the parts disassembled in the same step is t_2=20s. $t_2<t_1$, so, the two parts are suited to be disassembled in the same step.

3. The principle of compatibility of materials

The manufacturing of every electronic product needs different materials because of functional requirements. Some of the parts have high value can be recycled or reused after manufacturing, but the materials of some parts are maybe harmful to the environment. These components and parts should not be disassembled together (Guo et al., 2005). When classify the steps of product, it is conducive to recycle the materials if the components and parts with the same or compatible materials are disassembled in the same step.

The components and parts which are hard to separate should be manufactured using the same or compatible material to increase the recycling rate. The compatibility of materials has two meanings: the one is the compatibility that two or more materials exist to fulfill the function and have no interference between each other during the using process of materials; the other is the materials can be recycled together, such as Polycarbonate and ABS used in the computer can be recycled together and obtain another material PC/ABS copolymers. The compatibility of ordinary engineering plastics is shown in Table 3-1 (Qi & Xu, 2007).

	PE	PVC	PS	PC	PP	PA	POM	SAN	ABS
PE	√	×	×	×	√	×	×	×	×
PVC	×	√	×	×	×	×	×	√	√
PS	×	×	√	×	×	×	×	×	×
PC	×	O	×	√	×	×	×	√	√
PP	O	×	×	×	√	×	×	×	×
PA	×	×	O	×	×	√	×	×	×
POM	×	×	×	×	×	×	√	×	×

notes：√ good compatibility, O normal compatibility, ×bad compatibility.

Table 4. the compatibility of ordinary engineering plastics (Qi & Xu, 2007)

4. The principle of easy sorting

The parts disassembled in each step need to be sorted. It is more convenient to sort the parts if each step contains fewer parts. Choose the sort of the electronic and electrical components in circuit board as an example. The number of the electronic and electrical components in circuit board is about 277, and they are connected to the circuit board by the surface mount technology and through hole technology. They have different disassembly methods. If all of electronic and electrical components are disassembled together, it is hard to sort each of them. But if a few of them are disassembled once, it will be easy to sort. Thus it will save the sorting time and raise the sorting efficiency. One provision here is that the number of parts in each step is not more than 10.

There is little or no consideration of the principles of classifying steps during the design stage of products. Therefore, when determining the steps of active disassembly products, first of all, it needs to meet the principle of disassembly and the principle of the shortest disassembly time, and then meet the principle of compatibility of materials and the principle of easy sorting as much as possible. In addition, the parts which are prone to be damaged at the high temperature should be disassembled at first. And the parts need different trigger media should be disassembled in different steps, which also requires the designers to adopt the same trigger medium to parts in the same step when classify the steps of products. Thus put designers forward a higher request, which requires them to consider the disassembly and recycle problems after products are abandoned when design the products. This is also the basic requirement of the Green design concept based on the life cycle of products.

4.2 The process of division of multi-step active disassembly products

According to the principles of division of multi-step active disassembly, a flow chart about the process of division of multi-step active disassembly products is worked out, as shown in Fig.4 (Zhao et al, 2011a). At first, number the parts of a product from outside to inside. The more inside the part is, the greater the number is. Then according to the process in Fig.4, finish dividing the steps of multi-step active disassembly.

Where i values between 1 and n, n denotes the quantity of the parts. When dividing the disassembly steps of a product by the flow chart, it needs to make the quantity of the parts in each step meet the principle of easy sorting, which will be conducive to sort materials afterward.

When the parts contained in each step of a product is determined, the disassembly of each step is allowed. In order to achieve the multi-step active disassembly products, shape memory materials with different trigger strength are used in product design, which makes the trigger strength(such as temperature, magnetic field, etc.) form a gradient. Then install the active disassembly parts in each step according to the gradient. Finally the products will be disassembled in the work area of each step.

According to the flow chart of division of multi-step active disassembly, the steps of a product can be divided based on the following procedures (Zhao, 2011b):

Procedure 1: create the product model and analyze the connection way of the product.

Product model is the foundation and premise of disassembly research, which makes the disassembly process be simulated on computer. Industrial design software provides adequate possibility and convenience for creating product model. After the product is designed, the connection way of each part of a product needs to be analyzed.

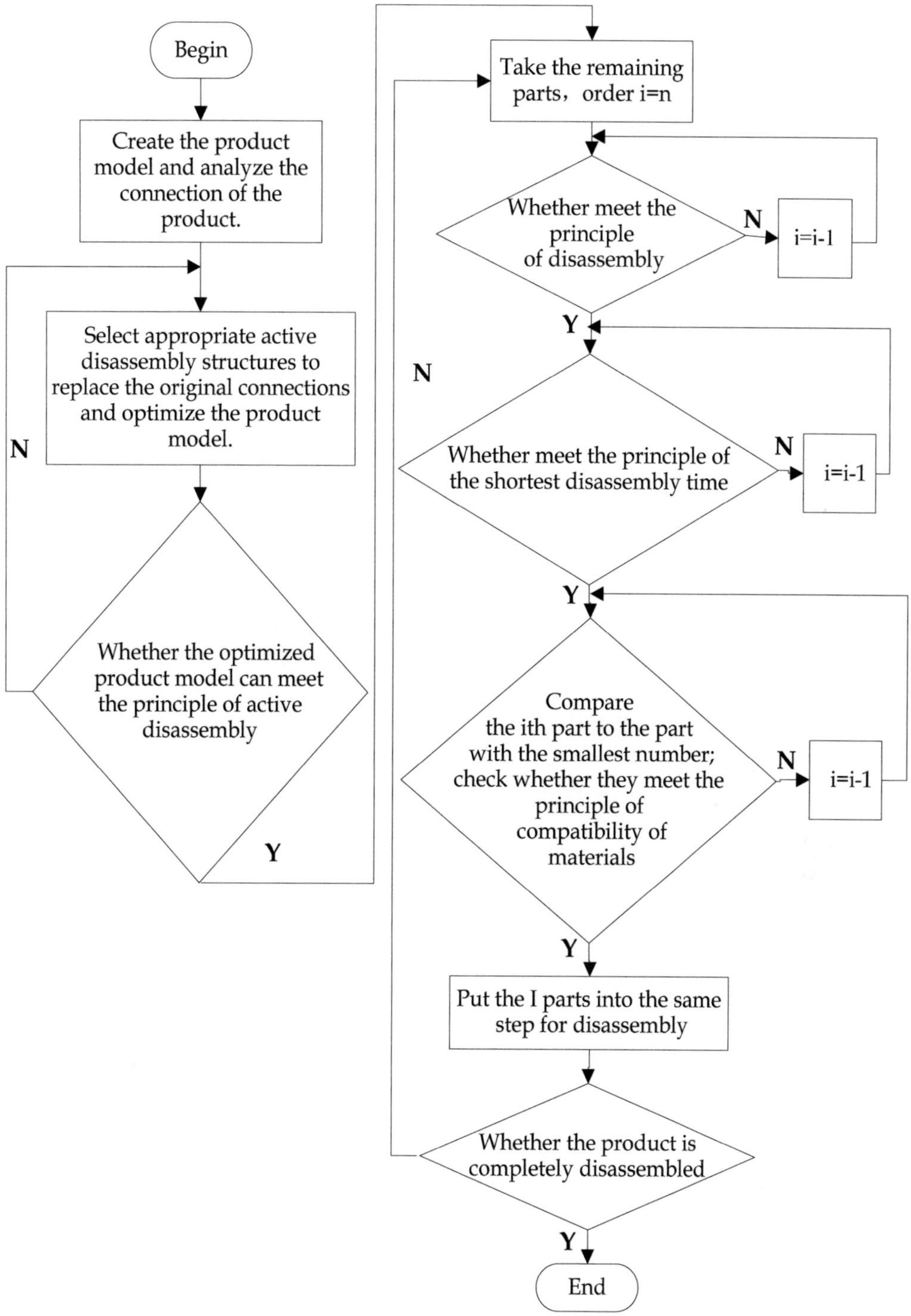

Fig. 4. The flow chart about the process of division of multi-step active disassembly (Zhao, 2011b)

Procedure 2: Select appropriate active disassembly devices to replace the original connections and optimize the product model.

When active disassembly product is designed, the original configuration should be designed based on traditional method, then select appropriate active disassembly devices and arrange them in the appropriate location to achieve active disassembly. Finally, optimize the configuration and appearance of active disassembly product.

Procedure 3: Check whether the optimized product model can meet the principle of active disassembly. If satisfied, transfer to step 4, otherwise, return to step 2.

Procedure 4: For the multi-step active disassembly products, take the remaining parts, and order i=n, then enter the next procedure.

Procedure 5: Check whether the i parts meet the principle of disassembly, if satisfied, transfer to procedure 6, otherwise, order i=i-1, that is, get rid of the component most inside, and continue to check until the principle of disassembly is satisfied.

Procedure 6: Check whether the i parts meet the principle of the shortest disassembly time, if satisfied, transfer to procedure 7, otherwise, order i=i-1, that is, get rid of the most inside component, and continue to check until the principle of the shortest disassembly time is satisfied.

Procedure 7: Compare part i to the part with the smallest number. Check whether they meet the principle of compatibility of materials. If satisfied, transfer to procedure 8, otherwise, order i=i-1, that is, get rid of the most inside component, and continue to check until the principle of disassembly is satisfied.

Procedure 8: Put the parts which have been checked through procedure 5, 6 and 7 into the same step for disassembly, then enter procedure 9. G1, G2… denotes the steps.

Procedure 9: Check whether the product is completely disassembled. If the parts are completely disassembled, end the procedure. Otherwise, transfer to procedure 5. Cycle this procedure until the product is completely disassembled.

All steps of product can be gotten through the above procedures, and the disassembly sequence is G1, G2….. When products are designed, it is convenient to achieve the purpose of multi-step active disassembly when the trigger strength is increasing in each step.

5. Conclusions

1. Aiming to the multi-step active disassembly, the chapter proposes the theory and method of the multi-step active disassembly, and analyzes the principles of multi-step active disassembly of products, then works out the process of classifying the steps of multi-step active disassembly and gives specific procedures of dividing steps according to the flow chart.
2. The theory and method discussed and researched in this chapter has the important academic value to replenish and improve the design method of dismountable products and the design theory system of active disassembly products.

3. The disassembly efficiency of products is greatly improved using the multi-step active disassembly method provided in this chapter. This method is in favor of parts and components sorting after the end-of-life (EOL) products disassembled, which can improve the recovery performance of EOL products and increase utilization rate of resource. To research this method, it is forming the foundation for the active assembly design of products and the industrialized application of this technology.

6. References

Chiodo J. D.,. Billett E. H &. Harrison D. J (1999) Preliminary Investigations of Active Disassembly Using Shape Memory Polymers, Cleaner Electronics Research, Brunel University.

Chiodo J.D (2002). Shape memory alloy actuators for active disassembly using "smart" materials of consumer electronic products. Materials and Design, Vol.23, No.11, (2001 November) pp. 471-478, ISSN 0261-3069

Chiodo, J. D. et al, (1998). Active Disassembly, Journal of Sustainable Design, Refereed Journal

Chiodo, J. D. et al, Available from http://www.activedisassembly.com/

Environmentally Conscious Design and Inverse Manufacturing, Proceedings of Eco Design '99: First International Symposium On pp. 590-596

Guo W, Liu Z, Liu G and so on (2005). Disassembly Sequence Planning Based on Modularization. Journal of Computer Aided Design & Computer Graphics, Vol.17, No.3, (2003 December) pp. 498-504, ISSN 1003-9775

Li X. (2008) Research of Active Disassembly Devices' Design Theory and Methods Based on Smart Materials. Hefei: *School of Mechanical and Auto Engineering, Hefei University of Technology.*

Liu G, Li X, Liu Z, et al. (2008). Research on Detachability of Remotes Based on ADSM. Proceedings of 15th CIRP International Conference on Life Cycle Engineering.

Liu Z, Cheng H, Li X, Zhao L (2011). The Design Method of Active Disassembly Device Using Electro and Heat Stimulation. *Machine Design and Research,* Vol.27, No.3, (2010 December) pp. 12-15, ISSN 1006-2343

Liu Z, Zhao L, Li X, Zhang H (2010) Research on Multi-step Active Disassembly Method of products Based on ADSM. Advanced Materials Research, Vols. 139-141, pp. 1428-1432 ISSN 1662-8985

Qi H, Xu H, 2007. Design theory & means for mechanical and electrical productions facing disassembling & recycling. Modern Manufacturing Engineering, Vol.2007, No.6, (2007 March), pp. 134-138 ISSN 1671-3133

Suga, T & Hosada, N (2000) Active Disassembly and Reversible Interconnection, Proceedings of International Symposium on Electronics and the Environment. IEEE

Wang B, Wang N (2005). Automatic Generation and Optimization of the Disassembly Sequence of Assemblies. Journal of Huaihai Institute of Technology, Vol.14, No.1, (2004 December) pp. 14-17 ISSN 1672-6685

Wang B, WANG N (2005). The Initial Population of the Optimal Disassembling Sequence Generated Automatically. Machine Design and Research, Vol.21, No.2, (2004 July) pp. 59-60, 63 ISSN 1006-2343

Wang B, Wang N (2006). Automatic generation and combinatory optimization of disassembly sequence for mechanical-electric assembly. Journal of Shandong University (Engineering Science), Vol.36 No.2, (2005 January), pp. 52-57 ISSN 1001-9839

Willems B, Design for Active Disassembly (2005) Katholieke University, ISEE

Zhao L (2011) Research on Multi-step Active Disassembly Methods and Design Principle of Products Based on ADSM and Its Environmental Reliability Analysis. Hefei: *School of Mechanical and Auto Engineering, Hefei University of Technology.*

Zhao L, Liu Z, Li X, Zhang H, 2011. Cheng H. Research on Multi-step Active Disassembly Methods and Design Principle of products Based on ADSM[J]. *China Mechanical Engineering*, Vol.22, No.7, (2010 May) pp. 848-852 ISSN 1004-132X

13

Material Flow, Energy Flow and Energy Flow Network in Iron and Steel Enterprise

Wen-qiang Sun and Jiu-ju Cai
Institute of Thermal and Environmental Engineering, Northeastern University
China

1. Introduction

Based on the theory of Metallurgical Process Engineering (Yin, 2011) and methodology of Systems Energy Conversion (Lu & Cai, 2010), the physical essence of the steel production process is re-observed, and the steel production system is decomposed into two parts: materials transportation process and energy transportation process.

The research on material flow, energy flow and the relationship between them will benefit the construction of a resource saving and environmentally friendly enterprise. Recently, more and more researchers are paying their attentions to the study of material flow and energy flow of iron and steel enterprise. The modern iron & steel joint enterprise is a complicated iron-coal chemical system, and the steel production process is, in essence, the dynamic operation of the materials flow under the drive and effect from energy flow following the definite rules (Yin, 2000, 2007, 2008). Lu proposed the *e-p* analysis method (Lu et al., 2000) and the standard material flow diagram for blast furnace - converter process (Dai & Lu, 2004) which is used to analyze the Fe resource efficiency. Based on the standard process energy consumption, the influence of material flow of iron and steel enterprise on its energy consumption was studied (Chen et al., 2002). The relationship between materials flow and energy flow was established (Cai et al., 2006) and they were expressed by the item of energy consumption per ton of steel (Cai et al., 2008).

In the following sections material flow and energy flow model of an iron and steel enterprise will be developed. In each model the flow of a unit, a process and an enterprise will be represented in order. Network is the running type of energy flows. Then, the operating behaviour and effectiveness of energy flow network and the interacting mechanism among several energy flows are analyzed in order to find out the new direction of energy conservation in an iron and steel industry enterprise. And finally, the conclusions on the research will be given.

2. Material flow

The material flow is formed when the material runs along its life cycle trace in an enterprise. An iron and steel enterprise is composed of several processes, and each process is also composed of several units. For instance, blast furnace - basic oxygen furnace route consists of processes related to the production of iron and steel, such as coke making, sintering, pelletizing, iron making, steel making, and steel rolling. For iron making process, it consists of several blast furnaces with different volume. These blast furnaces are the units in iron making process. So, the units' material flow structure, flow rate, flow direction is varied. They together determine the structure, flow rate, flow direction and flow indices of the whole material flow of an enterprise. To analyze the material flow of an enterprise, the material flow model of a unit should be established.

2.1 Material flow model of a unit

For iron and steel enterprise, five material flows shown in Fig. 1 may emerge in *j*-th unit of *i*-th process, described as unit (*i*, *j*). The flow rates of five material flows are average values of a statistical period (one year, one month, one day, etc.). Actually, flow rates of all material flows through a unit changes when time is changed. So, flow rate of material flow is the assembly average of the material flow rate per unit time.

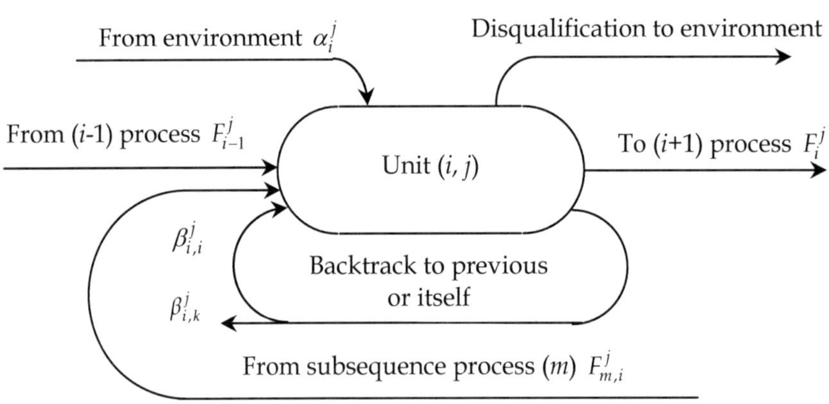

Fig. 1. Material flow model of a unit

i. Material flows from the anterior producing unit (F_{i-1}^j). They are products of (*i*-1)-th process served as materials of unit (*i*, *j*). e.g., hot metal of blast furnace flowed into a converter.

$$F_{i-1}^j = \int_{\tau_0}^{\tau_1} P_{i-1-F}^j(\tau)d\tau \Big/ \int_{\tau_0}^{\tau_1} P_n(\tau)d\tau \tag{1}$$

ii. Material flows from environment (α_i^j). They are materials of unit (*i*, *j*) from the outside of this unit. e.g., steel scrap flowed into a converter.

$$\alpha_i^j = \int_{\tau_0}^{\tau_1} P_{i-\alpha}^j(\tau)d\tau \Big/ \int_{\tau_0}^{\tau_1} P_n(\tau)d\tau \tag{2}$$

iii. Material flows to the environment (γ_i^j). They consist of emitted wastes ($\gamma_i'^j$) and saleable products ($\gamma_i''^j$), i.e., $\gamma_i^j = \gamma_i'^j + \gamma_i''^j$. e.g., the iron dust emitted from a sintering machine.

$$\gamma_i^j = \int_{\tau_0}^{\tau_1} P_{i-\gamma}^j(\tau)d\tau \Big/ \int_{\tau_0}^{\tau_1} P_n(\tau)d\tau \quad (3)$$

iv. Material flows recycled through unit (*i*, *j*) ($\beta_{i,i}^j$, $\beta_{i,k}^j$ and $\beta_{m,i}^j$). The waste products of unit (*i*, *j*) and its downstream processes recycled into unit (*i*, *j*) or other upstream processes as materials. e.g., iron dust from converter served as materials of sintering, crop ends of steel rolling served as materials of steel making.

$$\begin{cases} \beta_{i,i}^j = \int_{\tau_0}^{\tau_1} P_{i,i-\beta}^j(\tau)d\tau \Big/ \int_{\tau_0}^{\tau_1} P_n(\tau)d\tau \\ \beta_{i,k}^j = \int_{\tau_0}^{\tau_1} P_{i,k-\beta}^j(\tau)d\tau \Big/ \int_{\tau_0}^{\tau_1} P_n(\tau)d\tau \\ \beta_{m,i}^j = \int_{\tau_0}^{\tau_1} P_{m,i-\beta}^j(\tau)d\tau \Big/ \int_{\tau_0}^{\tau_1} P_n(\tau)d\tau \end{cases} \quad (4)$$

v. Material flows to the followed producing unit (F_i^j). They are qualified products of unit (*i*, *j*). e.g., hot metal from iron making process.

$$F_i^j = \int_{\tau_0}^{\tau_1} P_{i-F}^j(\tau)d\tau \Big/ \int_{\tau_0}^{\tau_1} P_n(\tau)d\tau \quad (5)$$

where, $P_{i-1-F}^j(\tau)$, $P_{i-\alpha}^j(\tau)$, $P_{i-\gamma}^j(\tau)$, $P_{i,k-\beta}^j(\tau)$ and $P_{i-F}^j(\tau)$ are flow rates of each material flow; $P_n(\tau)$ is the output flow rate of the last process; τ_0 and τ_1 are the initial time and end time of the statistical period, respectively.

Based on 1 ton of steel, the input-output relationship model for unit (*i*, *j*) is developed as

$$F_{i-1}^j + \alpha_i^j + \beta_{m,i}^j + \beta_{i,i}^j = F_i^j + \gamma_i^j + \beta_{i,k}^j + \beta_{i,i}^j \quad (6)$$

Therefore, the parameters and evaluating index of material flow of unit (*i*, *j*) are listed as

i. Waste rate (w_i^j). It is defined as the wastes produced by per unit of qualified product of unit (*i*, *j*) and is expressed as

$$w_i^j = \left(\gamma_i^j + \beta_{i,k}^j + \beta_{i,i}^j\right) \big/ F_i^j \quad (7)$$

ii. Recycle rate of waste (r_{w-i}^j). It is defined as the rate of recycled wastes to total wastes produced and is expressed as

$$r_{w-i}^j = \left(\beta_{i,k}^j + \beta_{i,i}^j\right) \big/ \left(\gamma_i^j + \beta_{i,k}^j + \beta_{i,i}^j\right) \quad (8)$$

From the view of reducing resource consumption, it is the demanded material flow per unit of product that is the most concerned for one producing unit. Thus, metal yield ratio of unit (*i*, *j*) is taken for evaluating the material flow of a producing unit.

iii. Metal yield ratio (R_i^j). It is defined as the output of products per unit of material and is expressed as

$$R_i^j = F_i^j \Big/ \left(F_i^j + \gamma_i^j + \beta_{i,k}^j + \beta_{i,i}^j\right) = F_i^j \Big/ \left(F_{i-1}^j + \alpha_i^j + \beta_{m,i}^j + \beta_{i,i}^j\right) = 1 \Big/ \left(1 + w_i^j\right) \quad (9)$$

It can be found from Eq. (9) that the metal yield ratio is related with the waste rate. To increase the metal yield ratio, waste rate of this unit should be reduced.

2.2 Material flow model of a process

Material flows through a producing process are shown in Fig. 2. Process i is consisted of k_i units. Then, the material flow of process i is the sum of material flows of the k_i units.

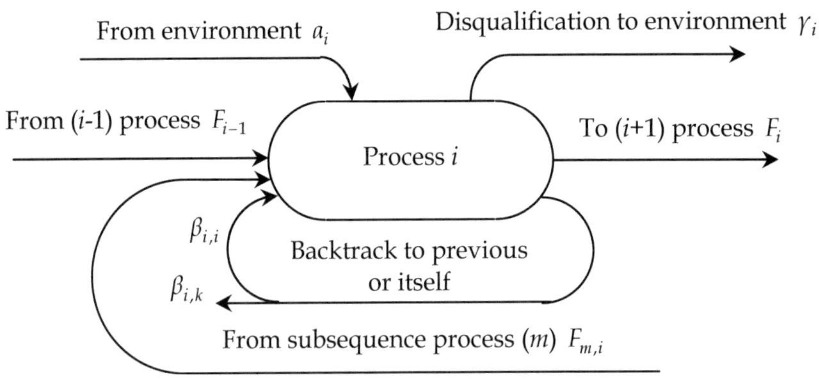

Fig. 2. Material flow model of a process

The material flows include: (i) Material flows from the anterior process (F_{i-1}); (ii) Material flows from environment (α_i); (iii) Material flows to the environment (γ_i); (iv) Material flows recycled through process i ($\beta_{i,i}$, $\beta_{i,k}$ and $\beta_{m,i}$); and (v) Material flows to the followed process (F_i). They are written as

$$\begin{cases} F_{i-1} = \sum_{j=1}^{k_i} F_{i-1}^j \\ \alpha_i = \sum_{j=1}^{k_i} \alpha_i^j \\ \gamma_i = \sum_{j=1}^{k_i} \gamma_i^j \\ \beta_{i,i} = \sum_{j=1}^{k_i} \beta_{i,i}^j, \beta_{i,k} = \sum_{j=1}^{k_i} \beta_{i,k}^j, \beta_{m,i} = \sum_{j=1}^{k_i} \beta_{m,i}^j \\ F_i = \sum_{j=1}^{k_i} F_i^j \end{cases} \quad (10)$$

So the input-output relationship model for process i is developed as

$$F_{i-1} + \alpha_i + \beta_{m,i} + \beta_{i,i} = F_i + \gamma_i + \beta_{i,k} + \beta_{i,i} \quad (11)$$

Likely, the parameters of material flow of process i are waste rate (w_i) and waste recycle rate (r_{w-i}); and metal yield ratio of process i (R_i) is taken for evaluating the material flow. They are expressed respectively as

$$w_i = (\gamma_i + \beta_{i,k} + \beta_{i,i}) / F_i \tag{12}$$

$$r_{w-i} = (\beta_{i,k} + \beta_{i,i}) / (\gamma_i + \beta_{i,k} + \beta_{i,i}) \tag{13}$$

and $R_i = F_i / (F_i + \gamma_i + \beta_{i,k} + \beta_{i,i}) = F_i / (F_{i-1} + \alpha_i + \beta_{m,i} + \beta_{i,i}) = 1/(1+w_i)$ (14)

From the view of reducing resource consumption, it is the demanded material flow per unit of product that is the most concerned for one process. To obtain a smaller demand of natural resource, it is better if waste is less for iron and steel enterprise. As for the waste produced, it is best to be recycled to reduce the consumption of natural resource. i.e., the objective of material flow recycling is to obtain a smaller w_i and a larger r_{w-i} mentioned in Eqs. (12) and (13).

2.3 Material flow model of an enterprise

Based on the material flow model of a process, it is considered that the iron and steel enterprise is composed of n processes. And the material model of an enterprise is shown in Fig. 3.

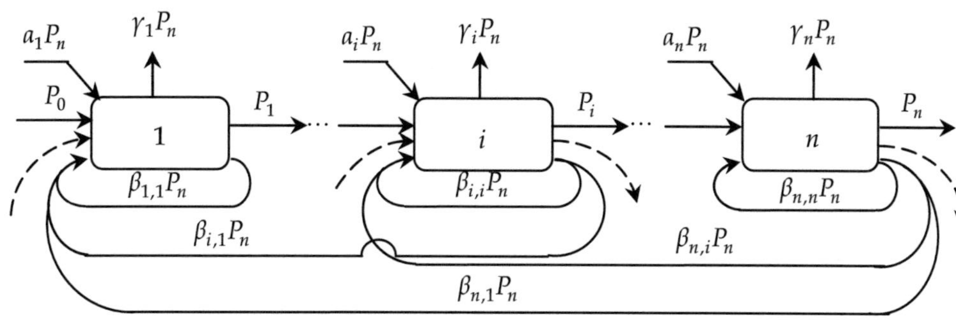

Fig. 3. Material flow model of an enterprise

P_i in Fig. 3 is the output of process i in statistical period and is expressed as

$$P_i = \left(1 + \sum_{l=i+1}^{n} \gamma_l + \sum_{l=i+1}^{n} \sum_{m=1}^{i} \beta_{l,m} - \sum_{l=i+1}^{n} \alpha_l \right) P_n \tag{15}$$

The material flow input-output balance of enterprise is

$$P_0 + \left(\sum_{i=1}^{n} \alpha_i \right) P_n = P_n + \left(\sum_{i=1}^{n} \gamma_i \right) P_n \tag{16}$$

Material flow per unit product can be gotten if both sides of Eq. (16) are divided by P_n. Then main parameters of material flow are:

i. External material flow per ton of steel.

$$\alpha = \sum_{i=1}^{n} \alpha_i \tag{17}$$

ii. Recycled material flow per ton of steel.

$$\beta = \sum_{i=1}^{n} \sum_{l=i}^{n} \beta_{l,i} \tag{18}$$

iii. Emitted material flow per ton of steel.

$$\gamma = \sum_{i=1}^{n} \gamma_i \tag{19}$$

iv. Waste rate of an enterprise.

$$w = \sum_{i=1}^{n} (p_i \cdot w_i) \text{ or } w = \gamma + \beta \tag{20}$$

v. Waste recycle rate of an enterprise.

$$r = \beta / (\gamma + \beta) \tag{21}$$

Resource efficiency is the output of final product per unit natural resource used. It is the evaluating index for material flow of an iron and steel enterprise with the expression as

$$R = 1/(1 + \gamma - \alpha) \tag{22}$$

Considering the waste rate and waste recycle rate of an enterprise, Eq. (22) can be rewritten as

$$R = 1/\left[1 + w(1-r) - \alpha\right] \tag{23}$$

It can be seen from Eq. (23) that resource efficiency (R) will be raised by reducing waste rate of an enterprise (w), increasing waste recycle rate of an enterprise (r), and/or increasing external material flow per ton of steel (α). To increase the resource efficiency of an iron and steel enterprise, for one thing the waste output of every process should be reduced, for another the formed waste should be utmost recycled and reused.

3. Energy flow

The energy flow is formed when the energy runs along the route of conversion, usage, recovery and emission. Similar as material flow, energy flow model of a unit will be established, based on which energy flow model of a process or an enterprise are developed.

3.1 Energy flow model of a unit

For iron and steel enterprise, six energy flows shown in Fig. 4 may emerge in *j*-th unit of *k*-th energy conversion process, described as unit (*k*, *j*). The flow rates of six material flows are average values of a statistical period (one year, one month, one day, etc.). Actually, flow rates of all energy flows through a unit changes when time is changed. So, flow rate of energy flow is the assembly average of the energy flow rate per unit time.

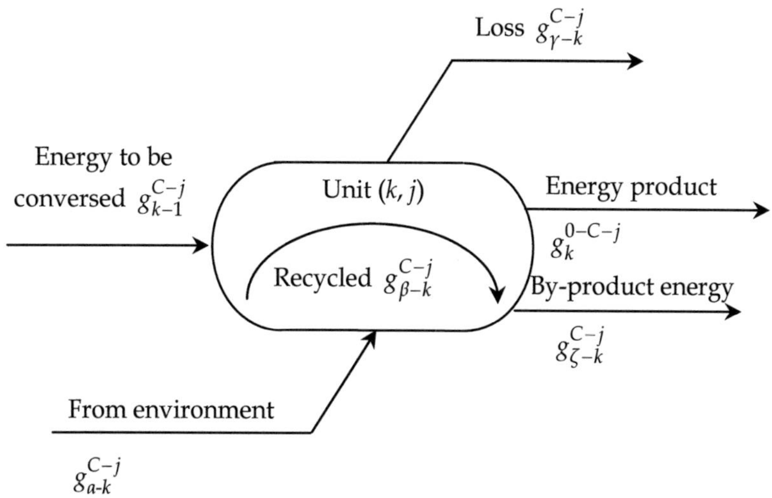

Fig. 4. Energy flow model of a unit

i. Energy flow of energy to be conversed.

$$g_{k-1}^{C-j} = \int_{\tau_0}^{\tau_1} g_{k-1}^{C-j}(\tau)d\tau \Big/ \int_{\tau_0}^{\tau_1} F_k^{0-C-j}(\tau)d\tau \tag{24}$$

ii. Energy flow of energy product.

$$g_k^{0-C-j} = \int_{\tau_0}^{\tau_1} g_k^{0-C-j}(\tau)d\tau \Big/ \int_{\tau_0}^{\tau_1} F_k^{0-C-j}(\tau)d\tau \tag{25}$$

iii. Energy flow from environment.

$$g_{a-k}^{C-j} = \int_{\tau_0}^{\tau_1} g_{a-k}^{C-j}(\tau)d\tau \Big/ \int_{\tau_0}^{\tau_1} F_k^{0-C-j}(\tau)d\tau \tag{26}$$

iv. Energy flow loss.

$$g_{\gamma-k}^{C-j} = \int_{\tau_0}^{\tau_1} g_{\gamma-k}^{C-j}(\tau)d\tau \Big/ \int_{\tau_0}^{\tau_1} F_k^{0-C-j}(\tau)d\tau \tag{27}$$

v. Energy flow recycled.

$$g_{\beta\text{-}k}^{C-j} = \int_{\tau_0}^{\tau_1} g_{\beta\text{-}k}^{C-j}(\tau)d\tau \Big/ \int_{\tau_0}^{\tau_1} F_k^{0-C-j}(\tau)d\tau \tag{28}$$

vi. Energy flow of by-product energy.

$$g_{\zeta-k}^{C-j} = \int_{\tau_0}^{\tau_1} g_{\zeta-k}^{C-j}(\tau)d\tau \Big/ \int_{\tau_0}^{\tau_1} F_k^{0-C-j}(\tau)d\tau \tag{29}$$

where, $g_{k-1}^{C-j}(\tau)$, $g_k^{0-C-j}(\tau)$, $g_{a-k}^{C-j}(\tau)$, $g_{\gamma-k}^{C-j}(\tau)$, $g_{\beta-k}^{C-j}(\tau)$ and $g_{\zeta-k}^{C-j}(\tau)$ are flow rates of each energy flow; $F_k^{0-C-j}(\tau)$ is the output flow rate of energy product; τ_0 and τ_1 are the initial time and end time of the statistical period, respectively.

Based on 1 ton of steel, the energy flow input-output relationship for unit (*k*, *j*) is

$$g_{k-1}^{C-j} + g_{\alpha-k}^{C-j} + g_{\beta-k}^{C-j} = g_k^{0-C-j} + g_{\beta-k}^{C-j} + g_{\zeta-k}^{C-j} + g_{\gamma-k}^{C-j} \tag{30}$$

Considering energy product is energy carrier, energy value of energy product is selected as the evaluating index of energy flow of unit (*k*, *j*). Its expression is

$$b_k^j = \left(g_{k-1}^{C-j} + g_{a-k}^{C-j} + g_{\beta-k}^{C-j}\right) - \left(g_{\beta-k}^{C-j} + g_{\zeta-k}^{C-j}\right) \tag{31}$$

where $\left(g_{k-1}^{C-j} + g_{\alpha-k}^{C-j} + g_{\beta-k}^{C-j}\right)$ is the sum of energy flows consumed in this unit, $\left(g_{\beta-k}^{C-j} + g_{\zeta-k}^{C-j}\right)$ is the sum of energy flows recycled in this unit.

It can be found from Eq. (31) that the energy value of energy product is determined by energy flows consumed and recycled.

3.2 Energy flow model of a process

Energy flows through a producing process are shown in Fig. 5. Process *k* is consisted of m_k units. Then, the energy flow of process *k* is the sum of energy flows of the m_k units.

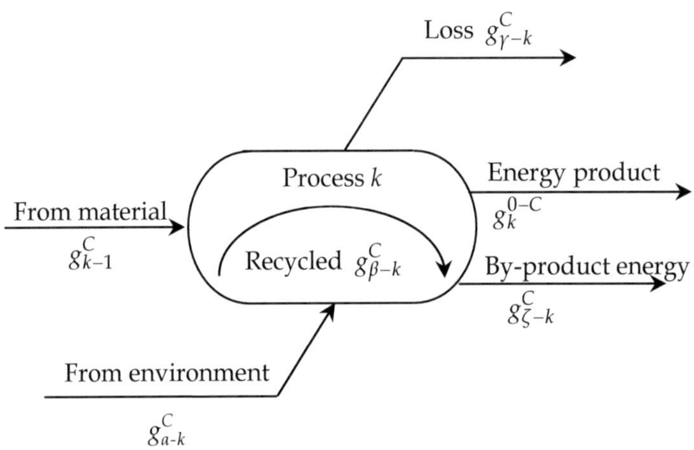

Fig. 5. Energy flow model of a process

The energy flows include: (i) Energy flows from material entering process (g_{k-1}^{C}); (ii) Energy flows of energy product after conversion (g_k^{0-C}); (iii) Energy flows consumed coming from outside process ($g_{a\text{-}k}^{C}$); (iv) Energy flows loss ($g_{\gamma-k}^{C}$); (v) ; Energy flows recycled ($g_{\beta-k}^{C}$); and (vi) Energy flows of by-product energy ($g_{\zeta-k}^{C}$). They are written as

$$\begin{cases} g_{k-1}^{C} = \sum_{j=1}^{m_k}\left(\omega_{k,j}\cdot g_{k-1}^{C-j}\right) \\ g_{k}^{0-C} = \sum_{j=1}^{m_k}\left(\omega_{k,j}\cdot g_{k}^{0-C-j}\right) \\ g_{\alpha\text{-}k}^{C} = \sum_{j=1}^{m_k}\left(\omega_{k,j}\cdot g_{\alpha\text{-}k}^{C-j}\right) \\ g_{\gamma\text{-}k}^{C} = \sum_{j=1}^{m_k}\left(\omega_{k,j}\cdot g_{\gamma\text{-}k}^{C-j}\right) \\ g_{\beta\text{-}k}^{C} = \sum_{j=1}^{m_k}\left(\omega_{k,j}\cdot g_{\beta\text{-}k}^{C-j}\right) \\ g_{\zeta\text{-}k}^{C} = \sum_{j=1}^{m_k}\left(\omega_{k,j}\cdot g_{\zeta\text{-}k}^{C-j}\right) \end{cases} \tag{32}$$

where $\omega_{k,j}$ is the product weights of each energy conversion unit with the expression as

$$\omega_{k,j} = F_{k}^{0-C-j}\Big/F_{k}^{0-C} \tag{33}$$

where F_{k}^{0-C-j} is the output of energy products of unit (k, j) and F_{k}^{0-C} is the output of energy products of process k.

The energy flow input-output relationship model for process k is developed as

$$g_{k-1}^{C} + g_{\alpha-k}^{C} + g_{\beta-k}^{C} = g_{k}^{0-C} + g_{\beta-k}^{C} + g_{\zeta-k}^{C} + g_{\gamma-k}^{C} \tag{34}$$

Energy value of energy product of a process is also the evaluating index of energy flow of energy conversion process, which is expressed as

$$b_k = \left(g_{k-1}^{C} + g_{a-k}^{C} + g_{\beta-k}^{C}\right) - \left(g_{\beta-k}^{C} + g_{\zeta-k}^{C}\right) \tag{35}$$

where $\left(g_{k-1}^{C} + g_{\alpha-k}^{C} + g_{\beta-k}^{C}\right)$ is the sum of energy flows consumed in process k, $\left(g_{\beta-k}^{C} + g_{\zeta-k}^{C}\right)$ is the sum of energy flows recycled in process k.

The generation of waste energy per unit energy product is

$$g_{w-k}^{C} = g_{\beta-k}^{C} + g_{\zeta-k}^{C} + g_{\gamma-k}^{C} \tag{36}$$

Define the recycle rate of waste energy of a process as

$$r_{e-k}^{C} = \left(g_{\beta-k}^{C} + g_{\zeta-k}^{C}\right)\Big/ g_{w-k}^{C} \tag{37}$$

Then the energy value of energy product can be rewritten as

$$b_k = \left(g_{k-1}^C + g_{a-k}^C + g_{\beta-k}^C\right) - r_{e-k}^C \cdot g_{w-k}^C = g_{k-1}^C + \left(1 - r_{e-k}^C\right) \cdot g_{w-k}^C \tag{38}$$

3.3 Energy flow model of an enterprise

It has several energy conversion ways for any type of energy. Each conversion needs several transportation and energy conversion devices. The energy flow model of an enterprise is shown in Fig. 6. Energy media k has l types of conversion. Anyway, the final output energy is g_k' after several conversions from the original energy g_k.

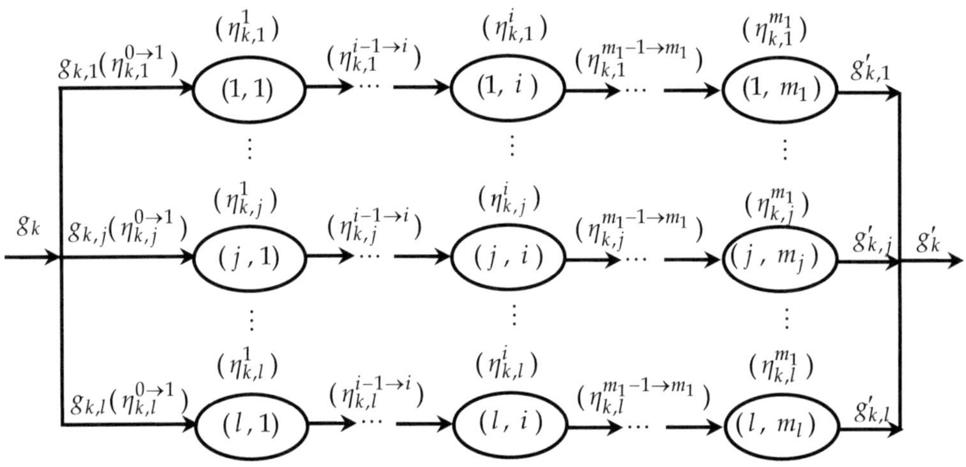

Fig. 6. Energy flow model of an enterprise

As for *j*-th energy conversion way of *k*-th energy media, supplied energy $g_{k,j}$ becomes $g'_{k,j}$ through m_j conversion devices and m_j times of transportation. The energy conversion efficiency of *j*-th energy conversion way is

$$\eta_{k,j} = g'_{k,j} / g_{k,j} = \prod_{i=0}^{m_j-1} \left(\eta_{k,j}^{i \to i+1} \cdot \eta_{k,j}^{i+1} \right) \tag{39}$$

where $\eta_{k,j}$ is the energy conversion efficiency of *j*-th energy conversion way, $\eta_{k,j}^{i \to i+1}$ is the transportation efficiency from unit *i* to unit (*i*+1) in *j*-th energy conversion way, and $\eta_{k,j}^{i+1}$ is the energy efficiency of unit (*i*+1) in *j*-th energy conversion way.

The energy conversion efficiency of media *k* is

$$\eta_k = \sum_{j=1}^{j=l} \left[\omega_{k,j} \cdot \prod_{i=0}^{i=m_j-1} \left(\eta_{k,j}^{i \to i+1} \cdot \eta_{k,j}^{i+1} \right) \right] \tag{40}$$

where $\omega_{k,j}$ is the ratio of energy input of *j*-th energy conversion way to total energy input, expressed as

$$\omega_{k,j} = g_{k,j} / g_k \tag{41}$$

It is indicated from Eq. (40) that some measures must be taken to improve the conversion efficiency and reduce the energy value of an energy product for an iron and steel enterprise. They are: increasing the weight factors of the energy conversion modes whose efficiency is higher; reducing the amount of the conversion units, optimizing the transportation mode of energy, and shortening the distance of transportation.

4. Energy flow network

Energy flow is dynamically operates in energy flow network following some rules. The energy flow network of iron and steel enterprise is composed of energy origin nodes (such as blast furnace, coke oven, converter), connectors (such as gas pipe), intermediate buffers (such as gas holder), and energy terminal nodes (such as sintering machine, blast furnace, heating furnace, boiler). The energy flow network of modern iron and steel enterprise is shown in Fig. 7.

According to the input-output feature of energy flow network, the energy flow network of iron and steel industry is divided into six regions. It includes energy transfer region (C), energy usage region (U), waste heat recycling region (R), surplus energy buffer region (B), energy storing region (S) and city energy system (CES).

In Fig. 7, the direction of arrow represents the flow direction of energy flow, parameters above flow lines represent the flow rates or allocation proportion of energy flows, and parameters below flow lines represent the temperatures or energy values of energy flows. C_F and C_G represent fuel conversion and power generation, respectively. U_H, U_M and U_L represent high-grade, medium-grade and low-grade energy consumers, respectively. R_H, R_P

and R_C represent heat recovery, power recovery and combined heat and power recovery, respectively. B_C and B_D represent centralized and distributed energy buffers, respectively. t is the temperatures of energy flows, according to which the energy values of energy flows can be gotten. λ in Fig. 7 is the allocation proportion of energy flow with

$$\lambda_0' + \lambda_0'' = \lambda_{CF} + \lambda_{CG} = \lambda_{CG_1} + \cdots + \lambda_{CG_n} = 1 \tag{42}$$

where λ_0' and λ_0'' represent the ratios of energy flow converted (G_0') and unconverted (G_0'') to external energy flow (G_0), respectively; λ_{CF} and λ_{CG} represent the ratios of energy flow of fuel conversion part (G_{CF}) and power generation part (G_{CG}) to converted energy flow (G_0), respectively; and so on.

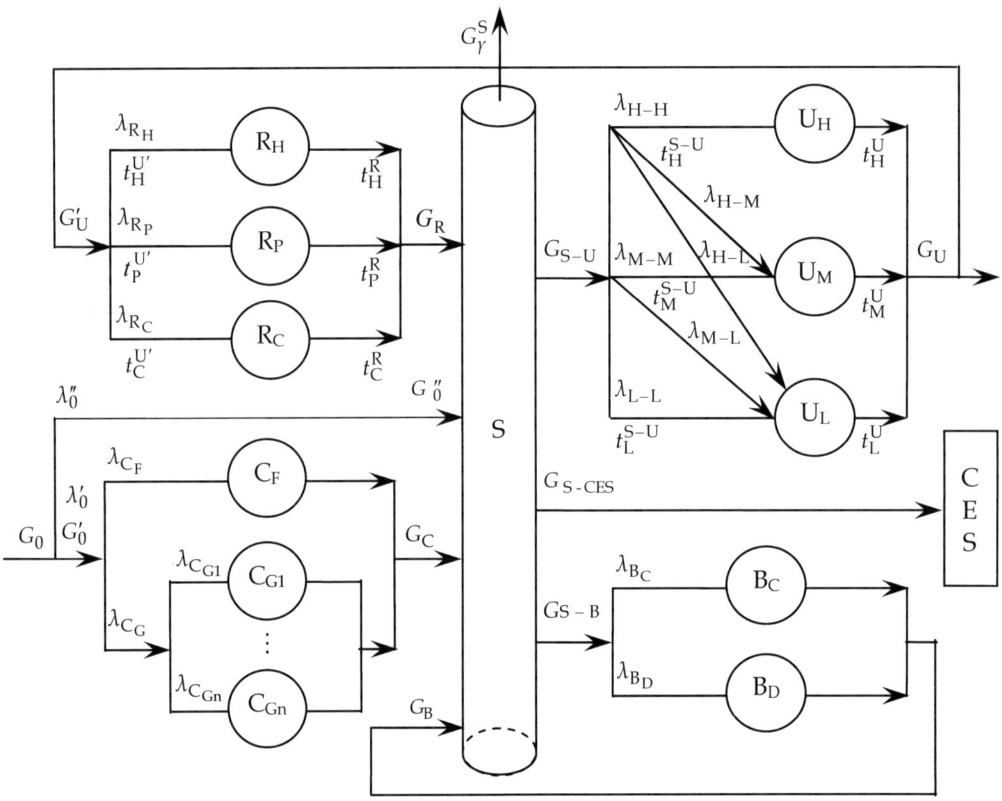

Fig. 7. Energy flow network model of an enterprise

Define η as the energy transportation efficiency of each process or user. $\eta_{\text{H-H}}^{\text{U}}$, $\eta_{\text{H-M}}^{\text{U}}$, and $\eta_{\text{H-M}}^{\text{U}}$ represent the conversion efficiencies of high-grade energy to high-grade energy, high-grade energy to medium-grade energy and high-grade energy to low-grade energy, respectively. $\eta_{\text{C-U}}^{T}$ represents the transportation efficiency from Region C to Region U. Then energy efficiencies of C, U, R and B are

$$\begin{cases} \eta_{\text{C}} = \sum_{i=\text{L}}^{\text{H}} \sum_{j=\text{L}}^{i} \left(\lambda_{i-j}^{\text{C}} \cdot \eta_{i-j}^{\text{C}} \right) \\ \eta_{\text{U}} = \sum_{i=\text{L}}^{\text{H}} \sum_{j=\text{L}}^{i} \left(\lambda_{i-j}^{\text{U}} \cdot \eta_{i-j}^{\text{U}} \right) \\ \eta_{\text{R}} = \sum_{i=\text{L}}^{\text{H}} \sum_{j=\text{L}}^{i} \left(\lambda_{i-j}^{\text{R}} \cdot \eta_{i-j}^{\text{R}} \right) \\ \eta_{\text{B}} = \sum_{i=\text{L}}^{\text{H}} \sum_{j=\text{L}}^{i} \left(\lambda_{i-j}^{\text{B}} \cdot \eta_{i-j}^{\text{B}} \right) \end{cases} \tag{43}$$

Systems energy efficiency is the evaluating index for energy flow network of iron and steel enterprise. It is defined as the output of steel product per unit of net energy input and is expressed as

$$\eta_{\text{S}} = \frac{\int_{\tau_0}^{\tau} \frac{1}{g_0(\tau) - g_{\text{B-CES}}(\tau)} d\tau}{\tau - \tau_0} \tag{44}$$

where g is the energy input per ton of steel.

Eq. (44) can be rewritten as

$$\eta_{\text{S}} = \frac{\int_{\tau_0}^{\tau} \frac{1}{g_{\gamma}^{\text{C}}(\tau) + g_{\gamma}^{\text{S}}(\tau) + g_{\gamma}^{\text{U/R}}(\tau) + g_{\gamma}^{\text{B}}(\tau)} d\tau}{\tau - \tau_0} \tag{45}$$

where $g_{\gamma}^{\text{C}}(\tau)$, $g_{\gamma}^{\text{S}}(\tau)$, $g_{\gamma}^{\text{U/R}}(\tau)$, and $g_{\gamma}^{\text{B}}(\tau)$ are the energy losses of Region C, S, U and R, and B, respectively.

i. Energy loss of Region C

$g_{\gamma}^{\text{C}}(\tau)$ is the energy loss of Region C and energy transportation process. It is expressed as

$$g_{\gamma}^{\text{C}}(\tau) = g_0'(\tau)\left(1 - \eta_{\text{C}}\eta_{\text{C-S}}^{T}\right) \tag{46}$$

ii. Energy loss of Region S

$g_{\gamma}^{\text{S}}(\tau)$ is the energy loss of Region S. It is related with the volume of energy storages and maximum energy consumptions of buffer users and CES. Its expression is

$$g_{\gamma}^{\mathrm{S}}(\tau) = g_{\mathrm{X-S}}(\tau) - g_{\mathrm{S}}^{\max}(\tau) - g_{\mathrm{S-B}}^{\max}(\tau) - g_{\mathrm{S-CES}}^{\max}(\tau) \tag{47}$$

where $g_{\mathrm{X-S}}(\tau)$ is the energy loss of energy storages, $g_{\mathrm{S}}^{\max}(\tau)$ is the surplus volume of energy storages, $g_{\mathrm{S-B}}^{\max}(\tau)$ is the maximum energy consumption of buffer users, and $g_{\mathrm{S-CES}}^{\max}(\tau)$ is the maximum energy consumption of CES.

iii. Energy loss of Regions U and R

$g_{\gamma}^{\mathrm{U/R}}(\tau)$ is the energy loss of Regions U and R. It is the sum of effective energy demanded for the production of iron and steel and energy losses of energy usage and recovery processes. Its expression is

$$g_{\gamma}^{\mathrm{U/R}}(\tau) = g_{\mathrm{C}}(\tau)\frac{\left(1 + \zeta - \zeta\eta_{\mathrm{U-R}}^{T}\eta_{\mathrm{R}}\eta_{\mathrm{R-S}}^{T}\eta_{\mathrm{S-U}}^{T}\eta_{\mathrm{U}}\right)}{\eta_{\mathrm{S-U}}^{T}\eta_{\mathrm{U}}} \tag{48}$$

iv. Energy loss of Region B

$g_{\gamma}^{\mathrm{B}}(\tau)$ is the energy loss of Region B and energy transportation process. It is expressed as

$$g_{\gamma}^{\mathrm{B}}(\tau) = g_{\mathrm{S-B}}(\tau)\left(1 - \eta_{\mathrm{S-B}}^{T}\eta_{\mathrm{B}}\eta_{\mathrm{B-S}}^{T}\right) \tag{49}$$

It can be achieved that from Eqs. (44) to (49) that systems energy efficiency of an iron and steel enterprise is related with energy conversion efficiency, energy usage efficiency, waste heat recovery efficiency and energy transportation efficiency.

With the continuous improvement of the technology and method of energy conservation, the energy conservation potential of the energy flow network of future steel enterprise lies in imbalance: the quantitative imbalance between the energy supply and energy demand of device, the qualitative imbalance of energy gradation between energy flow inlet and outlet, and the dynamic imbalance between the generation and consumption of secondary energy. So, the dynamic method of imbalance must be used to recognize the running regularity of energy flows, such as gas, oxygen and steam. And the imbalance between energy generation and utilization should be scientifically plan, design and schedule to seek for a new approach from non-equilibrium state to orderly structure. i.e., more attentions should be paid to energy conversion efficiency, energy usage efficiency, waste heat recovery efficiency and energy transportation efficiency.

5. Conclusion

Based on the five material flows including those from the anterior producing unit or process and those recycled through a unit or process, the material flow model for a production unit, a process and an enterprise are constructed in this chapter. The influence that the material flow of a process has on the metal yield ratio is described, and so is the influence of the parameters on the resource efficiency of an enterprise. For iron and steel enterprise, the produced waste material should be recycled to the process itself or others to reduce the consumption of natural resource.

In this chapter, energy flow through an energy conversion unit or a process, such as the converted energy and the energy of the product, is discomposed into six parts, and the energy flow model for a conversion unit, a process, and an enterprise are established. The influence that the flux of each energy flow, the largening extent of equipments and the recovery of residual heat (or energy) have on the energy value of the product is analyzed.

It is proposed in this chapter that the energy flow network is divided into six regions according to their functions: energy conversion region, energy usage region, waste heat recycling region, surplus energy buffer region, energy storing region and city energy system. The energy flow network model is established for evaluating the comprehensive energy level with the index of system energy efficiency.

Directions of next research include: (i) The forecasting of energy generation and energy consumption; (ii) The optimized allocation of surplus energy; and (iii) The optimized connection between energy flow network of an iron and steel enterprise and city energy system.

6. References

Cai, J.J., Wang, J.J., Lu, Z.W. & Yin, R.Y. (2006). Material Flow and Energy Flow in Iron & Steel Industry and Correlation between Them. *Journal of Northeastern University (Natural Science)*, Vol.27, No.9, (September 2002), pp. 979-982, ISSN 1005-3026

Cai, J.J., Wang, J.J., Zhang Q. & Li, G.S. (2008). Material Flows and Energy Flows in Iron & Steel Factory and Their Influence on CO_2 Emissions. *Research of Environmental Sciences*, Vol.21, No.1, (January 2008), pp. 196-200, ISSN 1001-6929

Chen, G., Cai, J.J., Yu, Q.B. & Lu, Z.W. (2002). The Analysis of the Influences of Materials Flows in Iron and Steel Corporation on Its Energy Consumption. *Journal of Northeastern University (Natural Science)*, Vol.23, No.5, (May 2002), pp. 459-462, ISSN 1005-3026

Dai, T.J. & Lu, Z.W. (2004). The Influence of Iron-Flow in Steel Production Process on Its Iron Resource Efficiency. *Acta Metallurgica Sinica*, Vol.40, No.11, (November 2004), pp. 1127-1132, ISSN 0412-1961

Lu, Z.W., Cai, J.J., Yu, Q.B. & Xie, A.G. (2000). The Influences of Materials Flows in Steel Manufacturing Process on Its Energy Intensity. *Acta Metallurgica Sinica*, Vol.36, No.4, (April 2000), pp. 370-378, ISSN 0412-1961

Lu, Z.W. & Cai, J.J. (2010). *The Fouondations of Systems Energy Conservation*, Northeastern University Press, ISBN 978-781102-846-1, Shenyang, China

Yin, R.Y. (2000). Analysis and Integration of Steel Manufacturing Process. *Acta Metallurgica Sinica*, Vol.36, No.10, (October 2000), pp. 1077-1084, ISSN 0412-1961

Yin, R.Y. (2007). Some Science Problems about Steel Manufacturing Process. *Acta Metallurgica Sinica*, Vol.43, No.11, (November 2007), pp. 1121-1128, ISSN 0412-1961

Yin, R.Y. (2008). The Essence, Functions, and Future Development Mode of Steel Manufacturing Process. *Scientia Sinica Technologica*, Vol.38, No.9, (August 2008), pp. 1365-1377, ISSN 1674-7259

Yin, R.Y. (2011). *Metallurgical Process Engineering*, Springer Heidelberg, ISBN 978-3-642-13955-0, New York, USA

14

Evaluation of the Energy Consumption of Recycling Process

Toshiharu Goto
Shizuoka University
Japan

1. Introduction

Human activities have always consumed energy. In ancient times, human beings started cultivation to increase the amount of energy taken in via food consumption, as a result, humans had time to invent culture and technology. That is to say, the surplus of energy made for an affluent society. This trend was dramatically accelerated by the Industrial Revolution. This era initiated the consumption of fossil fuels in every aspect of our life, even in cultivation. This means that our life is supported by energy consumption.

Fossil fuels are formed from dead, buried organisms and it takes millions of years to generate fossil fuels, meaning that our consumption rate is too fast as a sustainable energy resource. One of the discussions about this issue was written by the Club of Rome (Meadows et al., 1972). Evaluation of the technology from the perspective of energy consumption is one of the most important discussions within engineering.

Because fossil fuels are limited, the amount of energy which we can use will have to peak at some point (Hubbert, 1956). Predicted oil peaks range between 2010 and 2040 according to the review by Richard (Kerr, 2007). The International Energy Agency (IEA) reports in the World Energy Outlook 2010 that the peak of conventional oil came on 2007. On the other hand, rapid economic growth of developing nations means that global demand for energy is growing. These situations indicate that the energy issue will become more serious in the near future. So, reduce, reuse and recycling (3R's) are more and more important for the resource problem.

Supercritical fluid (SCF) is expected to be one of the technologies used for the recycling of plastics because of its reactivity. On the other hand, there are general discussions on the energy consumption of the recycling of plastics, which indicate that the recycling can be wasteful of energy in some cases (Takeda, 2008). If much more energy is consumed for recycling than to make virgin material, recycling cannot save resources. This is especially the case when using SCF; energy consumption should be considered because large amounts of supercritical water or methanol require large amounts of energy. A phase diagram of methanol is shown in Fig. 1. It indicates that with SCF, the temperature and the pressure of the materials are higher than critical points. Critical points of water and methanol are shown in Table 1. Critical temperatures of the alcohol and water are higher than 250°C. Figure 2 shows the internal energy of methanol (Reuck & Craven, 1993) which increases with temperature at 10 MPa, which means that making such a high temperature and pressure state requires large amounts of energy.

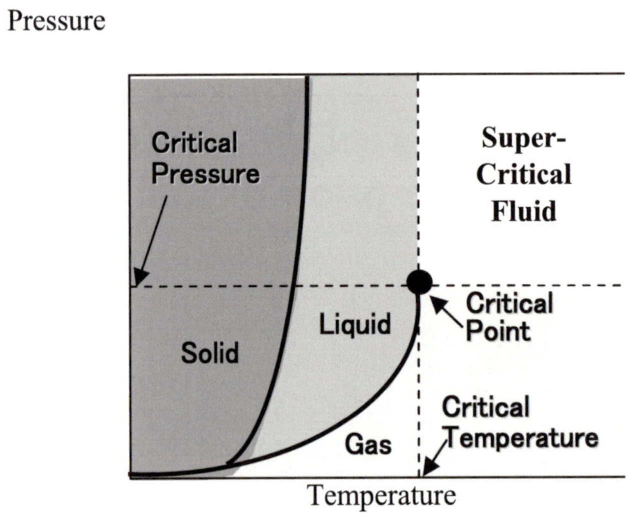

Fig. 1 Schematic phase diagram

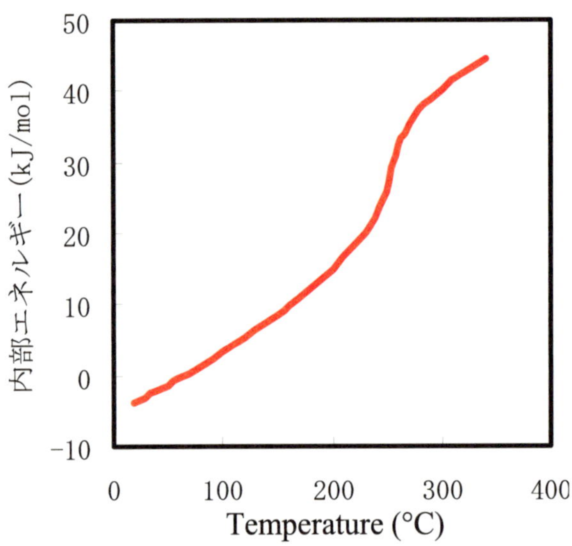

Fig. 2 Internal energy of methanol at 10 MPa

Material	Methanol	Water
Temperature(°C)	240	374
Pressure(MPa)	8.0	21.8

Table 1. Critical points

So, a technology which can minimize the amount of SCF was required for the recycling process using SCF to avoid wasting energy.

It was reported that the extruder can be used as a feeder and a reactor for the supercritical alcohol (Goto et al., 2005). Supercritical fluid can be injected into the polymer because the extruder can pressurize the polymer, which means that such process can minimize the amount of supercritical fluid and energy consumption. Here, LCA (Life Cycle Assessment) and EPR (Energy Profit Ratio or EROI:Energy Return on Investment) is applied to evaluate the energy consumption of this technology.

General knowledge of the LCA and SCF was explained first. Then, chemical reaction for the recycling of Si-XLPE in the supercritical methanol (Goto et al., 2008; Goto et al., 2001) was shown. After that, the technology which uses the extruder for the chemical reaction in supercritical fluid to develop the continuous process for industrialization (Goto et al., 2011) was introduced. The obtained recycled PE properties satisfied the requirements of Japan's industrial standard of insulation for 600V cross-linked polyethylene cable (600V XLPE cable) (Goto et al., 2006). Then energy consumption to make recycled PE is compared with that of virgin PE by LCA as one example of the evaluation of energy consumption for recycling technology.

2. Life Cycle Assessment (LCA) for evaluation of recycling technology

The final aim of LCA is the evaluation of the influence of human activity on the environment and society to select the best combination of products, services or technology. So, it is useful for the development of technology to take an objective view. LCA is quantitative analysis from raw material extraction through to processing, distribution, use, maintenance, recycling and disposal (Ito et al., 2007). It is useful for engineers, who tend to take the seeds oriented approach, to take a wider view of the technology. The procedure is standardized by ISO 14040 and 14044 in four distinct phases as follows:

1. Goal and Scope

It is impossible to gather all of information which is required for LCA. It is required to consider which raw materials and process of works affect the goal, to determine the boundaries of investigation. In this phase, function unit, system boundaries, assumptions, the allocation methods are defined.

2. LCI

Next, the scenario of life cycle is assumed from and to nature system. The data which cannot be directly measured can be compensated for by the database of unit base for LCA.

3. LCIA

The effect of the obtained results are discussed with regard to the goal and scope, such as effect on global warming, acid rain or resource problem. Normalization by weighting is conducted depending on goal and scope. But weighting will also make the meaning of the obtained data obscure in some cases. So, when it comes to studying energy consumption, it is preferable not to use weighting.

4. Interpretation

The key purpose of this phase is checking the credibility and accuracy of data, and sensitivity of parameters. It is also evaluated if the goal and scope is achieved.

3. Energy Profit Ratio (EPR)

The concept of energy profit ratio is one of the methods for the interpretation of the results of LCA. EPR originates from the idea of the 'rabbit limit' which is based on the case of rabbit hunting shown in Figure 3 (Gunther).

Fig. 3. Rabbit hunting

When you have to chase a rabbit to obtain food, you need to consume energy for hunting. If the energy taken for hunting the rabbit is greater than that of the energy obtainable from eating the rabbit, rabbit hunting is not sustainable. EPR is determined using the following equation (Amano, 2008):

$$EPR = \text{yielded energy} / \text{consumed energy}$$

EPR must be more than 1 if the activity is sustainable. The larger the amount of EPR means the greater the efficiency of the energy resource. So, EPR is also convenient for consideration of sustainatbility. This method was originally used to evaluate the value of energy resource such as oil, coal and so on. Here, we adopt the concept of EPR for recycling. The final aim of the recycling technology is to contribute to the sustainability of human life, that is to say, evaluation of recycling technology requires the concept of the rabbit limit. In the case of recycling, it must be evaluated if consumption of energy decreases. Here, energy consumption ratio (ECR) is defined as follows (Goto et al., 2009):

$$ECR = \text{Consumed energy for virgin product} / \text{Consumed energy for recycled products}$$

In this case, it is assumed that the recycled product is as valuable as the virgin product, such as in the case of horizontal recycling. This concept can also be adopted in the case of the improvement of industrial processes, distribution of products or supply chain to compare before and after improvement.

4. Supercritical fluid for recycling of plastics

A comparison of properties between gas, liquid and SCF is shown in Table 2. The density of SCF is 1/10-1/2 of liquid. On the contrary, viscosity close to the gas and diffusion constant is in the middle of gas and liquid. As a result, SCF has strong chemical activity. So, SCF is expected to be the method for decomposition of harmful chemical materials (Sako et al., 1998). Moreover, dielectric constant decreases around the critical point, which means that the properties as solvent are change around critical points (Saito, S.,1996). For example, water can be mixed with oil above the critical point of water.

When you need to use the reaction between water and polymer which is not dissolved in water, supercritical water can be used without other solvents. Another merit of using supercritical fluid is the ability of diffusion. Cross-linked polymers are difficult to stir to accelerate the chemical reaction because cross-linking of polymers robs the liquidity of polymer. The high diffusion constant of SCF is positive effect for water or methanol to diffuse into the polymer without stirring. All these properties lead us to expect that SCF has advantages for the chemical recycling of cross-linked polymers.

Properties	Gass	Supercritical fluid	Liquid
Density (kg/m^3)	0.6～1	200～900	1000
Diffusion Constant (m^2/s)	10^{-5}	10^{-7}～10^{-8}	10^{-9}
Viscosity (Pa·s)	10^{-5}	10^{-5}～10^{-4}	$<10^{-3}$

Table 2. Properties of supercritical fluid, liquid and gas

5. Application of chemical reaction in supercritical fluid

Here, a general analysis of chemical reaction in supercritical fluid is given. Chemical reaction in SCF is basically the same as chemical reaction at room temperature. SCF is useful if you need to accelerate the chemical reaction kinetics. Most of the research of chemical reaction of supercritical water and alcohol is transesterification. For example, decomposition of carbonate (Arai, 2002), and also applied for the siloxane bond such as decomposition of silica gel (Kitahara, 1969; Kitahara et al., 1970) or silicon rubber. All of this research takes advantage of acceleration of chemical reaction in supercritical fluid.

On the other hand, the disadvantages of SCF are difficulties in the development of the process for commercialization because it is difficult to put solid material into the SCF and take materials from the SCF. This difficulty is mainly derived from the properties of SCF which is close to the highly pressurised gas. So, development of a continuous process for SCF is discussed in the following section.

6. Recycling technology for Si-XLPE using supercritical fluid

Silane cross-linked polyethylene (silane-XLPE) is widely used as an insulation material for wires and cables (Maruyama, 2003). Recycling of cross-linked polymers poses one of the most difficult problems in the recycling of polymer wastes. Most of industrial waste silane-XLPE is buried in landfills or burned as fuel, because silane-XLPE is difficult to process due to its low fluidity derived by cross-linking elements formed by the siloxane bond. The research for recycling of silane-XLPE as polymer is as follows:

1. Clashed XLPE pellet is mixed with virgin polyethylene using a twin screw extruder (Tokuda et al., 2003),
2. Powdered XLPE is used as a filler of polymer (White, et al., 2000; Voldner, 2000).

These technologies require further investigation of the application for recycled products and the development of the market due to the apparent difficulty in using directly the recycled product as insulation of wire or cables (horizontal recycling).

6.1 Chemical reaction for recycling of Si-XLPE

Chemical reaction in supercritical alcohol is shown in Figure 4. Siloxane bond (Si-O-Si) can be decomposed into the alkoxide group (-OR) or the hydroxyl group (-OH) by supercritical alcohol selectively.

The condition for selective decomposition of cross-linking bond is investigated. The gel fraction and molecular weight of products against decomposition temperature is shown in Figure 5. The gel fraction vanished at over 300°C when methanol was used as supercritical fluid. Using water, a temperature over 370°C was necessary to decline the gel components to 0wt%. These results indicate that the cross-linking element was completely decomposed at over 300°C using methanol and at 370°C using water. Molecular weight was decreasing at over 340°C using water and methanol. This result agreed with the general knowledge about the thermal decomposition temperature of PE.

By using water, the molecular weight of the products would be lower than raw PE whenever the gel fraction of the products decreased to 0%. It is indicated that the reaction in supercritical water was not selective only for the cross-linking element, but also for the main chain of PE.

In the case of methanol, at temperatures between 300°C - 340°C and pressure more than critical pressure, the cross-linking element was completely decomposed and the molecular weight of the recycled PE was kept the same as virgin PE before being cross-linked. It means that using supercritical methanol is necessary to obtain thermoplastic PE which has the same molecular weight as virgin PE.

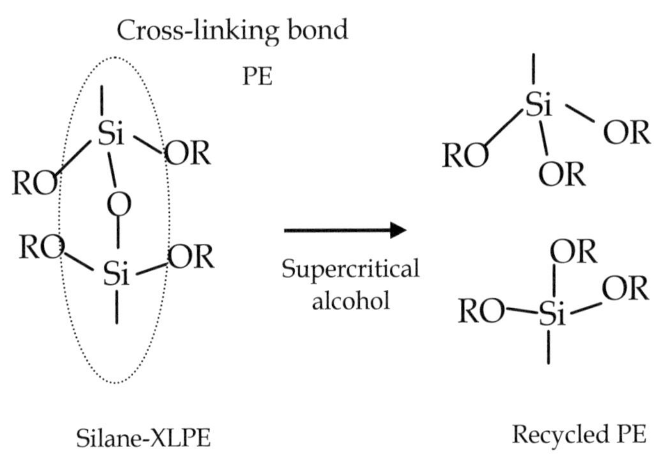

Fig. 4. Chemical reaction of crosslinking bond in supercritical alcohol

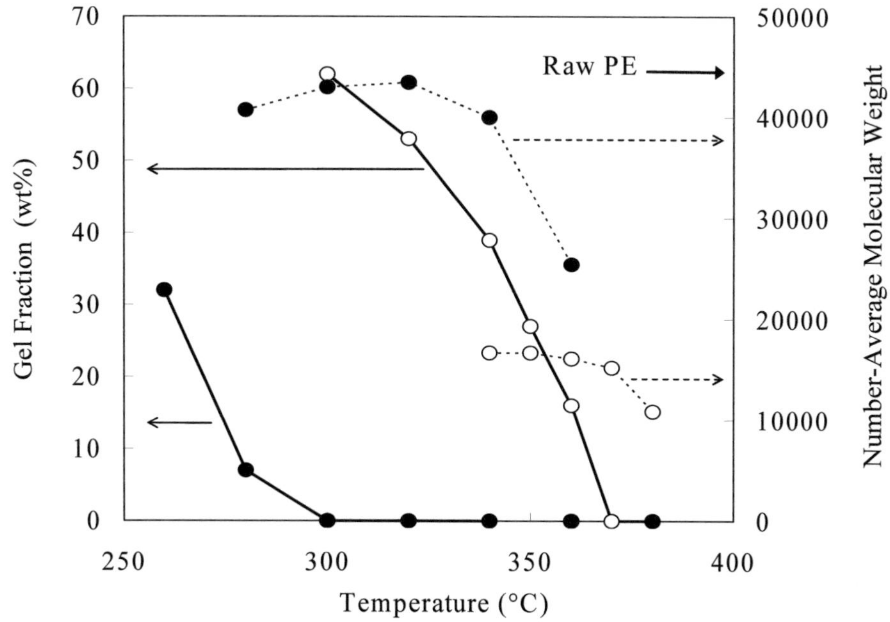

Fig. 5. Gel fraction and number molecular weight of recycled product made by supercritical methanol and water. Black circle:made by methanol, White circle: made by water

6.2 Process design

One of the most important issues with supercritical fluid technology is the development of a continuous industrial process for the solid material. The conventional process for SCF is compared with the process using an extruder that is developed for the recycling of cross-linking polyethylene in Table 3. The most fundamental method is the batch process, but in using this process, the reactor should be cooled to take out the product and put raw materials in the reactor. That is to say, efficiency of heat energy is low. In the case of a semi-continuous process, heat efficiency is better than that of the batch process, but the amount of water or alcohol required is about the same as the batch process because the pressure is applied by the materials used as SCF. The amount of SCF cannot be saved in this process, which means improvement on the efficiency of energy is small. In the general continuous process for the supercritical fluid, the solid material was crushed into small particles and dispersed in the solvent to make a slurry. Then, a liquid pump was used to feed the slurry, but this kind of process also has a problem for industrialization as follows. It is very expensive to crush the materials into particles to make powder for slurry. Moreover, the process requires much more solvent than the solid material itself. Ordinary, 10 of solvent required for one of solid material to make slurry. To solve these problems, we developed a new continuous process for the supercritical fluid using an extruder. Extruder is used as the processing machine for polymers. Pressure and heat can be applied to the polymer at 400°C and 50 MPa with the extruder. The polyethylene which is crushed into 5 mm in diameter can be fed to extruder. This means that the new process using extruder does not require making powder of solid material. Furthermore, a minimum amount of supercritical fluid is required for the chemical reaction into pressurized polymer.

	Nessesity of break into pieces	Heat Efficiency	Amount of Solvent	Continuous process	Productivity
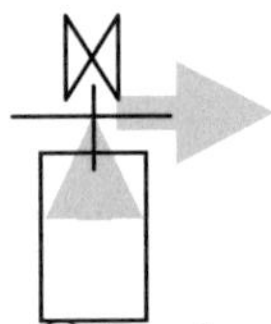 ① Batch	○ Not necessary	× Low	× Large	× Impossible	× Low
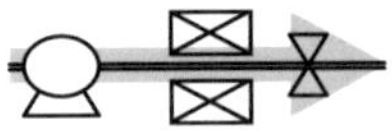 ② Pump	× Necessary	△	× Large	△ Possible	△
 ③ Semi Continuous	○ Not necessary	△	× Large	× Impossible	△
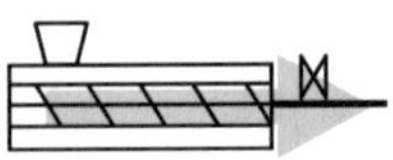 ④ Extruder	○ Not necessary	○ High	○ Small	○ Possible	○ High

Table 3. Comparison of the process for supercriticalfluid

The process was designed to make recycled PE from crushed Si-XLPE waste continuously and automatically. A scheme of the continuous process is shown in Figure 6. The twin screw extruders are made by Japan Steel Works and were used as equipment for the chemical reaction (Ext-Chem) and degassing (Ext-Degas). Methanol was fed using a high pressure pump from the methanol tank and heated to supercritical state by the heater before it was injected to the Ext-Chem. The tubular reactor was attached to the Ext-Chem to keep Si-XLPE and methanol for more than 15 min at 10 MPa, 335°C. The pressure control valve was connected to the tubular reactor. Ext-Degas was mounted to the outlet of the pressure control valve to separate the recycled PE and the gas. Pelletizer was prepared to cut the recycled PE strand into the pellets. The whole electric power demand for this process was measured using the attached ammeter to evaluate the energy consumption.

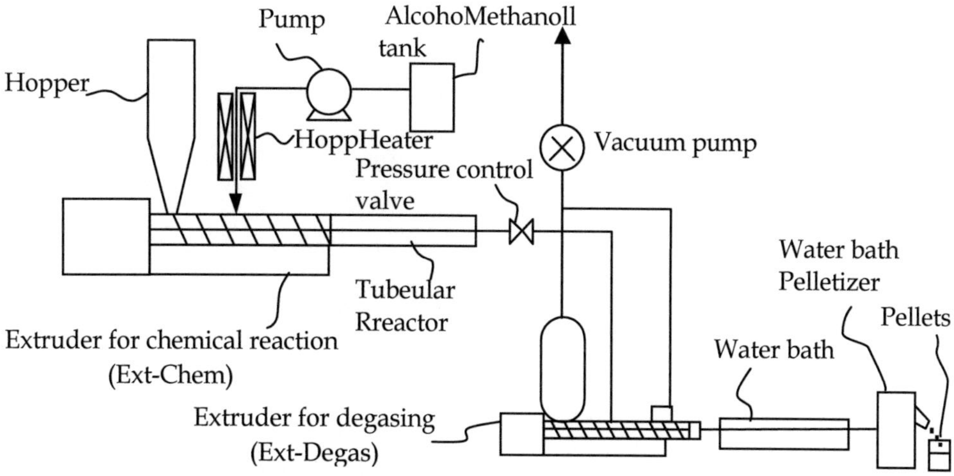

Fig. 6. Schematic diagram of the continuous process for recycling of silane-XLPE by supercritical alcohol using twin screw extruders.eCa SE

The Si-XLPE pellet with 30% in the degree of gel fraction was gathered from the factory for the production of 600V XLPE cable. The methanol used here was made by WAKO Chemicals.

40kg/hour of Si-XLPE was charged to the extruder from the hopper, then 10 phr of alcohol was injected to the Ext-Chem. The injected supercritical alcohol was kept at over 300°C at 10 MPa. The reactor was kept at 330°C and 10 MPa. The Ext-Degas was set at 200°C to extrude the recycled PE.

Electric power demand for this process is shown in Figure 7. Power demand is stable at around 160A 200V in 10 hours, which means that the process was well controlled. The average electric power demand in 10 hours is 159.3A, 200V.

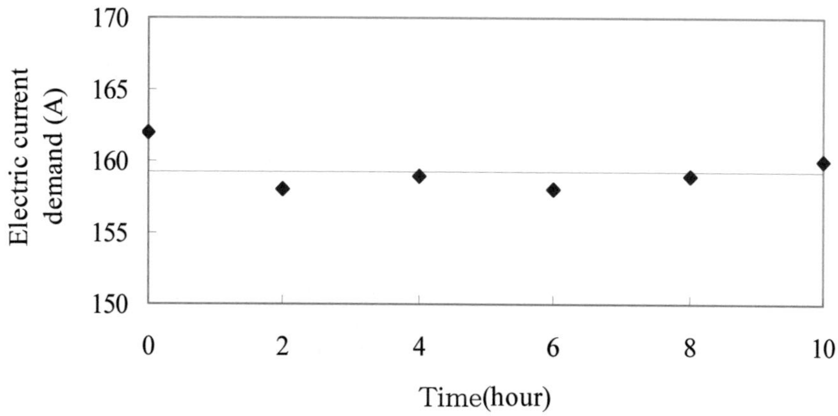

Fig. 7. Electric power demand for the recycling.

6.3 Properties of the productst

The appearance of cable extrusion of the recycled PE and the 600V rating XLPE cable (cross section of conductor 38mm2, thickness of insulation:1.2mm) is shown in Figure 8. The proccessability of the recycled PE was good enough to be used as cable insulation.

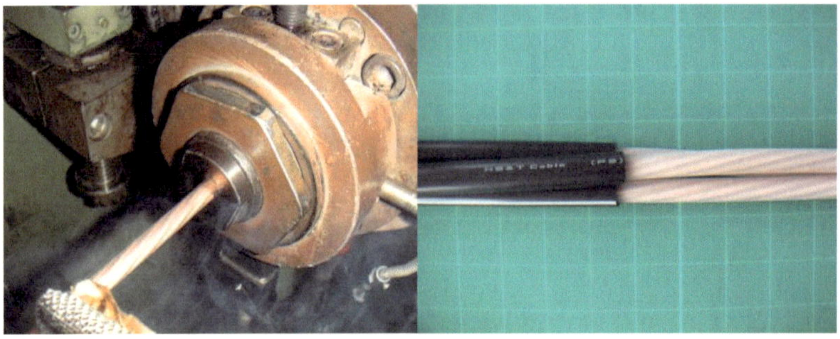

(a) Cable extrusion (b) Recycled 600V rating XLPE cable

Fig. 8. Trial for cable recycling

The properties of the recycled cable are shown in Table 4. The initial mechanical and electrical properties of the cable satisfied JIS C3605.

Items	Recycled Cable	Required
Tensile Strength at break(MPa)	20	>10
Elongation(%)	500	>200
Volume Resistance(MΩ/km)	10×10^5	$>1.5\times10^3$
Breakdown Voltage(kV)	28	-
Heat Deformation(%)*	0.9	<40

* : At 120°C under the force of 20N

Table 4. Properties of the recycled cable

7. Life cycle assessment of recycling of silane-XLPE

600V XLPE cable is the target application of this technology. The structure of 600V XLPE cable is shown in Figure 9. The conductor is covered by insulation and sheath.

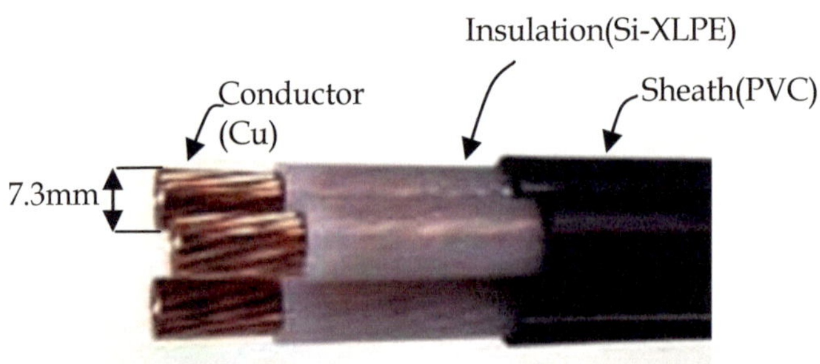

Fig. 9. Structure of 600V XLPE cable.

Figures 10 and 11 illustrate the process for production of 600V XLPE cable using virgin PE and recycled PE as insulation respectively. The same copper (Cu) and polyvinyl chloride (PVC) can be applied to the virgin and the recycled product as the conductor and sheath respectively.

It is reported that recycled PE can be used the same as the virgin PE (Goto et al., 2006), which means that energy used for the cable processing should be the same in each product. So, the system boundary was determined as the hatched box in Figures 10 and 11. Energy consumption to produce the recycled PE was compared with that of the virgin PE in this study. The life cycle energy assessment of the whole process for the production of cable is reported in another paper (Goto, 2004).

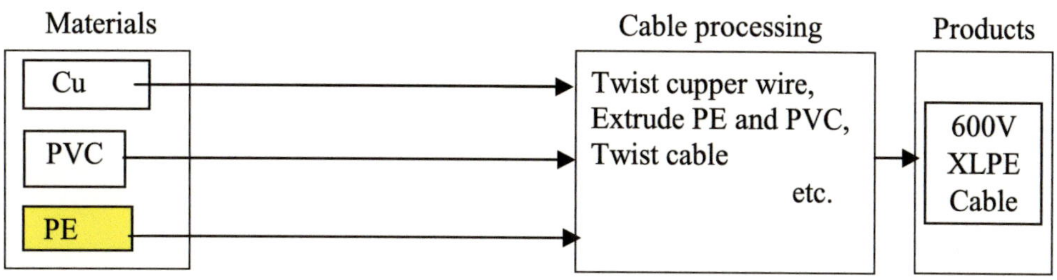

Fig. 10. Process for the the virgin products.

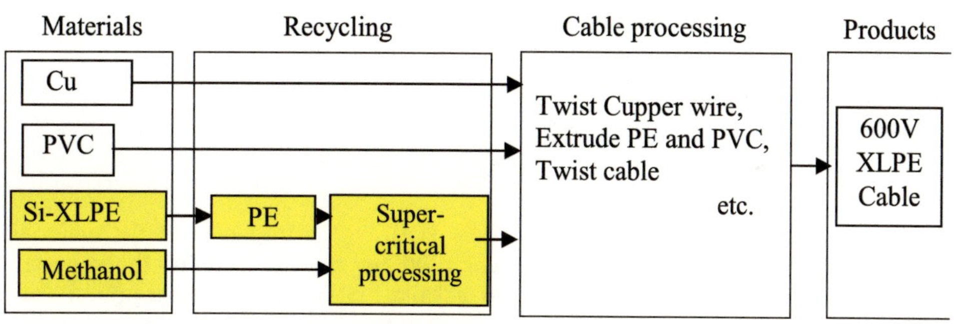

Fig. 11. Process for the recycling products.

Si-XLPE was assumed to be gathered from the cable factory and recycled at that site. Energy unit bases of virgin PE, methanol, electric power and steel were represented in Table 5 (Amano, 2006). These values were converted to the energy per 1kg of PE for calculation.

	Parameters	Unit base
PE	U_{PE}	7054 kcal/kg
Methanol	U_{MeOH}	5431 kcal/kg
Electric power	U_{ele}	2250 kcal/kWh
Steel	U_{Fe}	6125 kcal/kg

Table 5. Energy unit base

Energy for the recycled PE (Er)was represented as following equation.

$$E_r = U_{MeOH} + E_1 + E_2 \tag{1}$$

E_1 and E_2 are the energy consumption for crushing Si-XLPE and for the supercritical processing respectively for 1kg of PE. Energy consumption for each unit process is represented as follows:

$$E_n = E_n^{'} + E_n^{''} \tag{2}$$

$E_n^{'}$ is electric energy consumption per 1kg of PE, which was estimated equal to the volt-ampere(VA) per 1kg of PE. Measured electric demand shown in Figure 6 was used for evaluation of that of the supercritical processing. On the other hand, electric demand for the crusher was estimated as 30% of the electric capacity of crusher which can treat 40kg/hour of the Si-XLPE waste.

$E_n^{''}$ is energy requirement to make the equipment in the unit process per 1kg of PE, which is represented as following equation.

$$E_n^{''} = \frac{E_n^{'''}}{V_n \times D_n \times Y_n \times R_n} \tag{3}$$

- $E_n^{'''}$: Energy requirement to make the equipment (kcal)
- V_n: Productivity of PE (kg/hour)
- D_n: Operation time in the year (hour/year)
- Y_n: Life span of the equipment (year)
- R_n: Occupation rate of equipment for the product(-)

Equation (4) was used to estimate:

$$E_n^{'''} = \sum_l U_l W_l \tag{4}$$

W_l is weight of the part of equipment (kg). U_l (kcal/kg) is unit base of the materials of which part of equipment consist. Equipment used for crusher and for the supercritical treatment mainly consist of steel. So, we calculate the energy consumption for the production of the equipment as following equation instead of equation (4).

$$E_n^{'''} = U_{Fe} W_n \tag{5}$$

W_n is total weight of the equipment for the unit process (kg). U_{Fe} is unit base of steal (kcal/kg).

The found energy consumption is shown in Table 6. The required energy for the recycling of Si-XLPE was smaller than that of virgin PE. These results indicate that the recycling of Si-XLPE using supercritical alcohol can save energy consumption.

Moreover, further information about the recycling process is revealed as follows. Energy required for the construction of the equipment is smaller than that of the electric energy used for the operation of the equipment. Electric energy was mainly used as the energy source for the heaters and for the motors. So, these results indicate that the heat insulation and the efficiency of the motors are important matters in saving energy rather than the size reduction of the equipment.

The general aim of recycling is saving resources. Energy is the most fundamental resource because every human activity, including mining, requires energy. So, LCA on energy is a useful tool for the development and evaluation of the meaning of recycling technology.

		Recycled PE (kcal/kg)	Virgin PE (kcal/kg)
E_{MeOH}		244.4	-
E_{PE}		-	7054.0
E_1	E_1'	0.6	-
	E_1''	103.7	-
E_2	E_2'	17.0	-
	E_2''	2240.2	-
Total		2605.9	7054.0

Table 6. Energy analysis of the recycled PE.

ECR is calculated from the data shown in table as following equation.

$$ECR = \frac{7054.0}{2605.9} = 2.71$$

The obtained ECR is 2.71 which means that recycling of Si-XLPE is the technology that can contribute for saving energy. This example also indicates that the LCA and EPR methods are useful for evaluation of recycling technology. Moreover, these methods will be effective in the development of the recycling process to look at the final target.

8. Conclusion

The energy consumption for the recycling of Si-XLPE was studied. It is indicated that energy can be saved if the extruder is applied as a feeder and a reactor for the reaction in supercritical methanol. A life cycle energy assessment can give an answer to the criticism of wasting of energy because of the high pressure and temperature of supercritical methanol. Moreover, the results give us a principle in the development of a more efficient process.

9. References

Amano, O.(2006) "JHFC report about result of the study on total efficiency", Japan Automobile Research Institute

Amano, O.(2008) Energy after Oil Peak

Amano, O.(2010) How to live after Oil Peak

Arai, Y(2002) Supercritical Fluid, Techno System

Donella, H Meadows, Dennis L. Meadows, Jorgen Randers, and William W. Behrens III. (1972).The Limits to Growth. New York: Universe Books. ISBN 0-87663-165-0

Gunther, F, The rabbit limit,

Goto T., Ashihara S., Yamazaki T., Sugeta T., Okajima I., Sako T., Yoshihiko Iwamoto, Ishibashi M., (2011) Continuous Process for Recycling Silane Cross-linked Polyethylene Using Supercritical Alcohol and Extruders. Industrial & Engineering Chemistry Research, 50, 5661-5666

Goto T., Ashihara S., Yamazaki T., Watanabe K.(2006) The transactions of the Institute of Electrical Engineers of Japan. B, A publication of Power and Energy Society 126,4, 400-406

Goto T., Yamazaki T., Sugeta T., Okajima I., Iwamoto Y., Kakizaki J., Ohtake K., Sako T., Kagaku Kogaku (2005) Research on Continuous Recycling Process for Cross-linked Polyethyelne by Supercritical Alcohol. Ronbunshu Vol.31, No.6, p.411

Goto T., Yamazaki T., Sugeta T., Okajima I., Sako T., (2008) Selective Decomposition of the siloxane bond constituting element of silane-crosslinked polyethylene by supercritical alcohol, J. Applied Polymer Sci., Vol.109, p.144

Goto T., Yamazaki T., Sugeta T., Okajima I., Sugeta T., Miyoshi T., Shigenobu, H., Ohtake, K., Sako T. (2001) Recycling of Cross-linked Polyethylene Using Supercritical Fluid, Kobunsi Ronbunsyu Vol.58, No.12, p.703

Goto, T.; Amano, O.; Okajima, I.; Sako T.(2009) Evaluation Method for Production Technology using Concept of Energy Profit Ratio, Journal of Mottainai gakkai ,3,1-6

Goto, T.; Yamazaki, T. (2004) Recycling of Silane Cross-linked Polyethylene for Insulation of Cables using Supercritical Alcohol. Hitachi Cable Review, 23, 24-27

International Energy Agency(2010) World Energy Outlook 2010

ISO 14040 (2006): Environmental management - Life cycle assessment - Principles and framework, International Organisation for Standardisation (ISO), Geneve

ISO14044(2006):Environmental management–Life cycle assessment - Requirements and guidelines, International Organisation for Standardisation (ISO), Geneve

Ito, T; Tahara, H; Narita, N(2007) An outline of LCA, Sangyo Kankyo Kanri Kyoukai Co.

Kerr, R (2007) The Looming Oil Crisis. Could Arrive Uncomfortably Soon, Science, 316, 20, 351

Kitahara, S. (1969) Dissolution and Alternation of Surfaceof Silica Gel Powder in Ethanol between 150°C and 250°C

Kitahara, S., Asano, T., Hirowatari, T (1970) Dissolution and Alternation of Surfaceof Silica Gel Powder in Ethanol between 150°C and 250°C

M.K. Hubbert (1956) Nuclear Energy and the Fossil Fuels, Presented before the Spring Meeting of the Southern District, American Petroleum Institute, Plaza Hotel, San Antonio, Texas, March 7–8-9

Maruyama, M.(2003) Research on the Life Cycle Model of the Waste Materials from the Recycling System of Cable and Wire. JECTEC NEWS, 7, 18-23

Reuck, K.M.; Craven, R.J.B (1993) Methanol, International thermodynamic tables of the fluid state - 12, IUPAC, Blackwell Scientific Publications, London

Saito, S.,(1996) Science and Engineering of Supercritical Fluid, Sankyo Buisiness

Sako, T., Sugeta, T., Okajima, I.(1998) Countermeasure for resource and enviromental problem, 34, 31

Takeda K.(2008) Society and Economic Aspect of Polymer Recycling, the Society of Rubber Industry Japan Vol.81, p.5

Tokuda, S.; Horikawa, S.; Negishi, K.; Uesugi, K.; Hirukawa, H. (2003) Material Recycling Technology of Crosslinked Polyethylene. Furukawa Denkou Jiho, 111, 1, 90-94

Voldner, E.(2000) Crosslinked Polyethylene Scrap Can Be Recycled. Innov. Rotational Molding Year Beyond, Ohio U.S.A. 1999, 73-82.

White, C. C.; Wagenblast, J.; and Shaw, M. T.(2000) Separation Size Reduction and Processing of XLPE from Electrical Transmission and Distribution Cable. Polym. Eng. Sci., 40, 4, 863-879

15

Size Reduction by Grinding as an Important Stage in Recycling

Marek Macko
Casimir the Great University in Bydgoszcz
Poland

1. Introduction

Today, the issue of size reduction methods are one of the important fields of research as regards processing and recycling problems as growing post-use waste volume is becoming more widespread. This concerns, among others, polymer materials. The annual volume of polymer materials manufactured all over the world reaches nearly 260 million Mg, out of which about 60 million Mg in Europe [18]. Taking into account short life (which particularly applies to packages), necessity to meet requirements of the sustainable development policy as well as criteria of environmental procedures being implemented (e.g. LCA), effective management of that waste shall have significant consequences in the near future. There are many waste management methods, among which material or energy oriented recycling are particularly interesting [5, 29]. Comminution (grinding) is the process of key importance in every scenario. It can be assumed that energy potential of polymer materials (recyclates) shall be of increasing importance as new clean technologies of waste incineration and energy utilization are implemented. Analysts estimate that up to 20% of the global plastic production is subject to recycling in the first place, due to poor quality of a product, and 100% of it is recycled by comminution after use [45]. Due to visco-elastic properties of the materials subject to grinding and as the grinding machines are usually of general-purpose, grinding is a relatively energy consuming process. The power demand for grinding generally amounts to (20 ÷ 500) kW·h/Mg, which makes 25 ÷ 50% of the total electric power needed for material processing [13, 45]. It is estimated that up to 8% of the global power demand is allocated for comminution, granulation and agglomeration, while in case of minerals processing, even up to 70 % of the total power is used for comminution [19]. Currently a lot of attention is paid to the efficiency of grinding, particularly as regards relationship between efficiency of machine grinding and size reduction of separate grain (as

Grinder type	Efficiency
Multi-edge cutting mill	25 ÷ 40%
Beater mills	17 ÷ 25%
Roller mills	7 ÷ 15%
Ball mills	6 ÷ 9%
Jet mills	1 ÷ 2%

Table 1. Range of achieved efficiency for selected grinders

a relationship of power demand between physical and model conditions). Table 1 shows estimated efficiency values of selected design solutions. The range of grinding application can also be relatively broad depending on the expected grain size level for different materials and power demand range (fig. 1).

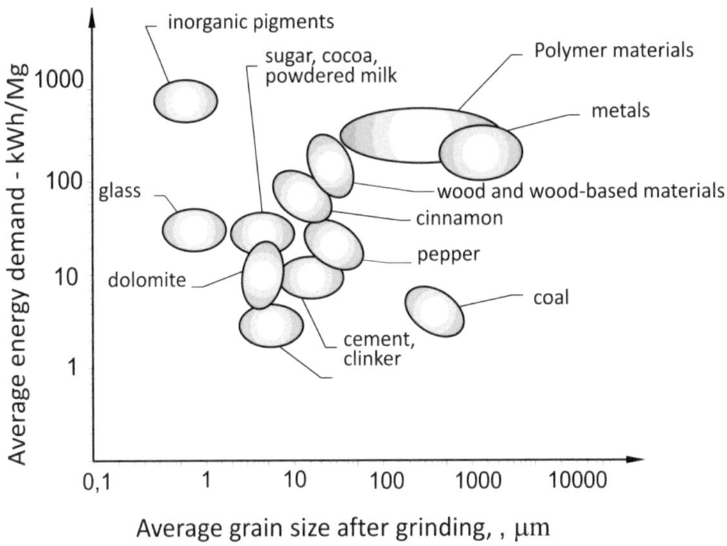

Fig. 1. Range of energy-related and dimensional requirements for wide spectrum of materials (based on studies by M. Pahl [35], J. Sidor [43], W. Peukert [36], J. Flizikowski [13], G. Schubert [42], V. Reppelin [39] and others)

The process efficiency issue is associated with the parameter of the unit energy consumption as the measure of energy needed to grind 1 kg of plastic. Energy demand for rough and medium grinding is 20 kW·h/Mg ÷ 130 kW·h/Mg, while for very fine grinding it reaches even 800 kW·h/Mg (for currently used solutions).

2. The state of the art: New and previous design solutions of comminution systems for non-brittle recycles new and former design solutions of comminution systems for non-brittle recyclates

The basic comminution methods (also known as mechanical or phenomenological models), are usually based on three main load types, i.e.: compression, tension and shear. Depending on physical properties, strength properties and the type of the material subject to grinding, to ensure high efficiency of the process, adequate reduction of shape and size of the input material must be used in the design stage.

According to H. Rumpf [40] and M.H. Pahl [34], there are four types of the grinding process used to produce elementary particles of the expected size and shape. They were briefly described and explained in fig. 2.

In case of polymer materials and cross-linked elastomers, the prevailing part in the grinding process is played by shearing (cutting) between two edges (marked in fig. 2 as grinding type I). This type of cutting is known as so-called „quasi-technological" cutting and it is used in, for example, multi-edge grinders characterized by low energy consumption, low noise level and good repeatability of grain classes [4, 13].

Cutting between two cutting edges is widely used in cutting mills, in particular for grinding of plastics and cross-linked elastomers [48, 3].

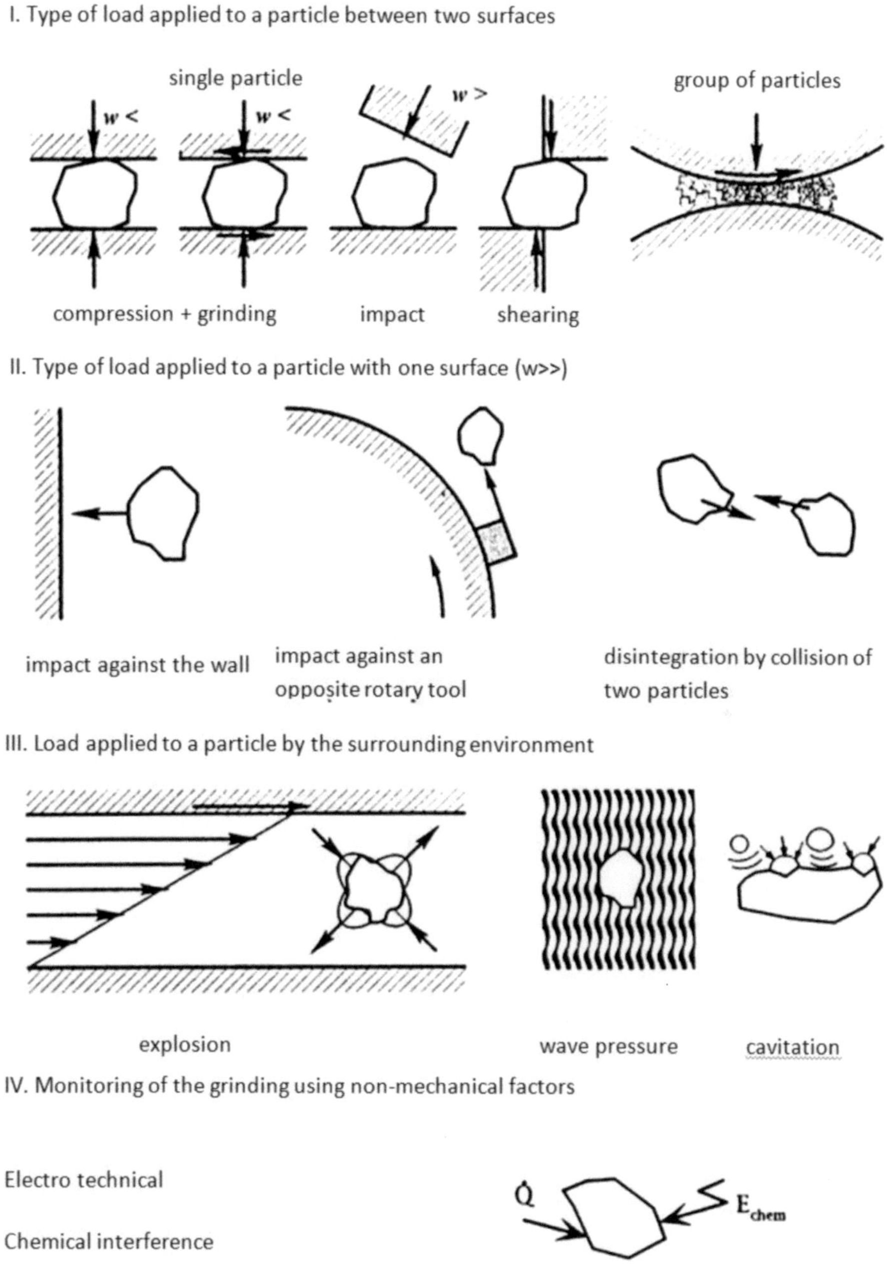

Fig. 2. Mechanical (fenomenological) models of grinding according to H. Rumph and M.H. Pahl [35, 40]

According to K. Höffl, comminution can be effected by two main division types only: breaking and shearing [17]. This is due to the increase of the distance between the particles caused by normal and shear stresses. However, main decohesion types are most often analysed based on the shear cutting theory (fig. 3) [9, 10, 13]:

- exclusively under the influence of normal force N, without lateral displacement of the tool (cutter) and without the impact of the lateral force; the slide cutting angle $\tau = 0$, (collisions); cutting is of dynamic nature (chopping),
- under the influence of the normal force N, and simultaneous influence of the lateral (tangent) force T without the slip; as the tool operation angle (slip cutting) is smaller than the friction angle $\tau < \varphi$, the tool (cutter) movement causes lateral displacement of the plastic particles being cut so that, as a result, cutting process takes place under influence of the normal force (beater-and-shear type),
- under the influence of the normal force N and the lateral force T, including the slide; the slide cutting angle is higher than the friction angle $\tau > \varphi$, so the friction resistance between the cutting edge and object being cut is smaller than the lateral component, which causes the cutter slide with reference to the material layer and thus it facilitates material cutting (shear cutting).

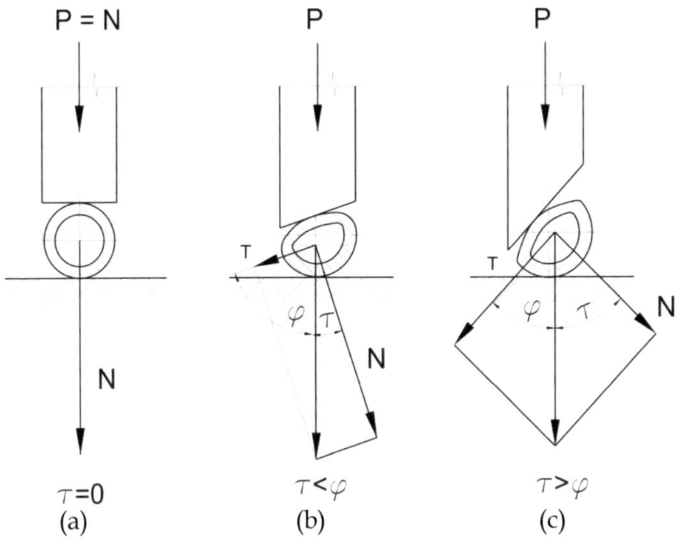

Fig. 3. Cutting types: a) cutting under the influence of the normal force only – beater type, b) cutting also under the influence of the lateral component, without the slide, c) cutting also under the influence of the lateral component, with the slide

Material and energy oriented recycling of relatively large parts (e.g. boxes, barrels) is performed in two stages. The first stage usually includes processing in roller and beater mills and then products are further ground (which is usually called as re-granulation) in order to obtain the form of the plastic that allows re-use. The main group of machines used at that stage includes granulators – cutting mills [13].

Current technical knowledge includes studies on grinding of selected materials using specific group of grinders. Polymer materials are ground using grinders of different design solutions. Some of them have been adapted to other fields of industry, e.g. agricultural engineering [9]. In the author's opinion, the most numerous group among polymer material grinding machines are beater and cutting grinders characterised by very diverse shape of the working space (fig. 4.).

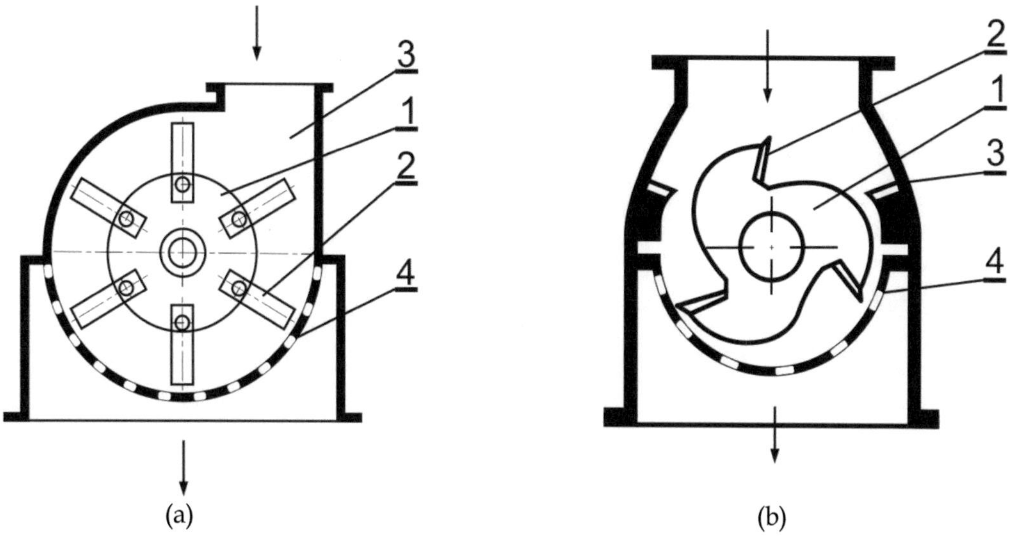

Fig. 4. Sample design solutions: a) beater mill: 1-rotor, 2-beater, 3-hopper, 4-screen, b) cutting mill: 1-rotor, 2-cutter, 3-counter-cutter, 4-screen.

The following types of grinders and mills described in the literature are currently designed and manufactured based on available theories and patents: beater mills, cutting mills, hyperboloidal mills, chopper mills, quasi-cutting mills, disk mills, roller mills and special (ball mills, vibration mills, rotary-and-vibration mills, rolling mills) and other design solutions developed as the modification of the above mentioned ones. One can distinguish main ways of load application in those solutions (strain, torque, bending, cutting, squeezing, breaking, grinding), but all those processes usually occur together where one of them is the prevailing process, depending on the mill design.

Cutting mills with cutters spaced over the drum circumference are widely used for rough and medium grinding of, among others, non-ferromagnetic light metals, polymer material waste, tyres and papers waste [46].

Cutting mills belonging to the group called „block knife shears" are equipped with several cuboidal cutters. Due to high durability of the components and heavy weight of the flywheel (fig. 5) those mills are capable of grinding thick-walled polymer and metal parts (e.g. of electric motors and transformers) at the speed of 5 ÷ 20 m/s.

Different design solutions include many components that intensify grinding process, such as, fixed cutters, rotating cutters or screens with specially shaped rebounding surface. Figure no. 6. shows the design of the mill manufactured by Alpine [35] company, equipped with rotating cutters fastened over the drum and two fixed cutters including bumping blocks fastened to the housing, where the process can be controlled by screwing in (letting in) the parts towards the chamber of the mill.

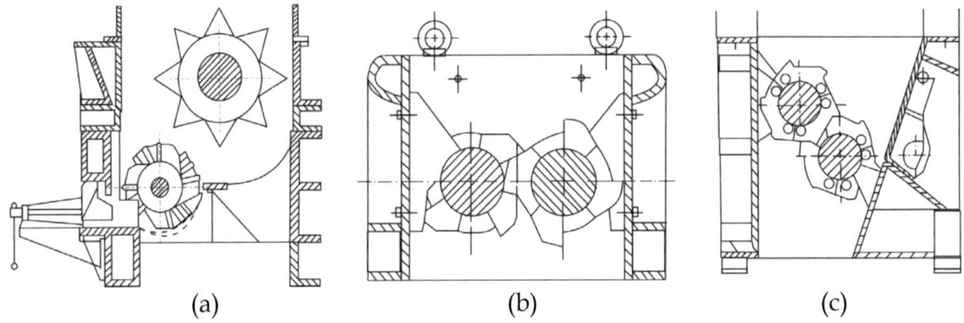

Fig. 5. Selected design solutions of mills equipped with double disk and roller systems: a) single-rotor system with angle gap (Alpine/Germany) [32], b) double-rotor system with radial gap (Hoger/Germany) [30], c) four-rotor system with the angle gap (SID/Switzerland) [31]

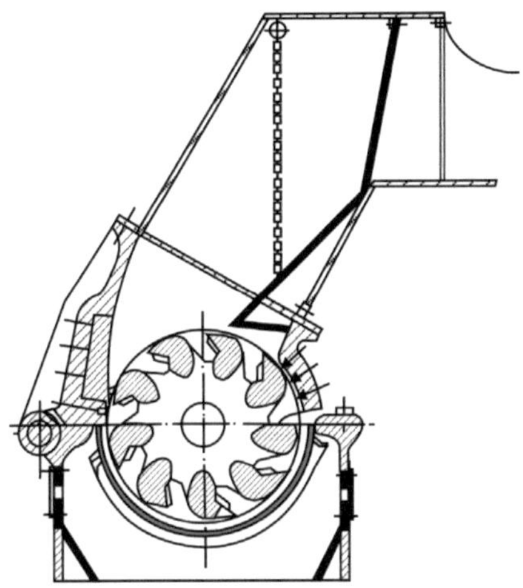

Rys. 6. The design of the mill manufactured by Alpine company, equipped with rotating cutters.

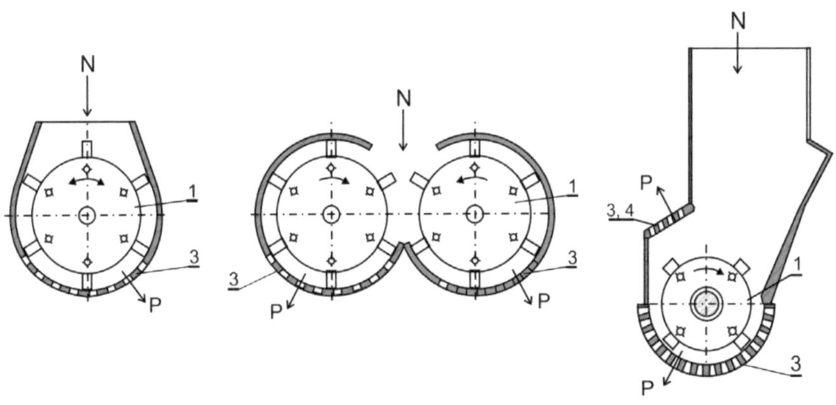

Fig. 7. Design solutions of mills with horizontal drive shaft: 1-rotor with rotating cutters, 2-fixed cutters, 3-screens, 4 swinging parts of plates with holes playing the part of screens [46]

Mills with horizontal drive shaft are equipped with disks with rotating and fixed parts and with screens spaced over the circumference (fig. 7).

An interesting design group is also constituted by mills in which the impact operation of elements spaced over the rotor circumference is used and which generate tangent stresses. The mill shown on the figure 8 can serve as the example of such a design solution.

An interesting design group is also constituted by mills in which impact operation of elements spaced over the rotor circumference is used and which generate tangent stresses. In the figure 8 can serve as the example of such a design solution.

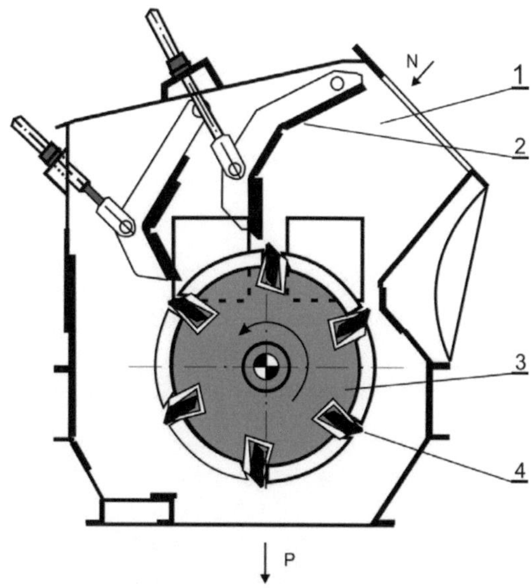

Fig. 8. View of the mill with beaters spaced over the rotor circumference, which also shear the material processed in the grinding chamber:
1-inlet of the fed material, 2-intensifying plates, 3-rotor, 4-beater.

Operating parameters of cutting mills depend on the number and layout of fixed and rotating cutters over the circumference as well as on the components intensifying the process (fig. 9).

The research works on new mill designs for vegetable and polymer materials has been conducted in the research and development centre of the City of Bydgoszcz (University of Technology and Life Sciences and Kazimierz Wielki University) for over 25 years. Among many design solutions developed there, it is worth to mention hyperboloidal mills invented by R. Konieczka [21, 22] and multi-edge mills developed and patented by J. Flizikowski and M. Bieliński [11]. The grinding process based on rotary cutting is used in knife mills that cut the material as a result of the cooperation between cutting edges of movable knives moving rotationally and fixed knives mounted on the mill housing. Fig. 10 shows a schematic diagram of a typical mill and its main parts.The grinding process based on rotary cutting is used in cutting mills equipped with hyperboloidal working chamber that cut the material as a result of the mating between cutting edges of rotating cutters moving rotationally, and fixed cutters mounted on the mill housing.

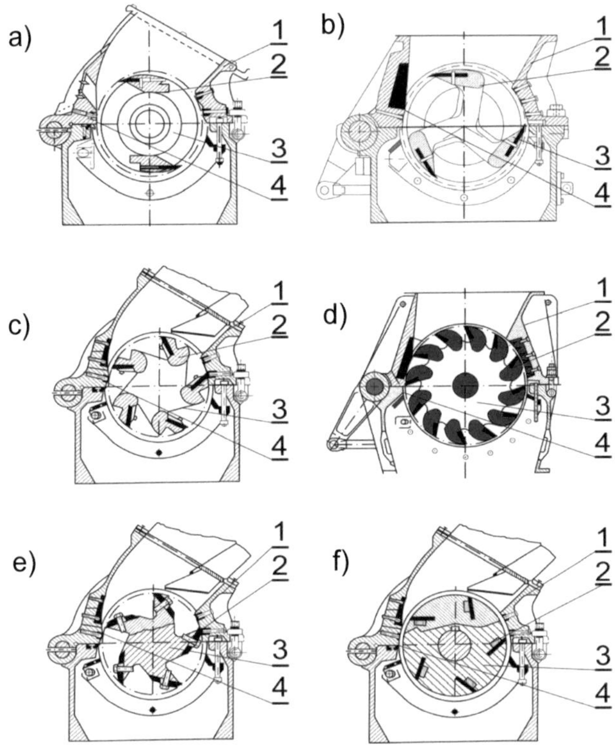

Fig. 9. Design solutions of cutting mills (based on the paper by M. Pahl) [35]:
a) with 2 rotating cutters and 1 fixed cutter, b) with 3 rotating cutters and 2 fixed cutters, c) with 5 rotating cutters spaced over the circumference and 2 fixed cutters, d) with 12 rotating cutters and 2 fixed cutters, e) with 5 rotating cutters and 2 fixed cutters, f) with milling cutters spaced over the circumference of the monolithic rotor and 2 fixed cutters; 1 - housing, 2 - rotating cutter, 3 - rotor, 4 - fixed cutter.

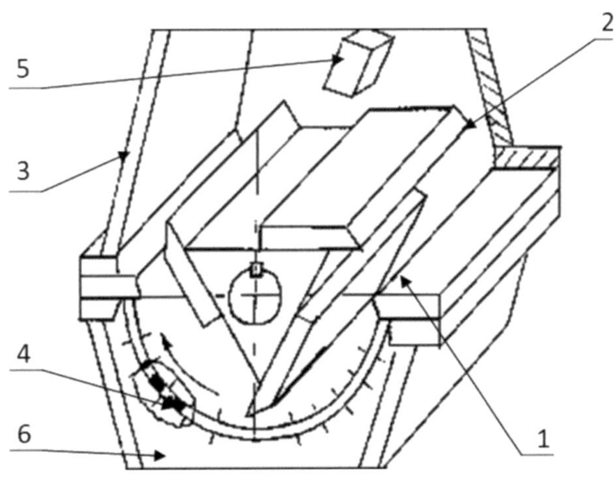

Fig. 10. Schematic diagram of a cutting mill [2, 8]: 1 - fixed knife, 2 - rotating knife, 3 - housing, 4 - sieve, 5 - fed material, 6 - product (elementary grain) [21]

The essence of the multi-edge grinder operation (fig. 11) is that it consists of the set of working disks mounted coaxially on the shaft and the disks are provided with holes with grinding edges spaced over diameters increasing from the input to the output of the material from the grinding space, while the speed between the neighbouring disks is the grinding speed [12]. Multi-edge mills are members of the group of equipment, where the material is processed at the linear speed of the grinding edge of about 1 m/s (fig. 5).

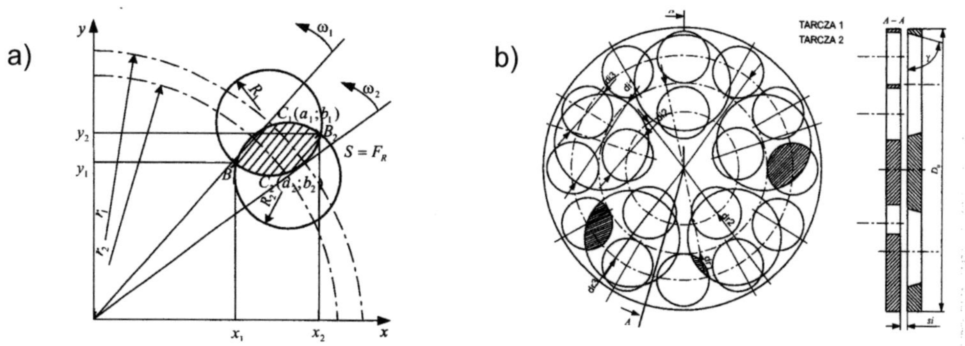

Fig. 11. The principle of the multi-edge grinder operation: a) geometric relationship between two holes of the neighbouring disks, b) view and section of two mating disks [13]

3. Expected quality range of the comminution product: Dimensions and a shape of a granule, particle size distribution (PSD)

Quality of comminution product is commonly assessed using screen analysis, i.e. by determining quantitative share of individual grain sizes in the sample of the polymer material using vibration of pneumatic screening through the set of analytic screens of specified rectangle mesh dimensions. The screen analysis is performed for grain size over 40 μm, because this is the minimum clearance of the screen mesh. The particle size distribution analysis for smaller grain is performed by sedimentation, which allows to separate grain of 1 ÷ 40 μm. Using optical and scanning microscope, it is possible to measure 0,6 ÷ 150 μm grain.

Grain size is a very important issue in most of operations and processes. Each grain is characterized by such parameters like: length, width, height. For the process analysis purposes, the grain size is determined by relationship between dimensions at three axes: a, b, c (as a cuboid).

Apart from the size, also qualitative characteristics of the granulated product and its form are important. The ground product should be of regular, usually cylindrical or cubical shape,

R. Koch proposed more precise classification of the process as he assumed the following criterion: specific category of the ground product should include over 50% of the grain of specific size [20]:

- rough - 0,5 ÷ 5 mm,
- medium - 50 ÷ 500 mm,
- fine - 5 ÷ 50 mm,
- colloidal - < 5 mm.

For process purposes, such as use of granular products obtained as a result of polymer and vegetable material grinding process, i.e. issues associated with their generation and future use, comparative indices are used. Those indices can be divided into the following three groups:

1. *Physical* indices characterizing geometric features and physical properties of the granular material include: grain sizes and shapes, grain collection distributions, size of control, outer upper or bottom particles as well as medium particles of specific weight percentage share of the control grain, specific surface area (kinetic), specific surface area (static), pycnometric density, micro-hardness, compressive strength, tensile strength, resistance to brittle fracture, hygroscopicity, adsorption and absorption properties, pyrophoric properties, colour, gloss, total volume of pores, average pore size and distribution, pore shape, internal friction angle, static and kinematical friction angle against metals etc.
2. *Chemical* indices specifying: content of the main component, content of the remaining components, content of impurities in individual phases (solid, liquid and gaseous), occurrence of oxide layers, corrosion resistance, chemical activity, electrochemical activity, catalytic properties, toxicity and other.
3. *Process-related* indices, including: degree of material size reduction (critical, medium, n-percentage), area increase degree, grindability, susceptibility to agglomeration, water content, total efficiency, usable efficiency, unit energy demand, grinding process efficiency ratio, looseness, compactibility, compressibility, sinterability etc.

From the point of view of the granulated product user, other factors are also important, i.e. physical and chemical properties of the ground material, such as: crystal lattice structure of the material, fibrousnesses, isotropy or anisotropy, plasticity, content of gases inside grains and gases absorbed over the grain surface, reactivity, adsorptivity, type and number of lattice defects, micro-structure, grain surface condition.

Granulated material quality indices depend on the type of the original material and planned use of the granulated material. Thus completely different quality indices are used for: polymer materials, (repeated recycling, fillers, dyes, fibres, seals, insulations etc,), fuels (combustion in power generation boilers, gasification, conversion into liquid fuel), cosmetics, ceramic bio-materials, metals, composites, brittle plastic and other materials.

4. Grinding process research methods

Taking into account useful form of the resulting product and the need for the analysis of high energy-consuming stages of the recycling process, it seems purposeful to develop research methods of the grinding engineering. Several thousands attempts to formally define the following aspects of the grinding process have been made for the last one hundred and fifty years, phenomena, system, relationships, design and the process itself [2, 6, 7, 16, 23, 38, 44, 47]. The main aim of the research works is to improve the general mathematical description and to help grinding equipment researchers and designers to increase the equipment efficiency: to reduce energy demand of comminution process and of drives, to increase efficiency and the quality of the grinder design, the charge and the grinding product (the function, grain size distribution parameters and the specific surface). There are many comminution hypotheses, but they can be categorized in three groups [10, 13]:

1. classic	2.	3.
associated with mechanics of the charge grinding process and relationships between results of the grinding process, such as: properties and energy demand of the charge deformation	associated with relationships between parameters of ground materials and the standards (relating mainly to grain size distribution parameters)	describing thermodynamic aspects of the grinding processes

The only hypotheses that have been applied in practice were those of the first group, i.e. associated with the energy of the deformation, i.e. symptoms of grindability, good proportions between the type, form parameters and grain size distribution of the charge and the product as well as the energy demand.

Mechanical engineering and machine operation methodology in recycling and issues associated with the properties of the charge and the grinding product are subject of many studies in Poland and abroad. Some of them indicate to similarity to the problems relating to material removal processes. Those studies allow energy demand analysis in those processes (e.g. machining), so that selection of less energy consuming variant is possible [1, 8, 33, 37].

Unit grinding, cracking, load distribution issues etc. are also important for better understanding of the grinding process (fig. 2.3) [36]. Studies in that field are performed, among others, at the University of Sheffield [41].

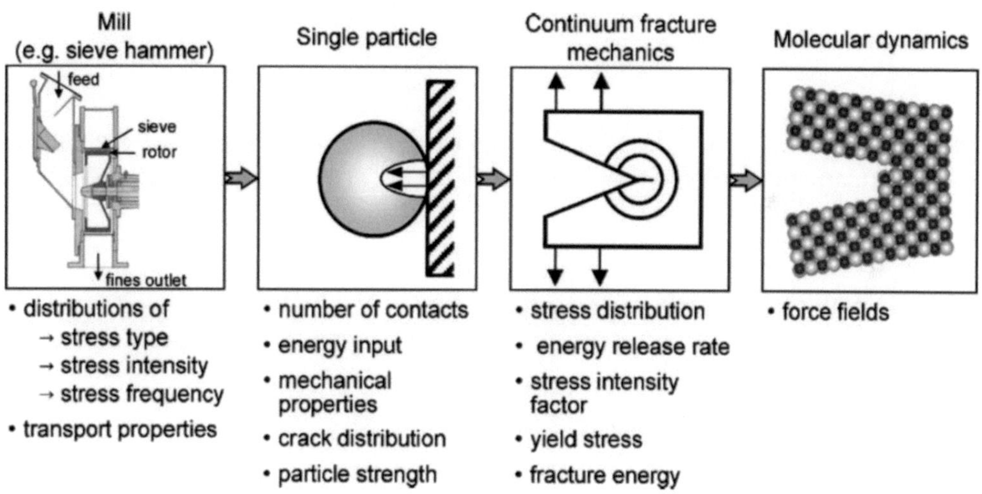

Fig. 12. The scope of research on grinding process [36]

5. Own research method of multi-edge grinding

Studies on the innovative design of multi-edge mills using author's own method are based on geometric relationships of the multi-edge mill for granulated and chunked materials patented in 1989 [11]. The multi-edge unit is a set of disks or drums with holes spaced over coaxial perimeters. Thanks to proper geometry of holes in their disks, drums or strips as well as due to appropriate relationship of the movement between neighbouring edges, multi-edge grinders are able to grind plastic using the neighbouring edges (fig. 13) [28].

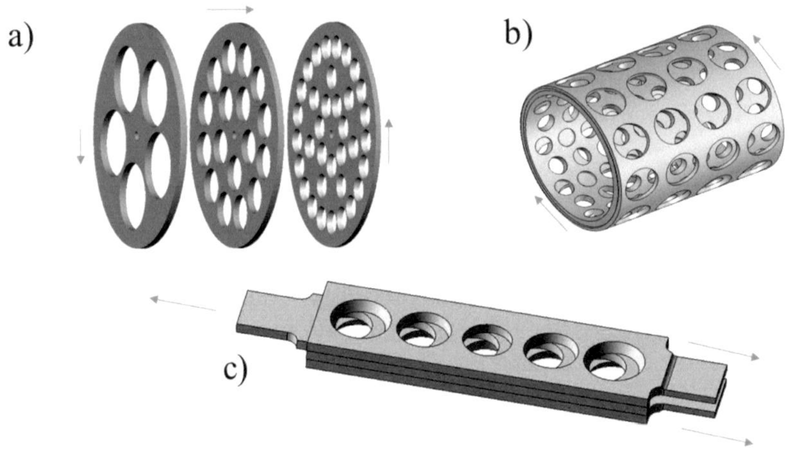

Fig. 13. Possible design implementatios of the disks – a), drums – b) and perforated plates – c)

Design solutions – from the allowable collection (1), (2) – that maximize or minimize values of selected operational characteristics were assumed as the general model of multi-edge grinding studies. The usability conditions are met by design features that maximize the function of: energy efficiency, overall efficiency, capacity, material size reduction degree for:

$$\left\{C_k^* \in \phi\right\} : \left\{ \bigwedge_{c_k \in \phi} H_u(c_k) < H_u(c_k^*) \right\} \text{ for: } H_u\text{,: } e_R \uparrow, W_u \uparrow, \eta_\rho \uparrow, \lambda \uparrow \tag{1}$$

While minimization of the power demand, unit energy consumption, energy dissipation, torque, angular, linear and rotational speed – as the measure of usability – can be achieved for effective design features:

$$\left\{C_k^* \in \phi\right\} : \left\{ \bigwedge_{c_k \in \phi} H_u(c_k) > H_u(c_k^*) \right\} \text{ for: } H_u\text{: } N_\upsilon \downarrow, E_T \downarrow, M_\chi \downarrow, \omega \downarrow, \varpi \downarrow, \nu \downarrow \tag{2}$$

where:

C_k^* – the solution to the problem,
ϕ – allowable space of the design features vector C_k^*,
H_u – operational characteristics,
e_R – energy efficiency function, - ,
W_u – mass capacity of the process, kg/h,
P_e – power demand, kW,
E_T – unit energy demand, J/g,
λ – break-up degree (product dimensions), mm,
M_c – torques, N·m,
ω, n, v – angular, rotational, linear speeds, rad/s, 1/s, m/s,
η_ρ – grinding efficiency, -.

The objectives of the study were fulfilled thanks to:

- assessment of the impact of material properties and the form of the sample on the function of forces and quasi-shear labour,
- the analysis of the impact of the disintegrating elements speed on the value of quasi-shear labour,
- determining the impact of the hole edge angle on the value of quasi-shear labour,
- implementation of the mechanisms of simulation research on quasi-shear resistance and sections for selected variants,
- assessment of the grinding process operational characteristics in physical conditions.

5.1 Description of a laboratory set-up of comminution, measure - possibilities

In this research, basic characteristics of a selected multi-edge grinding type were verified in laboratory set-up. These include: energy demand for the process, torques and output. Stress values were determined using mathematic relationships describing the area of grinding, taking into account design features of the grinding unit [14]:

$$2A_r = \frac{6,11}{\rho \cdot s_i}((0,043 \cdot s_i)^{-1,12D_{oi}+1,18}) \cdot v_r \cdot e^{1,88 \cdot n_w^2} \tag{3}$$

where:

A_r - the area of grinding, mm²,
ρ - density of materials subject to grinding, g/cm³,
s_i - the index of the gap between disks (drums) of the grinder, (-),
D_{oi} - diameter indicator of the i-th disk (drum) provided with holes, (-),
n_w -the factor dependent on the type of processed materials; n = 0,75÷1,00,
v_r relative speed between quasi-shearing edges, m/s.

The author performed laboratory research on multi-edge grinding for two design versions of the grinding unit, chosen based on the matrix of possible solutions [27], meeting preliminary criteria as regards energy efficiency and the grain size curve [14]: i.e. one of the drum and the disk version, provided with polygonal openings (fig. 14).

5.1.1 Description of a laboratory station

The laboratory station was built according to the author's own concept as part of KBN (State Committee for Scientific Research) Research project performed in the years 2003-2005 [25], and then adapted to the purposes of grinding energy efficiency research as part of the MNiSW (the Ministry of Science and Higher Education) project [24] (fig. 15, 16). The frame and the drive were prepared to allow measurement of torques and angular speeds at any position of the drive shaft (from the vertical to the horizontal one). The proposed design of the station ensures that the drive system, together with the measuring sensors, is an integral base for research using different disintegrating (grinding) units. The shaft end with working systems of the grinder fixed to it are used for that purpose. This ensures that the measuring parameters are unchanged and guarantees possibility to assess the energy input needed to grind unit mass (volume) of the material using different configurations (disintegration methods). Research can be performed with stepless control of the motor rotational speed in the range 0 ÷ 2900 rpm.

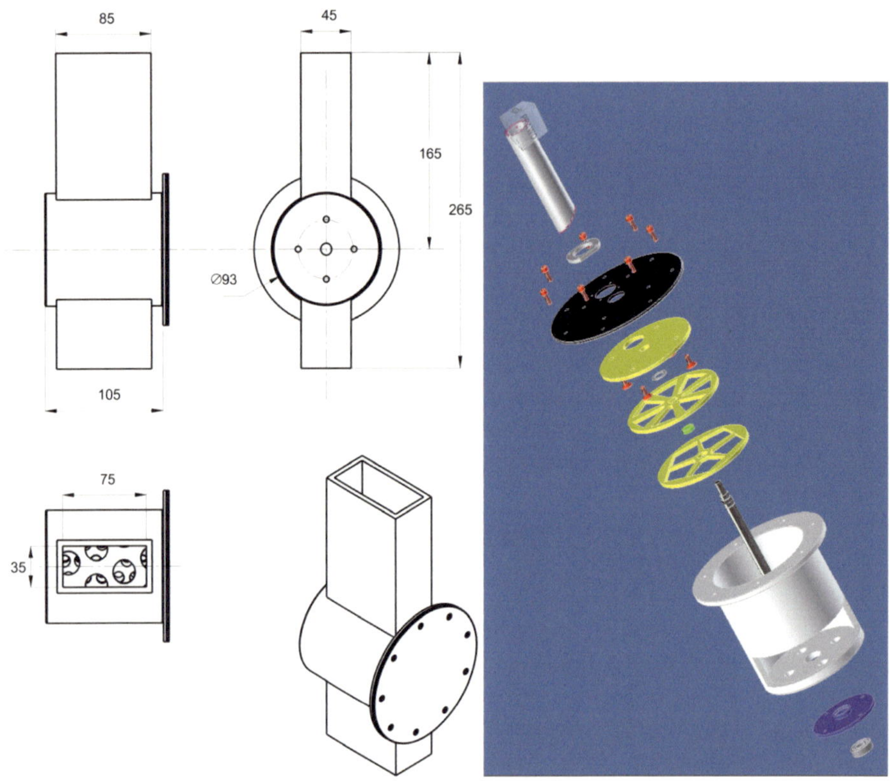

Fig. 14. The multi-edge grinder unit: a) drum type - orthogonal projection and axonometric view of the housing, b) disk type - exploded view

Fig. 15. The station of the general-purpose laboratory grinder ULR-2,0/2004 - with horizontal position of the drive shaft

Figure no. 16. shows mechanical diagram of the station of the all-purpose laboratory grinder ULR-2,0/2004.

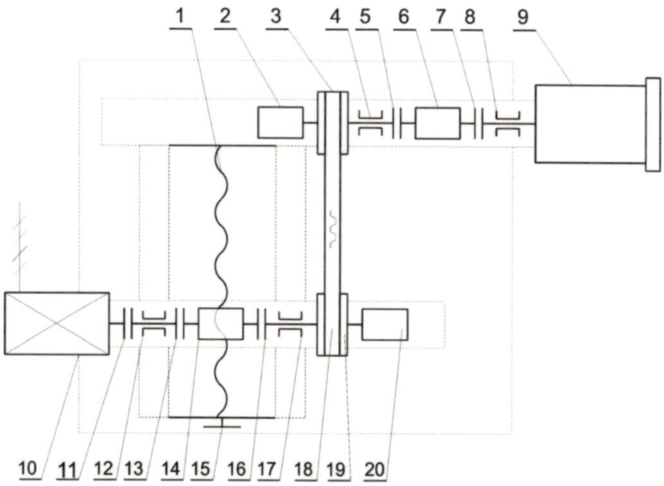

Fig. 16. Mechanical diagram of the station of the all-purpose laboratory grinder ULR-2,0/2004: 1 - the screw for adjustment of the distance between the motor shaft and the grinder shaft, 2 - rotary-pulse transducer, 3 - the pulley of the toothed belt transmission, 4 - journal bearing, 5 - clamp coupling, 6 - torque meter, 7 - clamp coupling, 8 - journal bearing, 9 - grinder chamber, 10 - motor, 11 - clamp coupling, 12 - journal bearing, 13 - clamp coupling, 15 - torque meter, 16 - clamp coupling, 17 - journal bearing, 18 - toothed belt of the transmission, 19 - the pulley of the toothed belt transmission, 20 - rotary-pulse transducer

Output signals from the torque meter are recorded using a 4 - channel measurement amplifier Spider 8 manufactured by HBM, while the study results were processed using the software CATMAN developed by HBM. Electric power was measured using the general-purpose power meter Hioki 3169 - 20/21 with the software (power measurement support software) Hioki 9625.

5.2 Results: Energy consumption for chosen modes of grinding, influence of design features on the process and the quality of grinding product

The study examined the energy and torque demand for different design versions of the grinders for processing polymer recyclates (fig. 17).

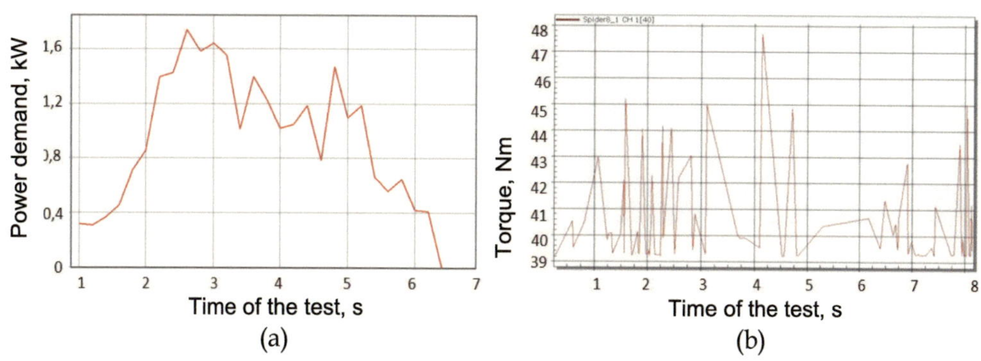

Fig. 17. Sample results of laboratory comminution tests: a) power demand - the drum mill, b) torque - the drum mill

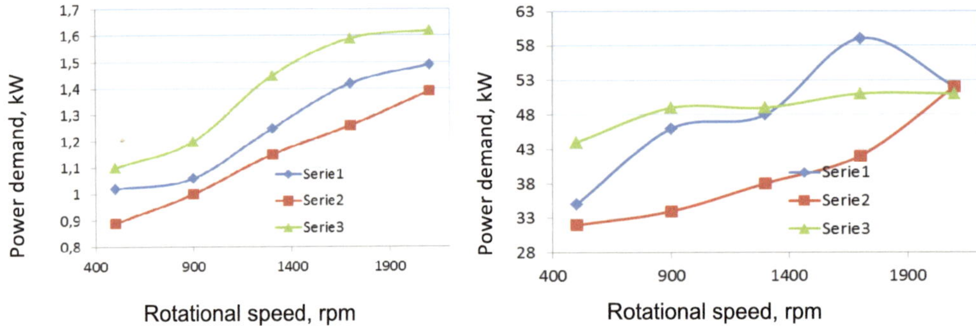

(a) (b)

Fig. 18. Influence of the rotational speed on the power demand during PP grinding process - a) Influence of the rotational speed on the torque value - b), series 1 - the disk mill with cylindrical holes, series 2 - the disk mill with polygon holes, series 3 - the drum mill

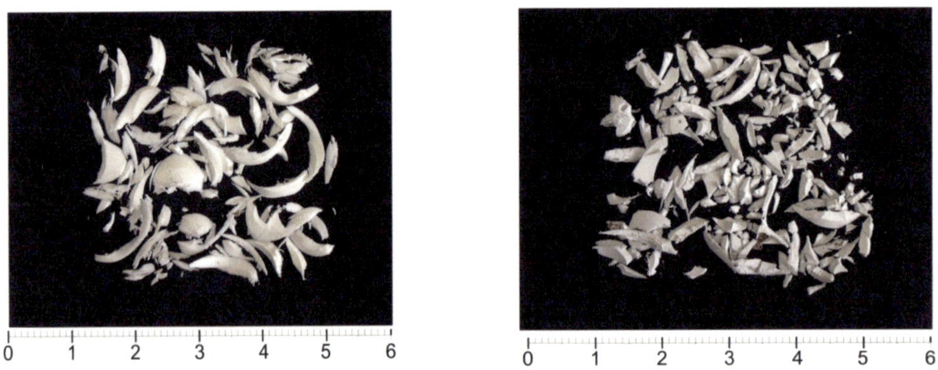

(a) (b)

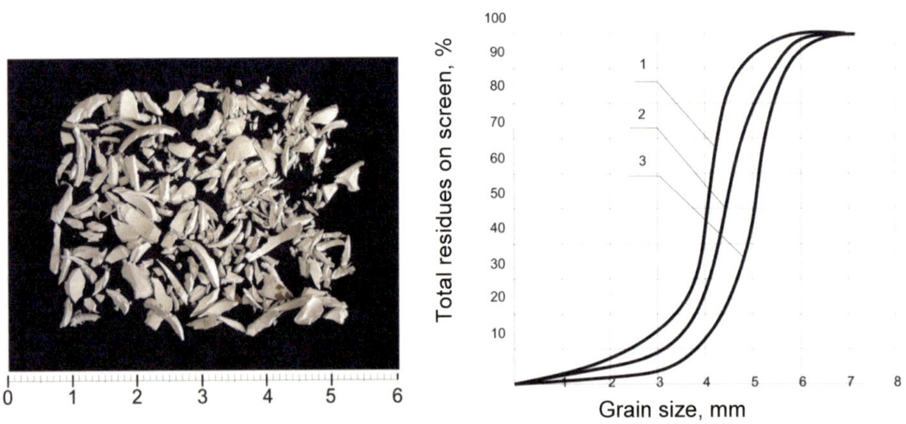

(a) (b)

Fig. 19. PVC obtained as a product of grinding using a disk mill with cylindrical holes at three different rotational speeds: a) 500 rpm, b) 1300 rpm, c) 2100 rpm, d) product grain size distribution curve: 1 - 500 rpm, 2 - 1300 rpm, 3 - 2100 rpm

Power demand and torque demand diagrams as the function of the rotational speed were prepared based on the average values of the obtained results (fig. 18).

The grinding product was assessed by means of the sieve analysis, performed using the vibration screen AS 200 basic manufactured by RETSCH (fig. 19). The product grain size distribution curve before the expected further processing ranges from 2 to 6 mm. With such range the product is classified as suitable for further processing.

5.3 New methods which open variety of new grinding research opportunities (CAD/CAE and AI – genetic algorithms)

5.3.1 Computer simulations of multi-edge grinding

Computer simulations are currently a necessary tool of machine and equipment design. Using those new technologies, it is possible to build virtual design models allowing efficient preparation of the drawing documentation and implementation of computer simulations (CAD/CAE). The purpose of grinding simulation results and analysis is to outline the range of design features for which assumed criteria are met. The application SolidWorks Motion is used to perform strength and kinematic simulations using virtual models. With the application it is also possible to verify future correct operation of the system at the prototype stage.

Grinding simulations were performed for two types of the general-purpose laboratory grinder ULR-2,0/2004: the drum and the disk one (fig. 20), and the load and stress simulations during single grinding were performed using the application SolidWorks Simulation.

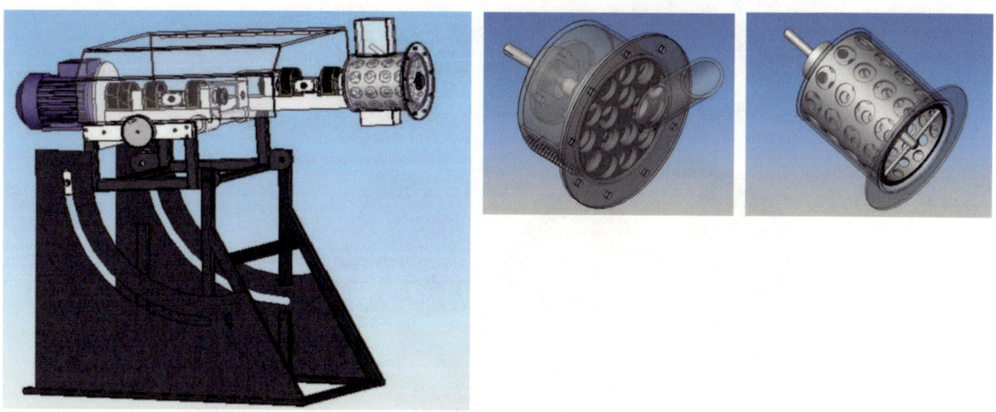

(a) (b) (c)

Fig. 20. Laboratory mill ULR-2,0/2004 designed in SolidWorks environment: a) general view, b) disk unit, c) drum unit.

As a result of simulation using SolidWorks Motion, it has bee determined that the value of the contact force and stress for both types of the grinder working units rises with the growth of the rotational speed within the range of 52,82 ÷ 203,02 MPa (for the drum type system) and 46,47 ÷ 142,40 MPa (for the disk type system).

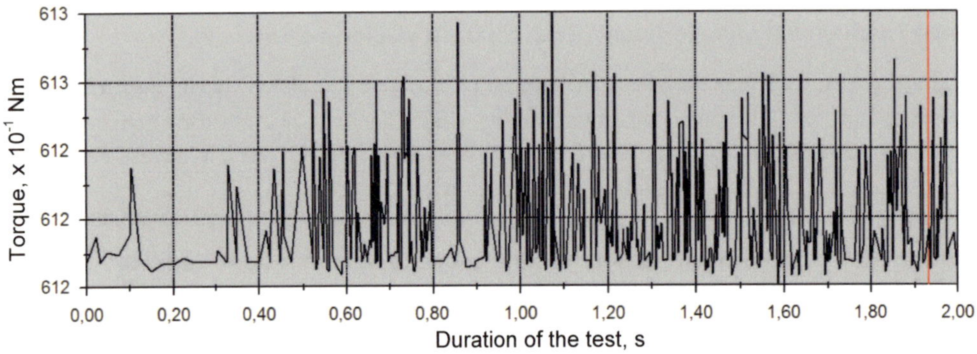

Fig. 21. Simulation of the torque function during grinding of PVC samples using the drum mill (SolidWorks Simulation).

Optimum design solutions of grinding disks and drums were found by implementation of genetic algorithms. The starting point in that method are design features of the grinder base disk. Figure no. 22 shows four consecutive stages of processing for the same disk (using the application tAG_Project1) [26].

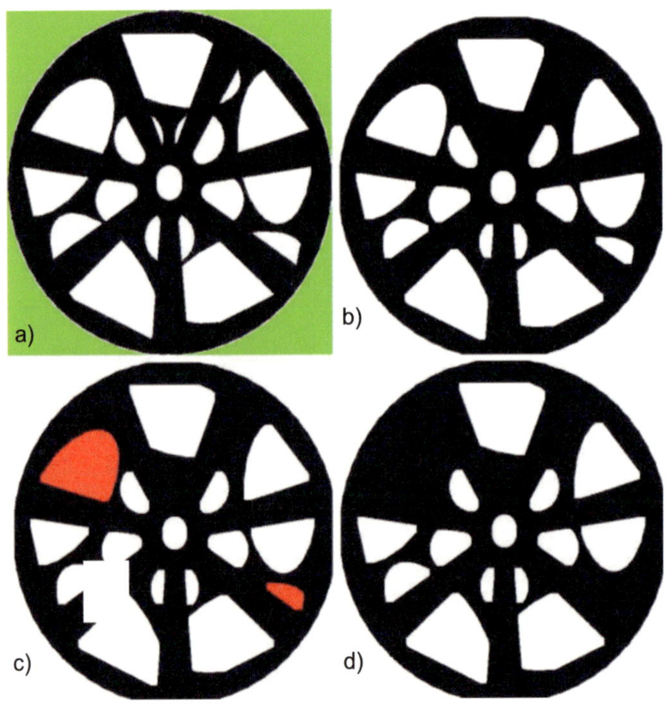

Fig. 22. Consecutive stages of the grinder disk modification: a) result of the crossing, b) the disk after morphologic optimisation, c) mutation phase, d) final result of the disk optimisation.

Based on geometric analysis performed using CAD applications, disks with the following hole geometry and layout were proposed (fig. 23):

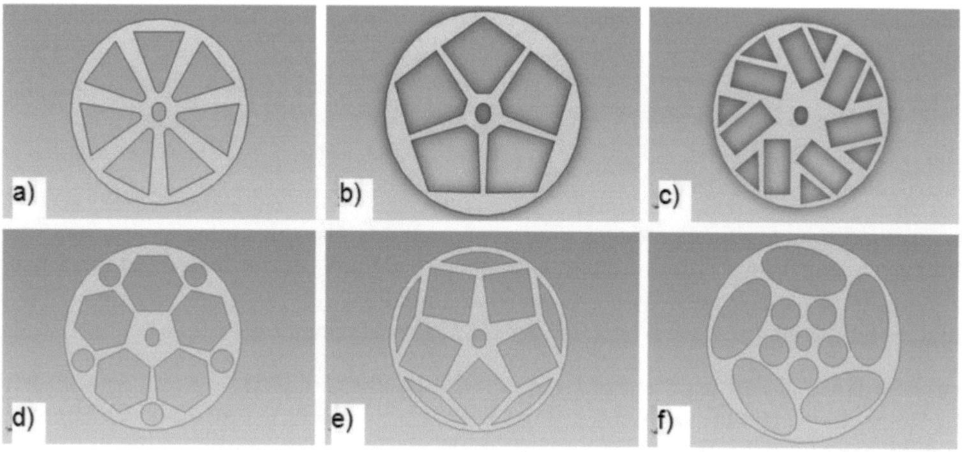

Fig. 23. Chosen disk design solutions as the area of design feature selection. a) the disk triangular holes, b) pentagonal holes, c) rectangular and triangular holes, d) hexagonal and cylindrical holes. e) quadrangular and triangular holes, f) circular and elliptical holes

5.4 Analysis of results

The following basic characteristics were verified in the study: energy demand for the grinding process, torque function and the efficiency. Laboratory tests of the multi-edge grinder were performed for three design versions of the grinding units: i.e. the drum and the disk version, provided with cylindrical and polygonal holes. The rotational speed values were included within the range (500 ÷ 2100) rpm. The study shows that the most advantageous solution as regards minimum power demand is the disk grinder with disk provided with polygonal holes. On average, it ensures 10 ÷ 15 % lower grinding power demand compared to the disk grinder with cylindrical holes and about 15 ÷ 20 % lower power demand compared to the drum mill. The average power demand ranges from 1 to 1,8 kW and the average torque value is included in the range 35 ÷ 60 Nm.

Power consumption for grinding of one mass unit to the size suitable for further processing was estimated on that base. In case of the drum unit, the power demand was 0,042 kWh/kg, while for the disk unit: 0,056 kWh/kg.

6. Conclusions and future work

It can be assumed that research works to improve grinder designs shall continue. This results from the necessity to reduce size of different materials characterized by wide scope of material properties. Thanks to introduction of innovative multi-edge grinder designs, original research procedures led to design and construction of high-grinding efficiency test station. With such test stations, it is possible to continue promotion of highly innovative grinding solutions. Genetic algorithms were successfully used to design the grinder disks and shapes and sizes of disk holes. Using modern technologies, it is possible to machine holes even of very complex shape in disks and drums.

The methodology of selecting multi-edge grinders for polymer materials processing includes state-of-the art IT tools, A general grinding database was built and developed using that methodology and it continues to be developed based on the backbone expert system. The highest usability within the research scope of the cutting hole edge angle (β = 60°, 75°, 90°, 105°, 120°) was achieved for β = 60° (2nd grade non-linear regression equations L = f (b). While minimum usability of quasi-shear labour was obtained for the angle β = 90° (samples with outer diameters $D_z \geq 0{,}5\ d_{otw.}$) and for the angle β = 120° - for samples with outer diameters of $D_z \leq 0{,}5\ d_{otw}$.

The precondition for further development of methods for selection of recycling-related grinding types is preparation of detailed mathematical models, improvement of recyclates recycling technologies and, as a result, manufacturing of recycled products that are characterised by quality level acceptable for further processing.

Recyclates with the average diameter of D_z < 50 mm and wall thickness of g < 6 mm which are planned to be ground using the grinder with three disks or drums should be processed using the machine equipped with minimum 5 kW electric drive motor and the transmission gear capable to reduce linear speed of the grinder edge to 0,5 ms^{-1}.

7. Acknowledgments

This work was financially supported by the Polish Ministry of Science and Higher Education

8. References

[1] Austin L.G., Klimpel R.R.: The theory of grinding operations. *Ind. Eng. Chem.* 56, p. 18-29 (1964).

[2] Austin L.G.: Introduction to the mathematical description of grinding as a rate process. *Powder Technology*, 5, p. 1-17 (1971-1972)

[3] Bauer W.: Untersuchung des Einzelzerkleinerungsvorganges in Schneidmühlen am Beispiel von Polypropylen. Reihe 3 - *Verfahrenstechnik*, Nr. 694. VDI Verlag, Düsseldorf 2001

[4] Bielinski M.: Materiałowa i przetwórcza charakterystyka wybranych tworzyw wtórnych. *Rozprawy nr 84*, Publishing House of ATR in Bydgoszcz 1998

[5] Błędzki A. K. (editing): Recykling materiałów polimerowych. WNT, Warsaw 1997

[6] Brożek M., Mączka W., Tumidajski T.: Modele matematyczne procesów rozdrabniania, Wydawnictwa AGH Kraków, 1995

[7] Butzen, G. A., Adams, G. A.: Size Reduction/Granulating. Van Nostrand Reinhold Co., *Injection Molding Handbook*, 1986

[8] Darlewski J., Gawlik J., Grzesik W., Jemielniak K., Ruszaj A., Weiss E., Zebrowski H.: Trendy w ubytkowych metodach obróbki, *Automatyzacja produkcji '97 Innowacje w technice i zarządzaniu*, Wrocław 1997, Scientific research works at the Institute of the Machine Engineering, Automation and Robotics of Wrocław University of Technology, no. 67, Conference series no. 29, volume 1, 175-212

[9] Dmitrewski J.: Teoria i konstrukcja maszyn rolniczych. T.3, PWRiL, Warszawa 1978

[10] Drzymała Z. (et al.): Badania i podstawy konstrukcji młynów specjalnych. PWN. Warszawa, 1992.

[11] Flizikowski J., Bielinski M.: Rozdrabniacz wielotarczowy zwłaszcza do materiałów ziarnistych i kawałkowych. Polish patent, P - 144566, UP Warsaw 1989

[12] Flizikowski J.: Badania i podstawy konstrukcyjne rozdrabniaczy wielotarczowych. Rozprawy ATR no. 42, WMN - ATR Bydgoszcz 1990
[13] Flizikowski J.: Rozdrabnianie tworzyw sztucznych. WMN - ATR Bydgoszcz, 1998
[14] Flizikowski J.: Sprawozdania z badań. Grant (KBN)MNiI, PB 622/T08/2002, AGH Kraków - ATR Bydgoszcz, 2002-2005
[15] Flizikowski J. : Micro- and Nano-energy Grinding. ISBN-10 9814303534. Pan Stanford Publishing. 2011
[16] Giersemehl, M., G. Plihal: Fine Grinding System with Impact Classifier Mill and Cyclone Classifier. Company Affiliation: Neuman & Esser GmbH Mahl- und Sichtsysteme Published in Journal: powder handling & processing Year: 1999 Volume: 11 Number: 3 Page: 269
[17] Höffl K.: Zerkleinerungs und Klassiermaschinen. Springer-Verlag, Berlin 1986
[18] http://www.plasticseurope.org/documents/document/20101028135906-final_plasticsthefacts_26102010_lr.pdf (December 2010)
[19] Kalman H., Hubert M., Grant E., Petukhov Y., Haim M.: Fatigue behaviour of impact comminution and attrition units. Powder Technology 146, 2004
[20] Koch R., Noworyta A.: Procesy mechaniczne w inżynierii chemicznej. WNT, Warsaw 1998
[21] Konieczka R.: Podstawy mechanicznych procesów recyrkulacji folii z polietylenu małej gęstości. Rozprawy ATR no. 74, Bydgoszcz 1996
[22] Konieczka R.: Urządzenie rozdrabniające. Polish patent no. 130061, UP Warsaw 1987
[23] Lowrison G. Ch.: Crushing and grinding. Verlag Butterworth, London, 1974
[24] Macko M. : Badania procesu rozdrabniania tworzyw polimerowych i materiałów biologicznych w kierunku podwyższenia efektywności energetycznej - MNiSW research project no. N N508 440536 (2009-2011)
[25] Macko M. : Tworzywowo zintegrowane podstawy inteligentnego rozdrabniania i granulowania w recyklingu - KBN research project (MNiI) no. 978/T08/2003/24 (2003-2005)
[26] Macko M., Czerniak J.: The evolutionary method for optimising disk design of multi-edge grinders. *Journal of Theoretical and Applied Mechanics* - the paper accepted for publishing in 2012.
[27] Macko M.: Metodyka badao procesu rozdrabniania wielokrawędziowego w recyklingu. *Inżynieria i Aparatura Chemiczna*, Gliwice, 3/2005, p. 51-52
[28] Macko M.: Wpływ cech konstrukcyjnych zespołu wielotarczowego na charakterystyki użytkowe procesu rozdrabniania rurowych recyklatów tworzyw sztucznych. Thesis. WM-ATR, Bydgoszcz 2000
[29] Marsh G.: Facing up to the recycling challenge, Reinforced Plastics, Vol. 45, Issue 6 (2001), p. 22-26
[30] N. N.: Doppelrotor-Shredder. Prospektmaterial, Fa. Hoger Maschinenbau GmbH & Co. KG, Bergkirchen
[31] N. N.: Rotorscheren Typ S, Typ I. Prospektmaterial, Fa. Societe Industrielle de la Doux SA, Saint-Sulpice (Schweiz), 2001
[32] N. N.: Universal-Zerkleinerer ZU. Alpine Aktuell Nr. 14, Prospektmaterial, Fa. Alpine AG, Augsburg
[33] Osman M.O.M., Mueller G.S.: Reference systems for cutting tool geometry and their transformation matrices. Int. J. Prod. Res. 1973, vol.11, nr 2, p. 113-124

[34] Pahl M. H.: Zerkleinerungstechnik. Praxiswissen Verfahrenstechnik. Mechanische Verfahrenstechnik: Auflage, Leipzig, 1993

[35] Pahl. M. H.: Zerkleinerungstechnik. Verlag TÜV, Rheinland 1994

[36] Peukert W.: Material properties in fine grinding. *Int. J. Miner. Process.* 74S (2004) p.3–17

[37] Pienkowski G.: Ocena energochłonności operacji obróbki ubytkowej. Thesis, Wrocław University of Technology, 2006

[38] Prasher C.L.: Crushing and Grinding Process Handbook. Verlag Wiley & Son, Chichester, 1987

[39] Repelin V., Govin A., Rolland M., Guyonnet R.: Energy requirement for fine grinding of torrefied wood. *Biomass & Energy* 34 (2010) p. 923-930

[40] Rumpf H.: Die Einzelkornzerkleinerung als Grundlage einer technischen Zerkleinerungswissenschaft. *Chemie Ingenieur Technik* 37, No.. 3/1965, p. 187-202

[41] Salman A.D., Biggs C.A., Fu J., Angyal I., Szabo M., Hounslow M.J.: An experimental investigation of particle fragmentation using single particle impact studies. *Powder Technology*, Volume 128, Number 1, 4 December 2002, p. 36-46

[42] Schubert G., Bernotat S.: Comminution of non-brittle materials. *Int. J. Miner. Process.*, 74S (2004), p. 19-30

[43] Sidor J.: Badania, modele i metody projektowania młynów wibracyjnych. *Rozprawy 150.* AGH Publishing House, Kraków 2005

[44] Siwiec A.: Związek pomiędzy opisem procesu rozdrabniania procesami Markowa a hipotezą Rittingera. Górnictwo. 1999 z. 4 (p. 237 - 244)

[45] Wang Y., Forssberg E.: Enhancement of energy efficiency for mechanical production of fine and ultra-fine particles in comminution. *China Particuology* 5 (2007), p. 193–201

[46] Woldt D., Schubert G., Jackel H.-G.: Size reduction by means of low-speed rotary shears. *Int. J. Miner Process.* 74S (2004) S405-S415

[47] Zawada J.: Wstęp do mechaniki procesów kruszenia. ITEE Radom, 1998

[48] Zimniak J., Konieczka R.: Zerkleinerungs- und Mischprozesse für Verbundwerkstoffe. *XV. Fachtagung über Verarbeitung und Anwendungen von Polymeren TECHNOMER'97*, November 1997, Chemnitz P36, p. 1- 8